Reconfigurable Cryptographic Processor

Leibo Liu · Bo Wang · Shaojun Wei

Reconfigurable Cryptographic Processor

Leibo Liu
Institute of Microelectronics
Tsinghua University
Beijing
China

Shaojun Wei
Institute of Microelectronics
Tsinghua University
Beijing
China

Bo Wang
Institute of Microelectronics
Tsinghua University
Beijing
China

ISBN 978-981-10-8898-8 ISBN 978-981-10-8899-5 (eBook)
https://doi.org/10.1007/978-981-10-8899-5

Jointly published with Science Press, Beijing, China

The print edition is not for sale in China Mainland. Customers from China Mainland please order the print book from: Science Press.

Library of Congress Control Number: 2018936628

Printed on acid-free paper

This Springer imprint is published by the registered company Springer Nature Singapore Pte Ltd.
part of Springer Nature
The registered company address is: 152 Beach Road, #21-01/04 Gateway East, Singapore 189721, Singapore

Foreword

The reconfigurable cryptographic processor is one of the ideal hardware implementations for encryption and decryption algorithms. Compared with traditional cryptographic processors, the reconfigurable cryptographic processor can meet the requirements of cryptographic application for high security, high energy efficiency, and high flexibility. In terms of security, the function of the dynamically reconfigurable computing architecture is not determined by either the hardware or the software alone, but it is dynamically defined by both the software programming and the hardware programming. The reconfigurable computing architecture has the "blank chip" feature after the power supply is cut off. Therefore, it is difficult to obtain the cryptographic algorithms through invasive attacks. Meanwhile, the execution model of the dynamically reconfigurable computing architecture has specialization and is indeterministic and it is very difficult to conduct side-channel attacks on it such as power analysis attacks, fault attacks, and electromagnetic attacks through behavioral modeling. Therefore, the dynamically reconfigurable computing architecture has high security against physical attacks. In terms of performance and power consumption, the dynamically reconfigurable computing architecture performs operations mainly by using a spatially parallel method, which is very suitable for the feature of cryptographic algorithms, and its energy efficiency (i.e., performance per watt) is high. In terms of functional flexibility, the reconfigurable computing architecture can change its hardware functions at runtime to fit them for different cryptographic algorithms and execution modes, and it has excellent flexibility. With the constant development of physical attack means in recent years, even if the security of cryptographic algorithms is very good, the attacker can still steal key information by using invasive or noninvasive physical attacks on the bottom-level cryptographic processor. Because the reconfigurable cryptographic processor has the above-mentioned outstanding advantages in resisting physical attacks, it has gradually become a hot research direction of cryptographic chips and a lot of relevant achievements on it have been published at the top conferences of the cryptographic field and in periodicals of the cryptographic field in recent years.

Traditional cryptographic processors are mainly the application-specific integrated circuits (ASICs) and the instruction set architecture processors (ISAPs). The ASIC has obvious disadvantages in security and flexibility. By reversely analyzing ASIC chips, attackers can crack the cryptographic algorithms by the circuit implementation and steal the secret information being processed. In addition, the ASIC chip can only implement the specific cryptographic algorithms and it almost has no functional flexibility. Therefore, it cannot meet the rapidly developing demands of applications. The ISAP also has some security problems. Because it is easy to model the execution process of the system, the ISAP is likely to suffer noninvasive side-channel attacks. Meanwhile, the energy efficiency of the ISAP has been unable to meet the demands of current practical applications. The reconfigurable cryptographic processor has solved these difficult problems very well, and it has gradually been applied in practice. Therefore, it is likely that the reconfigurable cryptographic processor will be both the research and application directions of the future cryptographic processor.

The research team headed by Prof. Leibo Liu, the author of the book, is one of the best domestic teams in the reconfigurable cryptographic processor research field because it has worked in the field for a long time and accumulated a lot of experiences. The Institute of Microelectronics, Tsinghua University, in which Dr. Liu works completed two key projects of the National 863 Program in the reconfigurable computing direction in the 11th and the 12th Five-Year Plan periods, and the institute was awarded a second-class prize of the 2015 National Technological Invention Award. Acting as a leader in the frontier research of the domestic reconfigurable computing field, the institute has published a series of influential academic papers at top-level meetings of the relevant fields and in relevant periodicals, published an academic monograph *Reconfigurable Computing*, obtained relevant patents, and won gold medal in the 2015 National Patent Award. The institute has conducted some fruitful pioneering research in the application of reconfigurable computing technology to the cryptographic field in recent years and proposed the key technologies for improving the security, flexibility, and energy efficiency of cryptographic processors by making use of the dynamic reconfigurable features, including fault attack countermeasures and electromagnetic attack countermeasures based on the dynamic reconfiguration with spatial and temporal randomization, technologies of generating physical unclonable functions by using reconfigurable architectures, and fault attack countermeasures by making use of the improved Benes network. The novelty, advancement, and effectiveness of these technologies have impressed me very much.

In this book, the description is precise, the materials are well-organized, the contents are novel, and the viewpoints are original. This book not only describes in detail the basic knowledge and current research status of cryptographic processors, but also discusses the design methods and development direction of reconfigurable cryptographic processors. It is an extremely good monograph in the field. Therefore, I would love to recommend it to readers.

Beijing, China Jiren Cai

Preface

Reconfigurable cryptographic processor has special technological advantages in security, energy efficiency (performance per watt), and functional flexibility, compared with the other cryptographic processors such as instruction set architecture processors (ISAP), field-programmable gate arrays (FPGA), or application-specific integrated circuits (ASIC).

The first technological advantage is the advantage in the security of the cryptographic processor due to the following two reasons. Firstly, though the processing elements and interconnections of reconfigurable cryptographic processors may be heterogeneous, they are still very regular in terms of circuits as well as the placement and routing. It is difficult to obtain the information of the cryptographic algorithms by observing the hardware structure and circuit construction. Therefore, the information of cryptographic algorithms will not be leaked during the tape-out process or even after the chip is lost. This feature is called the "blank chip feature" of reconfigurable cryptographic processors, which is very concerned by a lot of users. Secondly, reconfigurable cryptographic processors have the features of dynamic reconfiguration and partial reconfiguration. The function of processing elements and the interconnection among them can be changed within several cycles. The reconfiguration time is from over 10 ns to dozens of nanoseconds (Note: The FPGA reconfiguration time is from several hundreds of milliseconds to several seconds). Therefore, its capability of physical attack resistance is far stronger than that of the traditional cryptographic processors, which is one of the reasons why the security of the reconfigurable cryptographic processor is high. The content of this part is one of the highlights of the book, and it will be described in detail.

The second technical advantage is that reconfigurable cryptographic processors have very high energy efficiency, while it can meet the functional flexibility demands for the diversification and constant evolution of cryptographic algorithms. This is mainly because the reconfigurable processor supports hardware programming as well as software programming, and it can meet the constantly changing software demands by dynamically changing hardware. As we all know, the energy efficiency of ASICs is the best. However, this type of circuits does not have any flexibility.

After the silicon implementation is finished, the functions cannot be changed and no new functions can be added to it unless it is redesigned and taped out again. Since now the chip research and development (R&D) costs are becoming increasingly high, such implementation mode that needs a lot of time and very high R&D costs will be gradually eliminated in the market. Cryptographic processors using the instruction set architecture are of the best functional flexibility. This type of cryptographic processors can be used to perform most of the cryptographic algorithms at present and even in future; however, its energy efficiency is very low. It can reach only $1/10^4$ of or even lower than the energy efficiency of the ASIC, which is far from meeting the application demands. In addition to the ASIC and the ISAP, there is a more common way to implement cryptographic processors, i.e., the way of using the programmable logic device. However, the performance of this implementation approach is still not ideal in energy efficiency and functional flexibility. The reasons for that can be explained in two aspects. Macroscopically speaking, the flexibility of programmable logic devices such as the FPGA are too flexible (Note: As long as the number of programmable units is large enough, digital logic in almost any form can be implemented), and the high flexibility is obtained at the cost of greatly reducing the energy efficiency and area efficiency (i.e., performance vs. area). Though cryptographic operation needs functional flexibility, it is not wise to obtain such high functional flexibility which is unnecessary and cannot be fully used by reducing the performance and increasing the power consumption and area consumption. Microscopically speaking, the programming granularity of programmable logic devices such as the FPGA is too fine (For example, the core processing element is the lookup table (LUT) of 1-bit granularity), which results in too much configuration information and too long configuration time, makes it impossible to implement dynamic and partial reconfiguration limiting the improvement of energy efficiency and area efficiency. Though some commercial FPGA advertisement claims that the product has such function, we do not think the FPGA can implement the same dynamic and partial reconfiguration as reconfigurable processors. This is determined by the architecture of FPGA. Cryptographic processors implemented in other types of technologies such as the system-on-a-chip (SoC), the system-on-a-programmable-chip (SoPC), the programmable system-on-a-chip (PSoC), and the application-specific instruction set processor (ASIP) are different combinations or variants of the above-mentioned three types. For example, a SoC is in fact a combination of an ISAP and an ASIC, while a SoPC is a combination of a programmable logic device and an ISAP and the ASIP is the customized ISAP for some specific fields. Though these cryptographic processors have inherited the advantages of ASICs, ISAPs, and the programmable logic devices, they have their inherent disadvantages. Therefore, their energy efficiency and functional flexibility are still not good enough and cannot be improved much in the future. Reconfigurable cryptographic processors are customized for cryptographic operations. Its functions can be dynamically reconfigured after silicon implementation. It is backward compatible with cryptographic algorithms. Its functional flexibility can meet the requirements for cryptographic algorithms, while its energy and area efficiencies are maintained with moderate flexibility. Our research results have shown that the energy and area efficiencies of reconfigurable cryptographic processors can reach 1–3 orders of magnitude or even higher than those of the ISAP and the programmable logic device on condition that the flexibility demand of cryptographic algorithms is met. Why and how we obtained such results will be described and analyzed in detail in the book.

This book consists of seven chapters: Chap. 1 describes the state-of-the-art researches on cryptographic processors, analyzes the advantages and disadvantages of traditional ASIC and ISAP cryptographic processors in terms of performance, power consumption, flexibility, and security, introduces the reconfigurable computing concept and the cutting-edge researches on reconfigurable cryptographic processors. In Chap. 2, the current mainstream cryptographic algorithms are presented using the reconfigurable computing architecture as the implementation hardware platform. The extraction of common logics of cipher algorithms and the features of data types, and the analysis of the parallelism of algorithms are described. This chapter also preliminarily discusses the hardware architecture design based on the implementation of cipher algorithms. Chapter 3 analyzes the hardware architecture design of reconfigurable cryptographic processors in two aspects: the datapath and the controller. The design methods of hardware architecture for cryptographic algorithms are also proposed. Chapter 4 introduces the compilation process of reconfigurable computing processors, discusses the special optimization methods for cryptographic algorithms, and demonstrates with instances of the compilation of specific cryptographic algorithms. Chapter 5 describes a reconfigurable cryptographic processor chip designed by our team. It includes the basic architecture, the key technologies, the integrated development tool, and the comparisons with other state-of-the-art designs. Chapter 6 describes several novel physical attack countermeasures for reconfigurable cryptographic processors, including countermeasures against physical attacks using random reconfiguration and the computing resources in the reconfigurable array. Compared with the applying of traditional countermeasures to a reconfigurable architecture, these new countermeasures based on the reconfigurable features can reduce the performance loss, area consumption, and power consumption caused by the security improvement through resource reuse and it is possible that it can resist new attacks in the future. Chapter 7 discusses the development trends of the reconfigurable cryptographic processor technology and focuses on exploring the hardware Trojan and the fully homomorphic encryption.

This book has embodied the collective wisdom of the reconfigurable cryptographic processor research team at the Institute of Microelectronics, Tsinghua University, for the past seven to eight years. We are very grateful to our colleagues and some students such as Bo Wang, Jianfeng Zhu, Hai Huang, Neng Zhang, Ao Li, Zhouquan Zhou, Dongxing Wang, Chenchen Deng, and Hanning Wang for their contribution. We really appreciate the great support and guidance from Prof. Shaojun Wei. We are very grateful to academician Jiren Cai, a famous expert in the information security field of China. He has read the book and written a preface though he was very busy at that time. Finally, I would like to thank my wife and children for their understanding. It is almost impossible for me to complete the work without their support, and they will be an important force that drives me forward and makes me continue to work hard in the future!

Beijing, China Leibo Liu

Contents

Contents

Chapter 1
Introduction

As a carrier to implement cryptographic algorithms, a cryptographic processor plays an important part in information security applications. With the development of network information technologies and integrated circuit technologies, the requirements for the cryptographic processors are no longer limited to pure computing performance. To support as many cryptographic algorithms and execution modes in the protocols as possible, a cryptographic processor should be flexible enough. To make a balance between performance and power consumption, energy efficiency (performance per watt) becomes a more reasonable metric compared with performance. To fight against increasingly intensive cipher-based physical attacks, security has gone beyond traditional metrics and become the most important one in cryptographic processors. Traditional cryptographic processors including application-specific integrated circuits (ASIC) and instruction set architecture processors (ISAP) cannot make a balance between the three metrics—flexibility, energy efficiency, and security. ASIC lacks flexibility, and ISAP has poor energy efficiency. Also, neither ASIC nor ISAP can meet the stringent requirements for a cryptographic processor in terms of security. ISAP uses the instruction-based software implementation mode. Because the generic form of instructions facilitates the modeling of physical attacks (e.g., establishment of the power consumption model in power consumption attacks). As a result, the ability of ISAP in physical attack resistance is generally 1–2 orders of magnitude lower than that implemented through hardware and is inapplicable to scenarios where high security is required. For ASIC, the security issue is mainly in leakage of algorithm information. In many application scenarios, the cipher algorithm itself needs to be secret. Typical examples are non-public algorithms in special fields such as military and aerospace. In case that the chip is lost when ASIC is used, the attacker may probably crack the secret cipher algorithm which ASIC uses by dissembling the chip through reverse engineering. In addition, in the context of global division of labor in the integrated circuit industry, the taping out process of chips is usually outsourced to foundries. Therefore, the secret cipher algorithm which ASIC uses may be probably leaked by untrusted foundries who may reversely parse the design file. As a new

L. Liu et al., *Reconfigurable Cryptographic Processor*,
https://doi.org/10.1007/978-981-10-8899-5_1

implementation method, a reconfigurable cryptographic processor can not only keep high energy efficiency while flexibly supporting various cryptographic algorithms, but also has its unique advantages in terms of security. On the one hand, a reconfigurable cryptographic processor is driven by a configuration context instead of instructions and thus has a much higher physical attack resistance than the software implementation of ISAP. On the other hand, a reconfigurable cryptographic processor has a basically similar array structure, and thus dissembling the chip after it is powered off will not cause leakage of the algorithm information. For this reason, a reconfigurable cryptographic processor is considered to have the feature of a blank chip. In a word, a reconfigurable cryptographic processor can reasonably balance the requirements for flexibility, energy efficiency, and security, and it represents an important and promising development trend in the field of cryptographic processor.

1.1 Information Security and Cryptographic Processor

With the rapid development and wide application of information technologies, great changes take place in the fields of information transmission, information storage, and information exchange. Information can be communicated, obtained, and utilized in much more ways. This greatly facilitates social development and people's life. However, as information technologies play a more and more important role in such fields as national economy, politics, diplomacy, national defense, and social management, information security has also become an important factor which affects such areas as personal privacy, property security, social finance and communication security, national political security, and defense security.

In March 2013, the largest hacker attack in history took place in South Korea. In this hacker attack, the computers in the mainstream banks and TV stations in South Korea were destroyed and thus could not provide services, and a large amount of data were stolen, which greatly influenced the social stability of South Korea. In January 2015, hackers claiming to be working on behalf of Islamic State militants intruded and seized control of the Twitter account and YouTube account of the US Central Command as long as 1 h. They replaced the logo of the US Central Command with "I love you ISIS." This event had negative influence on the national security and image of the USA. In April 2015, a hacker group named Cyber Caliphate intruded the television broadcast transmission channel of a French television station because of their unsatisfaction with French President Hollande's participation in the international counterterrorism action. This intrusion brought a large amount of network attacks to TV5 of the French TV station. In September 2016, the "iCloud leaks" event took place. In this event, the hackers stole photographs uploaded by users and a great amount of user privacy data were exposed on the Internet. This brought great threat to people's privacy security. Information security has become an important basis for normal and stable operation of today's society with abundant information. If anything goes wrong with information security, it will greatly influence individuals, groups, and nations. As such, America developed *International Strategy for Cyberspace* and *Strategy for Operating in Cyberspace in 2011*, and other countries including Russia, the UK, France, Germany, Canada, Australia, Japan, and Korea also developed their network security plans and built their network warfare forces to cope with various security pressures and challenges brought by information networking. China also established the Central Leading Group for Internet Security and Informatization in 2014.

Specifically, information security indicates that information and information systems are protected from being accessed, used, leaked, interfered, modified, and damaged without authorization, with the goal of ensuring the confidentiality, integrity, and availability of information [1]. Figure 1.1 shows different layers of information security. As basic carriers of information include devices, software, and communication, the underlying information security is to ensure the security of devices and software as well as the security of the communication process, so as to realize the security of personal information and then the security of national and social information. The cryptographic algorithm is the basis of realizing the information security target, because it is a necessary condition for ensuring the confidentiality and integrity of information.

(1) Information confidentiality indicates limiting the right to access and disclose information. The symmetric cryptographic algorithm is usually used to implement information confidentiality. For information containing only a small amount of data, the public-key cipher is also often used for information confidentiality.
(2) Information integrity indicates that the information is protected from being improperly modified or damaged, and it is ensured to be undeniable and true. Information integrity is mainly implemented through hash functions and public key ciphers.

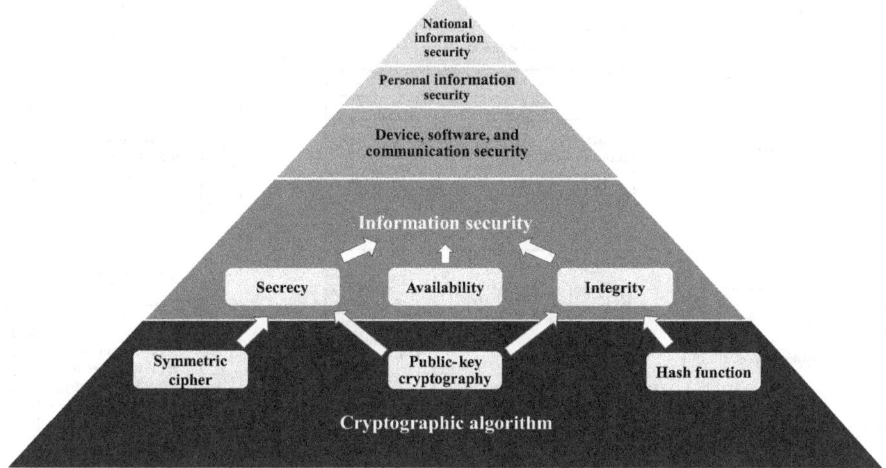

Fig. 1.1 Different layers of information security

Hash functions are often used to generate message verification codes and ensure that the information is not damaged. In addition, they can also be used to implement message authentication codes (MAC) and ensure that the messages have not been modified without any authorization. A public-key cipher is usually used to construct digital signatures and ensure that the information is undeniable.

Cryptographic algorithms are implemented using cryptographic processors, which act as the infrastructure of information security. It has been widely used in many different fields such as IoT, Internet finance security, secure communication, and transportation. It has received wide attention and has been deeply researched in academia and industry.

For example, the IoT system and architecture are usually divided into the perception layer, the network layer, and the application layer (Fig. 1.2), and information security issues exist in each layer. The perception layer is a feature of the IoT which distinguishes the IoT from the Internet. It collects data and monitors the system state through sensors. It mainly uses radio frequency identification (RFID) to communicate with the network layer. During this communication process, the communication may suffer such attacks as interference, shielding, theft, and impersonation and cause information security issues if sufficient security measures are not adopted. As such, any IoT design which places a high priority on security will add a cipher chip to the perception layer to encrypt/decrypt the data of the devices in the perception layer and provide an integrity checking mechanism to

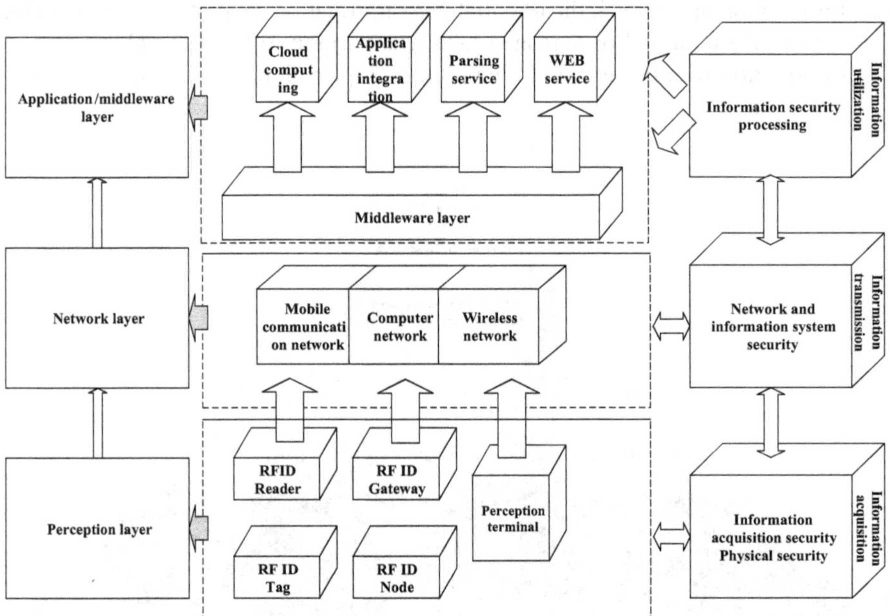

Fig. 1.2 Three-layer architecture of Internet of Things (IoT) [2]

ensure information security. Take a pacemaker as an example, many security vulnerabilities exist in the pacemaker and its peripheral systems installed in the human body according to the research result of the leading provider of WhiteScope, a security company. This may pose serious threats to the patients, and the security of the cipher chip and the system software need to be greatly improved. As such, the secure communication of the perception layer poses higher and higher requirements for the cryptographic processor. The network layer and application layer of the IoT are similar to those of the Internet. The existing information security and cryptographic processor technologies available in the TCP/IP network, wireless network, and mobile communication network can also be utilized in IoT. To sum up, cryptographic processors are hardware basis for the information security on each layer of the IoT.

Another example can be found in the financial industry. With increasingly higher requirements being raised for information security, China has stopped issuing traditional magnetic stripe bank cards since 2015, which are replaced with chip cards with a higher security level. The cryptographic processor for the financial transaction card supports various cryptographic algorithms. As shown in Fig. 1.3, the cryptographic processor for the financial transaction card is integrated with

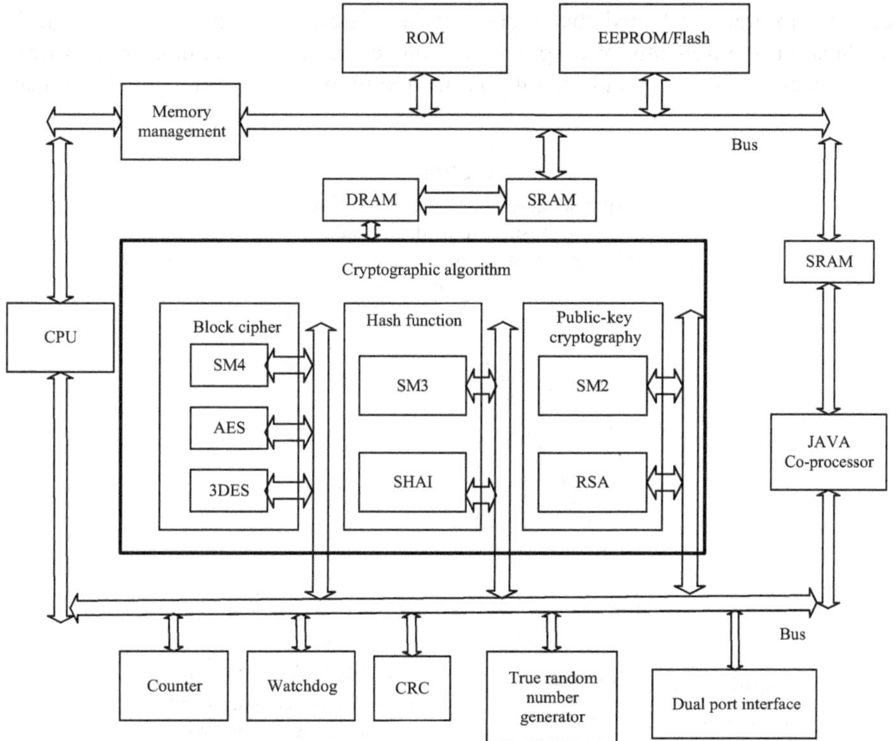

Fig. 1.3 Cryptographic algorithms available in the cryptographic processors of financial transaction cards [3]

block cipher algorithms SM4, AES, 3DES, hash functions SM3 and SHA1, and public-key algorithms SM2 and RSA. A cryptographic processor integrating many cryptographic algorithms can effectively ensure the security of financial transaction cards.

Cryptographic processors mentioned in this book refer to various integrated circuits that are used for implementing cipher applications. Customized optimizations are required to meet the requirements of cipher applications. To be specific, cryptographic processors can be divided into ASIC, ISAP, and reconfigurable cryptographic processor. What should be noted is that the ISAP mentioned in this document indicates all processors based on instruction set, including general-purpose processor (GPP), digital signal processor (DSP), graphic processing unit (GPU), and application-specific instruction-set processor (ASIP).

1.2 Challenges of Cryptographic Processor Application Requirements

As the information security application develops increasingly, cryptographic processors face unprecedented challenges in terms of design. As shown in Fig. 1.4, the application requirements of cryptographic processors can be summarized as flexibility, energy efficiency, and security. As the requirements are mutually affected and

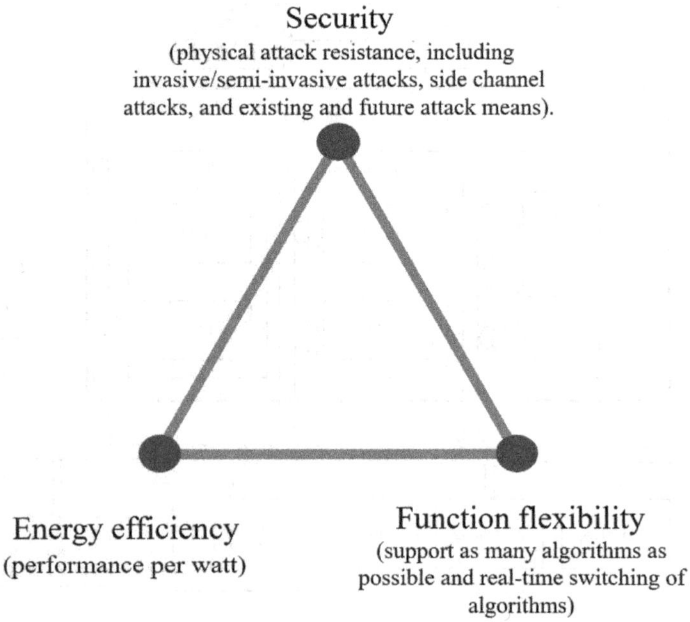

Fig. 1.4 Application requirements challenges of cryptographic processor

restricted, you need to balance the relationship between these demands reasonably when designing a cryptographic processor.

1. Flexibility

A cryptographic processor is mainly used to process various cryptographic algorithms. Cryptographic algorithms can be divided into symmetric cryptographic algorithm, public-key cryptographic algorithm (also called asymmetric cryptographic algorithm), and hash function. A symmetric cipher is mainly used in encryption of large data blocks or data streams, and it can be further divided into block cipher and stream cipher. A public-key cipher is mainly used in sharing of digital signatures and keys. Hash function is mainly used in data integrity check and message authentication. From the aspect of algorithms and standards, cryptographic algorithms and standards will be updated continuously as higher and higher requirements are raised for data security by computing systems. At present, there are at least dozens of symmetric cryptographic algorithms that have become standards, including AES, 3DES, Camellia, three categories of public-key cryptographic algorithms, including large number factorization (e.g., RSA), elliptic curve discrete logarithm (e.g., ECC), and finite field discrete logarithm (e.g., ElGamal); and there have been at least dozens of hash functions, including MD5, SHA-2, and SHA-3. In addition, each algorithm has many variants. For example, the symmetric cryptographic algorithm AES can be further divided into AES-128, AES-192, and AES-256 according to the key length, and the elliptic curve encryption algorithm in public-key cryptographic algorithm has many variants according to the selected number field and curve; the hash function can be further divided into SHA-3-224, SHA-3-256, SHA-3-384, and SHA-3-512 [1, 4] according to the digest length. What's more, there are many algorithm-independent execution modes can be selected for each cipher to meet different security demands in various applications. For example, each block cipher can run at least five modes [4], including electronic code book (ECB), cipher block chaining (CBC), and counter (CTR). All these factors can be combined to form a very large cryptographic algorithm space. From the aspect of application and protocol, there may be as many as hundreds of cryptographic algorithm used in a common information security solution. For example, the IP-layer security protocol IPsec supports at least 15 cryptographic algorithms, including symmetric cryptographic algorithms AES, DES, 3DES, Blowfish, CAST, IDEA, RC4, RC6, public-key cryptographic algorithms Diffie–Hellman and ECDH, and hash functions MD5, SHA-1, SHA-2, SHA-3, and SM3; the transport-layer security protocol SSL supports dozens of cryptographic algorithms [5, 6], including Fortezza, RC2-40, and DSS. What should be noted is that the specific cryptographic algorithm used by these security protocols in a scenario may vary according to the session negotiation result.

What have been discussed are only the cryptographic algorithms which have become standard cryptographic algorithms and are usually used in the business field. Actually, such special departments as National Defense and Military, Aerospace, and Energy and Electric Power impose even higher requirements (nearly harsh requirements) for the security of cryptographic algorithms. These special departments usually design special cryptographic algorithms according to their own demands. For example, they can construct sequence-specific cryptographic algorithms [7] which are different from standard algorithms and are in various forms by changing the number of stages and tap position of the linear feedback shift register (LFSR). There are various these specific and non-public cryptographic algorithms which are updated and upgraded much rapid and need quick switch at a higher frequency during usage.

According to the analysis above, you can see that there are various public cryptographic algorithms and the algorithm standards are also updated constantly. Old standards will expire, and new ones will be established subsequently. Cryptographic algorithms in security protocols are increasing constantly in number and are changing constantly. Existing algorithm may also be attacked and lose effect, and securer new algorithms will be proposed subsequently. Cryptographic algorithms used in special departments are quite different from general cryptographic algorithms in terms of variety, number, modification and upgrade frequency, and usage (e.g., frequency of dynamic function switching). All these application requirements pose great challenges to cryptographic processors in terms of function and flexibility.

2. Energy Efficiency (Performance per Watt)

Within a certain range, the security strength of a cryptographic algorithm is directly related to the amount of computation required. Adopting a long key and increasing the amount of computation can improve the security strength of a cryptographic algorithm to a certain extent. For example, 3DES has a higher security strength than DES, and the amount of computation is twice that of DES; RSA whose key is 2048 bits long has a security strength much higher than that of RSA whose key is 1024 bits long and the amount of computation is 4–8 times [8] that of RSA whose key is 1024 bits long. Cryptographic algorithms based on different principles are also quite different from each other in terms of the amount of computation. For example, the amount of computation of a public-key cipher is much greater than that of a block cipher. The amount of computation between them can differ by orders of magnitude. As constant increase of security strength, complexity, and computation amount of the cryptographic algorithm, many new security applications impose higher and higher requirements for cryptographic processors in terms of computation speed. For example, the throughput of a high-speed security network has improved from 1–10 to 100 Gb/s or even higher [9, 10]. A cryptographic processor in such a security network must meet such requirements in terms of speed. In addition to performance, another metric to evaluate a chip is power consumption. Power consumption is another big problem for cryptographic processors. For example, for a cryptographic processor used in a high-performance security server, excessive power consumption means more than energy waste. The main problem is that the subsequent increase of temperature will cause decrease of overall reliability, or even functional error [11], to the system; for a mobile cryptographic device which uses batteries, power consumption is essential for the endurance of the mobile cryptographic device. For example, affected by high computation complexity of cipher applications, it will take 75% of the battery energy [12] of the handheld device (e.g., for a typical laptop connected to a wireless network, the battery equipped in it can work for 3–5 h), if the Blowfish algorithm whose key is 32 bits long is used to encrypt 0.0136 MB data.

Performance and power consumption are always two core metrics in the chip design. Like cryptographic processors, other types of processors are also evaluated by these two metrics. The only difference is that they have different emphases due to different application fields. It is important to note that performance improvement usually means increase of power consumption. These two metrics are usually conflicting. Therefore, it is unnecessary to go into details about the importance of these two metrics and the challenge they pose to chip design. It is more appropriate to use energy efficiency (performance per watt) to give a comprehensive evaluation to these two core metrics of processor chips [13]. We take energy efficiency as an important metric for comprehensively evaluating the performance and power consumption of a cryptographic processor.

3. Security

For cipher applications, security is always the core design element. For a cryptographic processor, security mainly refers to physical attack resistance. As a new attack method emerging at the end of the twentieth century, physical attacks are more threatening than traditional math attacks targeting cryptographic algorithms and protocols [14]. Targeting the implementation of cryptographic algorithms, physical attack steels secret information such as encryption keys from power consumption, electromagnetic radiation, sound, fault propagation pattern leaked from the circuit. Physical attacks can be divided into invasive attacks, semi-invasive attacks, and non-invasive attacks according to the way the attacker acts on the circuits [15].

Invasive attack needs to damage the package and remove the passivation layer of the chip, and obtains the circuit function or the key information by imaging the circuit structure using an optical microscope, or direct electrical contact with the circuit by using such tools as a microprobe workstation and a focused ion beam workstation [16]. Invasive attack is a powerful attack, and the attacker can even implement attacks by directly modifying the physical wiring of the circuit. However, it requires a long time and a high cost, and may cause permanent and irreversible damage to the target of attack. Such attacks as reverse engineering and microprobing are invasive attacks. Reverse engineering falls into two categories according to its purpose. One is to obtain the structure and function of the circuit and the other is to obtain the secret information in the memory. In the first category of reverse engineering, the attacker images the circuit structure and recreates the layout by using a device like a confocal microscopy. To separate different physical layers, the attacker needs to conduct a series of reverse process for the chip, for example, material removal based on corrosion or polishing. After obtaining the pattern of each physical layer, the attacker can obtain the structure and function of the attacked chip through combination and analysis. In this way, the design information of the circuit is obtained. The second category of reverse engineering mainly targets mask read only memories of some certain categories. After removing the metal layer on the top or conducting a selective etching, the attacker can directly observe the data information in the memory by using optics method. Microprobing makes direct electrical contact with the circuit by using a microprobe workstation. The attacker can get access to sensitive data by establishing connection with the bus or other data channels, or interact with the circuit by injecting test signals and observe the feedback.

Compared to invasive attack, semi-invasive attack also needs to remove the package of the chip. However, it does not need to further damage the passivation layer of the chip. That is, it does not need to establish direct electrical contact with the circuit, and thus will not cause mechanical damage to the circuit [17]. Both backside imaging and fault attacks (FA) based on optical injection are semi-invasive attacks. With the improvement of precision of modern laser, FA based on optical injection has become a very threatening attack. By using the photoelectric effect generated due to acting of laser on the circuit to generate specific fault in specified execution step of the algorithm, and analyzing the information like the fault propagation pattern in the cipher text, the attack can crack the key of the algorithm with only few faulty cipher text. Figure 1.5 shows injection of laser faults into a platform [18]. The attacked chip is fixed to a micropositioner, and the laser probe will firstly conduct large-scale point-by-point scanning for the chip to determine the effective area of attack, and then will continuously inject faults into the specific attack area to obtain enough required faults.

The non-invasive attack does not damage the original circuit or its package. It usually makes an attack by analyzing such information as power consumption and electromagnetic radiation which is leaked during chip operation. Therefore, it takes much lower cost than the other two attacks. The non-invasive attack is more secret

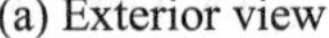

(a) Exterior view

(b) Interior view

Fig. 1.5 Laser fault injection platform [18]

because the chip owner cannot determine whether the chip has been attacked. The non-invasive attack mainly involves the information leaked from the side channel of the chip, and thus it is also called side-channel attacks (SCA).[1]

The earliest non-invasive attack starts with power attacks [19]. It is discovered that the power consumption generated during chip running is closely related to the data and operation. If power consumption traces recorded during operation are analyzed mathematically and matched with the power consumption model established by the attacker, information on the secret key might be obtained by removing the effect from the noise (including electronic noise and noise by irrelevant operations) in the signals. Experiments show that such an attack can easily crack an electronic chip without any protection measures. In addition, power attacks are also of low cost, because it mainly relies on such simple devices as an oscilloscope and a differential probe. Figures 1.6 shows the result of a simple power attack (SPA). In such an attack, the attacker can directly obtain the information about the algorithm based on the shape of the power consumption curve. Figure 1.6a shows the power consumption traces during running of the DES algorithm on an intelligent chip, with which you can clearly distinguish the 16 rounds of operations of the algorithm. In Fig. 1.6b, the power traces corresponding to the second and third rounds of operations in Fig. 1.6a are enlarged, and more details about the power consumption curve are given. For example, in the key register of DES, there is one rotation (the left arrow) during the second round of operation, and two rotations (the right arrow) during the third round of operation. These slight differences between different rounds of operation are caused by the conditional jumps based on secret keys or intermediate data, and will greatly increase the risk of simple power attacks. Compared to simple power attacks, differential power attack (DPA) is more threatening. It can make an attack in case where there is no obvious difference for the trace shape (e.g., the base noise is high or some countermeasures have been taken) through statistical analysis of large number of power traces. Figure 1.7 shows the results of a group of differential power attacks targeting the AES algorithm input in the look-up table. From the enlarged view on the right, you can see that when the guessed key is the correct key, the correlation coefficient between the power consumption model and the actually measured power consumption curve is much higher than other scenarios where the guessed key is incorrect. This means that the attacker can successfully crack the secret key of the algorithm. The electromagnetic attack (EMA) makes an attack by using the electromagnetic radiation signals which are generated during circuit running and are collected by an electromagnetic probe [20]. Except that it uses a different way to collect signals, the EMA uses an attack method similar to that of the power attack. Compared to power attack, the ambient noise of the data collected in an electromagnetic attack is much higher. The spatial precision of the electromagnetic attack, however, is much higher. That is, you can use a smaller electromagnetic probe to collect the signals in

[1]In some documents, SCA is more strictly limited to the range of timing attack, power attack, and electromagnetic attack. In general, SCA is a vague definition.

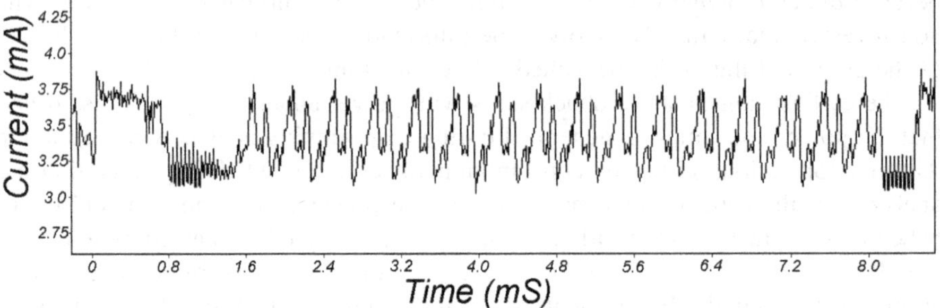

(a) Power traces of the 16 rounds of operation of the algorithm

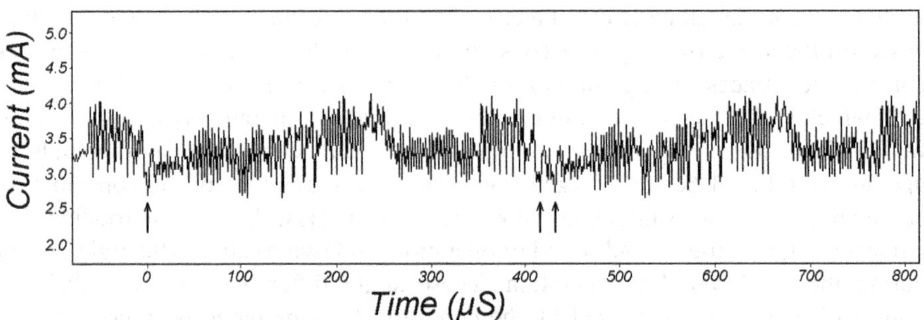

(b) Enlarged view of the curve of the second and third rounds of operation

Fig. 1.6 Result of the simple power attack targeting the DES algorithm [19]

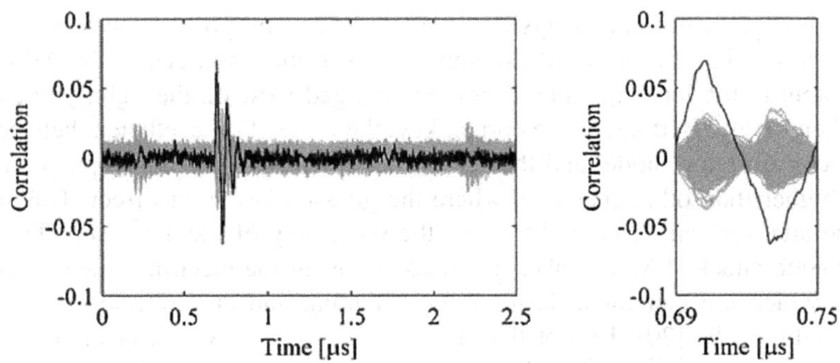

Fig. 1.7 Result of the differential power attack targeting the AES algorithm [14]

a specific area of the chip. Figure 1.8 shows the basic equipment to implement an electromagnetic attack. The core devices are an electromagnetic probe and an oscilloscope [21].

In addition to such traditional non-invasive attacks as power attack and electromagnetic attack, many new attacks have emerged in recent years. The attack

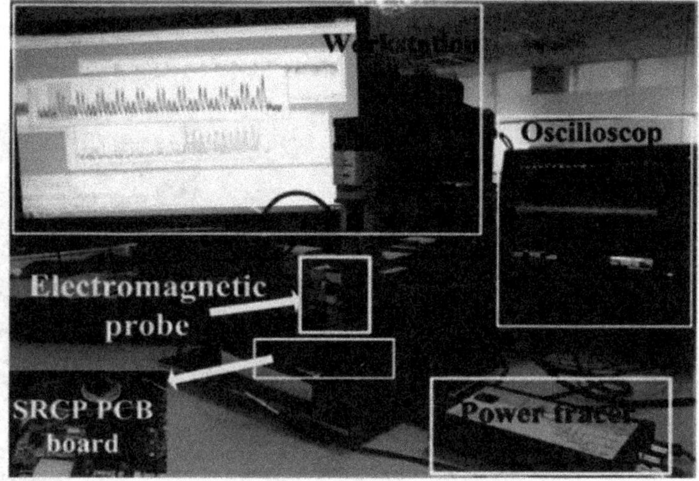

Fig. 1.8 Basic equipment for electromagnetic attack [21]

based on sound signals is one of these new attacks. This attack cracks the key by collecting the low-frequency acoustic wave signals leaked during the processor operation. Figure 1.9 shows an example of the sound attack, in which the attacker can complete the whole attack process just by using a simple sound receiving device to collect signals beyond several meters [22]. In addition, there are many interesting attack methods targeting specific cases. The attack shown in Fig. 1.10 is one example [23]. The attacker can obtain the key information just by touching the laptop with his bare hand for several seconds. This attack utilizes the side-channel

Fig. 1.9 An example of the sound attack [22]

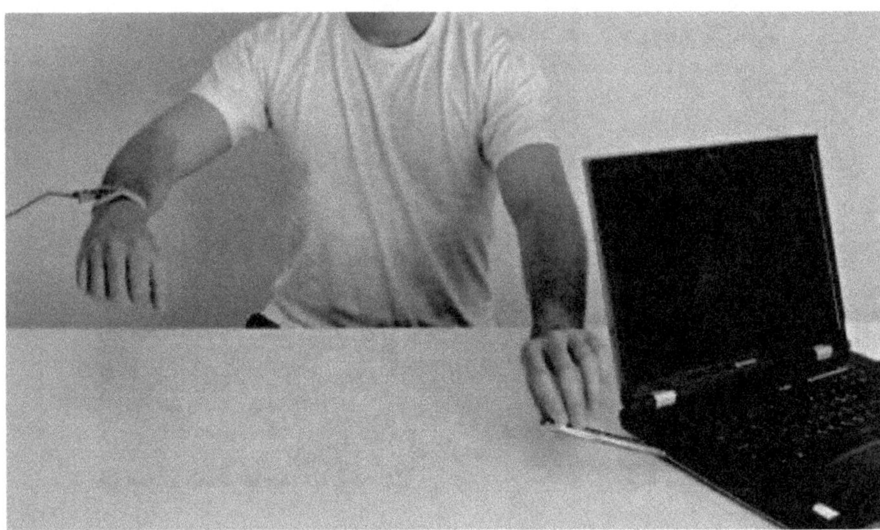

Fig. 1.10 New attack based on the earth potential [23]

information—the fluctuation of earth potential. The metal shell of a laptop is the ideal earth potential by default. During the actual computation, however, the earth potential can generate fluctuations related to data and operation. As a result, the signals are sent out when the human body touches the shell, and an attack is made unconsciously.

Under threat of various physical attacks, many countermeasures are developed to protect cryptoprocessors against them. For invasive attacks which need to damage the integrity of the chip, people cover many layers of metal shield and sensors on the top of the chip, thus making it more difficult for the attacker to establish direct electrical contact with the internal circuit or to conduct indirect optical detection [24]. For fault attacks, traditional countermeasures mainly resist them by means of redundancy and comparison. For example, time redundancy or hardware redundancy is added by repeatedly executing the circuit for many times or duplication [25]. The results of redundant computation are compared. If the comparison result is different, the output is null or a random value to avoid a fault attack. When there are more than two redundant paths, the circuit can not only detect errors, but also realize fault tolerance by voting. In addition to such traditional countermeasures against fault attacks as error detection and fault tolerance, infection countermeasures are also proposed [26]. Instead of preventing the attacker from obtaining the faulty ciphertexts, the fault propagation pattern is scrambled in a random way so that the attacker cannot retrieve the secret key even though faulty ciphertexts are obtained. For power attacks and electromagnetic attacks, the main purpose of the countermeasures is to reduce the signal-to-noise ratio of the attacked target, that is, reduce the ratio of the power consumption or electromagnetic radiation generated by the attacked operations related to the key in the total power consumption or electromagnetic radiation. Hiding and masking are two commonly used methods to reduce the signal-to-noise ratio. Hiding is to make the measured power consumption or electromagnetic radiation and data independent of operation. This can be implemented in two modes: randomize the power consumption or electromagnetic radiation generated by circuit in each clock period, or keep the power consumption or electromagnetic radiation of each clock period constant. The first mode can be implemented by randomly inserting dummy operations or shuffling of operations [27]. The second mode can be implemented by modifying low-level circuit structure to keep the power consumption and electromagnetic radiation generated during operation relatively constant such as dual-rail precharge (DRP) [28]. Masking is adding (e.g., exclusive OR operation and multiplication) a random value to the plaintext. In this way, the intermediate values that occur during the computation are masked, and power consumption traces are independent of ciphertexts [29]. The theoretical principle of masking is secret sharing. That is, the value of the original intermediate data is determined by both the masked value and the mask itself. Therefore, any information about the original intermediate data cannot be recovered if only the masked value or the mask itself is known. Similar to secret sharing, masking can also use multiple independent random numbers to protect the intermediate data. This method is also called higher-order masking.

However, although numerous countermeasures against physical attacks have been proposed and implemented, physical attack resistance is still a key element for the design of cryptographic processors for two main reasons.

(1) All countermeasures will inevitably introduce overheads in performance, area, and power consumption. Therefore, it is still essential to consider how to reduce the overheads of the countermeasures against physical attack while maintaining certain security level, and how to make a balance between the security and the overheads. For example, when a fault attack is resisted, the time redundancy measures based on repeated computation will inevitably cause a performance degradation, and the hardware redundancy measures based on duplication will inevitably cause an area waste. For countermeasure against power attacks and electromagnetic attacks, inserting random delay will cause an obvious increase to the total computation time. For random masking, it is required to continuously track the masked data to finally restore the data for the correct output without any mask. This will obviously increase the delay and circuit scale. When the higher-order masking is used, such overhead is especially obvious. Circuit-level method has similar overheads. For example, dual-rail precharge keeps the power consumption constant and independent of with the data. However, it tends to maintain the power consumption in each period to be maximum, which will obviously increase the power consumption.
(2) As the attackers continuously update the attack measures, many traditional countermeasures are no longer effective. Meanwhile, many new attack methods are emerging. This makes it necessary to constantly develop new countermeasures against physical attacks. In recent years, such new physical attacks as multiple fault attacks and sound/electromagnetic-based low-frequency attacks

have posed serious security threat to cryptographic processors. Multiple fault attacks will probably affect the effectiveness of the current countermeasures which are based on redundancy computation [30]. When same faults are injected into the main processing path and redundant path, the original comparison operation will become ineffective. As a result, the attacker can still obtain the faulty ciphertexts of the fault and make an attack. In addition, the precision of optical injection, such as laser injection, becomes higher and higher. Meanwhile, a series of biased fault attacks aiming at raising the fault collision probability, such as differential fault intensity analysis (DFIA), are proposed. It has become possible to generate two or more identical faults. This means that multiple fault attacks have become real threat but not just in theory [31, 32]. Recently, a low-frequency physical attack method which is based on sound or electromagnetic has emerged [11, 33]. This attack method can make attacks by using the specific MHz or even KHz signals leaked by the attack target even it works at GHz level. This means that the previous conclusion that a high-frequency device is more secure in terms of physical attack resistance becomes no longer valid. This also exposes the cryptographic processors which adopt only lightweight countermeasures with an aim to improve the processing performance to serious threats.

1.3 Traditional Cryptographic Processors

Traditional cryptographic processors can be divided into two types: ASIC and ISAP. These two different architectures will be presented and analyzed in terms of capability of handling application challenges.

1.3.1 ASIC Cryptographic Processors

ASIC is the integrated circuits which are designed and manufactured according to the requirements of specific users and demands of specific electronic systems. Compared to general-purpose ICs, ASIC has such advantages as small area, low power consumption, high reliability, high performance, and low cost. The research topics of ASIC-based cryptographic processors mainly focus on high-speed and high-throughput design and low power consumption and high area efficiency design. State-of-the-art ASIC-based cryptographic processors with different design requirements are presented and analyzed in terms of physical attack resistance.

1. High-Speed and High-Throughput Design

The key technologies for high-speed and high-throughput design include pipeline and retiming [34–36]. Lin and Huang propose a high-throughput AES architecture supporting 128-bit, 192-bit, and 256-bit keys, and four modes of operation including ECB, CBC, CTR, and counter with CBC message authentication code (CCM) [34]. The overall architecture of the proposed AES cipher is shown in Fig. 1.11a. It mainly consists of IO interface, first in, first out (FIFO), and AES core. The datapath of the AES core is designed with two-stage pipeline, as shown in Fig. 1.11b. To fit the timing of the datapath, the key is also generated using pipeline. This architecture can process two separate data streams on a single datapath. In CCM mode, this architecture can process two different data streams in parallel because it only needs the encryption function in CCM mode. This improves the throughput effectively. In addition, XORs and multiplexers are optimized using the retiming technology to further improve the critical path. The AES cipher is implemented using 0.13 μm CMOS process with a maximum frequency of 333 MHz and a maximum throughput of 4.27 Gb/s.

Ueno et al. present a new round-function-based pipeline architecture for the CBC encryption of AES [35]. Figure 1.12a shows the overall framework of the proposed architecture, which consists of round function and key scheduling. The function framework of round function is shown in Fig. 1.12b. Different function modules of the round function adopt the unification technique, so that affine transformation and linear mapping (isomorphism and constant multiplication) can use the same architecture, and only one 128-bit 4-to-1 multiplexer is required for the whole round function (multiple multiplexers are usually required for a similar architecture). This effectively reduces the false critical paths. In addition, this architecture also adopts the technologies of operation-reordering and register-retiming. As a result, the inversion operations of encryption and decryption can share the same architecture without extra delay overheads. Reordering of internal function modules of the round function and retiming of the register during encryption and decryption are shown in Fig. 1.12c, d. For encryption, the affine transformation in the ShiftRows and substitution box (S-Box)[2] is exchanged, and then the affine transformation, MixColumns, and AddRoundKey are merged together. For decryption, the inversion and InvShiftRows in the S-Box are exchanged, so that the inversion transformation in the S-Box is at the beginning of the round. This architecture is implemented with 65, 45, and 15 nm CMOS processes. Compared to other works, a higher throughput per unit area (53–72% higher than other architectures), a lower critical path delay (because of fewer series gates in critical paths), and a smaller area are achieved.

Liu et al. develop a dual-field ECC processor with high performance and high flexibility, which provides a maximum key length of 576 bits [36]. By initializing curve parameters and instruction codes stored in ROM, the proposed processor can perform arbitrary ECC operations over dual-field, various elliptic curve scalar multiplication (ECSM) algorithms (such as binary method and Montgomery ladder algorithm), and various ECC standards such as FIPS 186-2, IEEE P1363, and ANSI

---

[2]Substitution box is abbreviated to S-Box in this book althougth it has different abbrevaions in previous literatures.

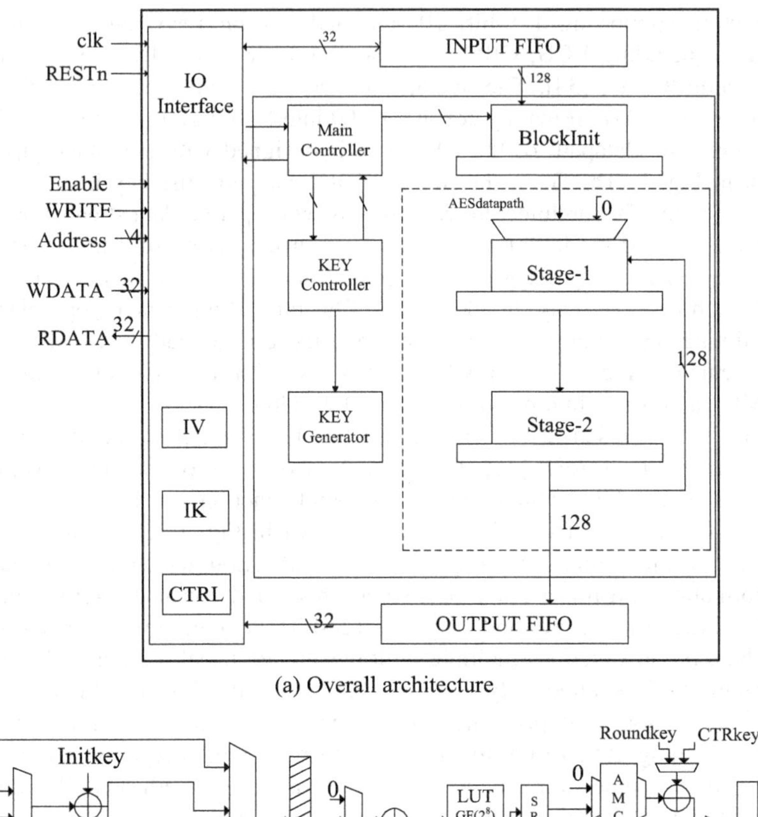

(a) Overall architecture

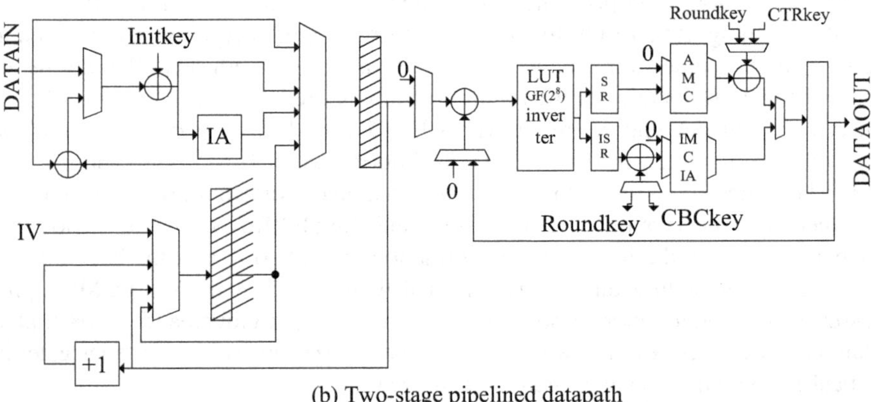

(b) Two-stage pipelined datapath

Fig. 1.11 AES architecture and its datapaths [34]

X9.62. As shown in Fig. 1.13a, the proposed ECC processor consists of ECC controller, modular arithmetic logic unit (MALU), ROM, register file, and advanced high-performance bus (AHB) interface. To achieve high flexibility, the MALU is integrated with various modular operations, including modular adder and subtractor (MAS), modular multiplication, and modular division, as shown in Fig. 1.13b. In order to reduce the delay in the datapath, carry save adder (CSA) and carry

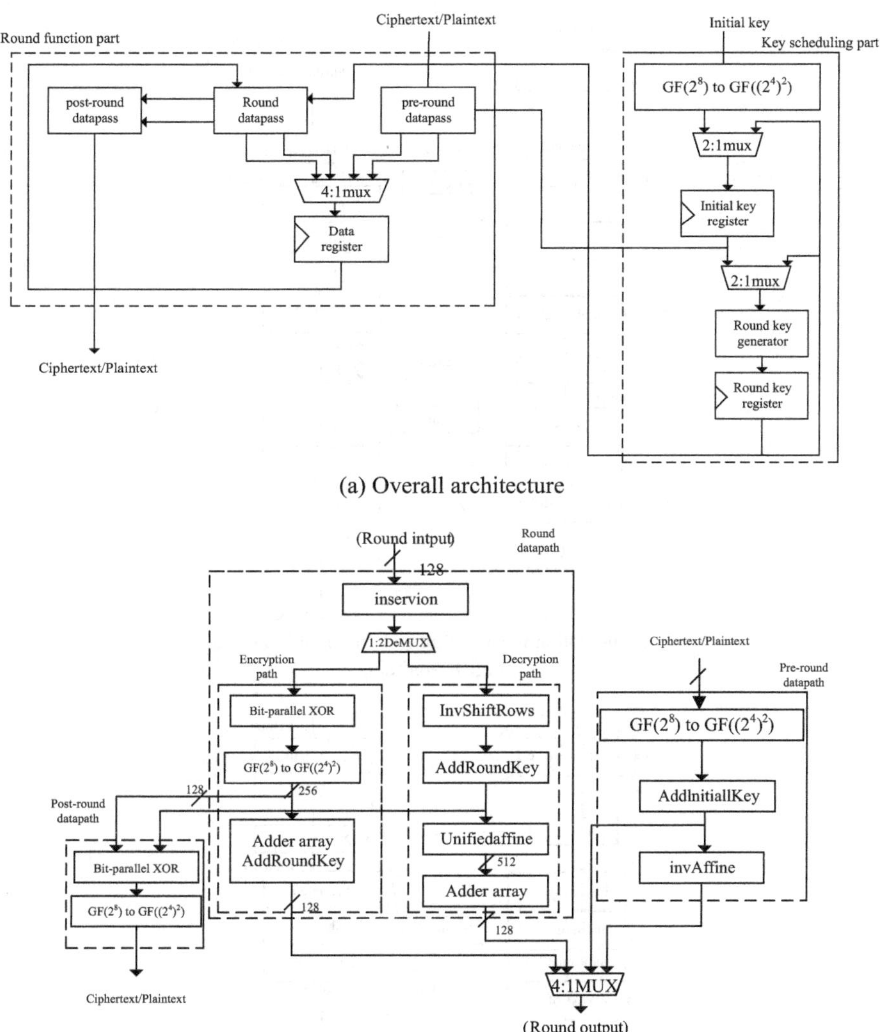

(a) Overall architecture

(b) Functional diagram of round function

Fig. 1.12 AES architecture and its hardware implementation [35]

propagation adder (CPA) are used to implement the radix-4 interleaved multiplication, modular doubling, and modular quadrupling. The hardware utilization of this processor can be improved by reusing some units of the MALU. For example, CSA2 and CPA2 can be used in MAS as well as modular division including modular halving and modular quartering operations; CSA0 can be used in both B $\pm$ A and modular addition over GF(2^m). The proposed processor is implemented in XMC 55 nm CMOS process with an equivalent gate of 189 K. It takes 0.60 ms to

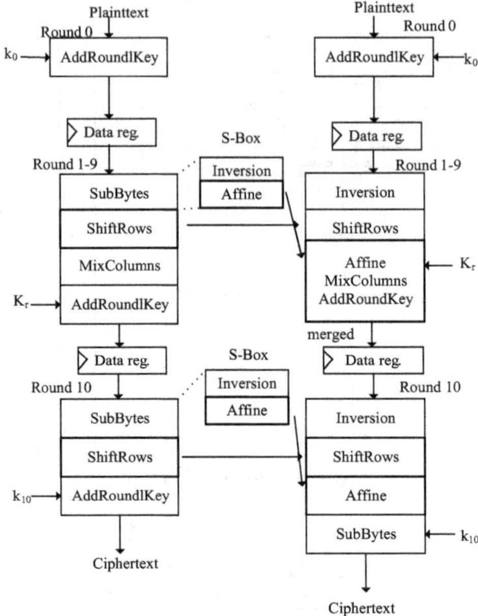

(c) Encryption flows before and after reordering and register-retiming

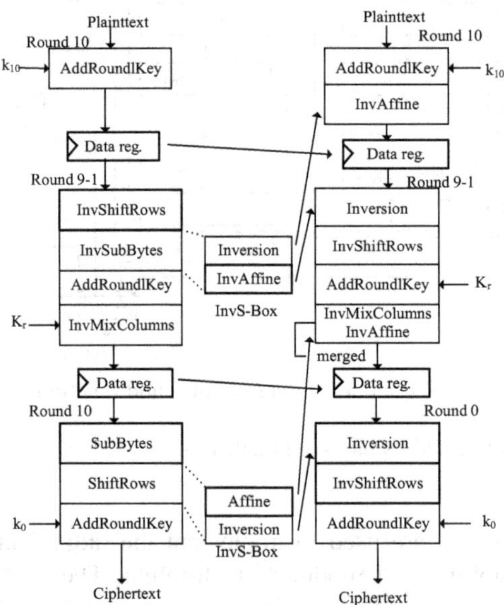

(d) Decryption flows before and after reordering and register retiming

Fig. 1.12 (continued)

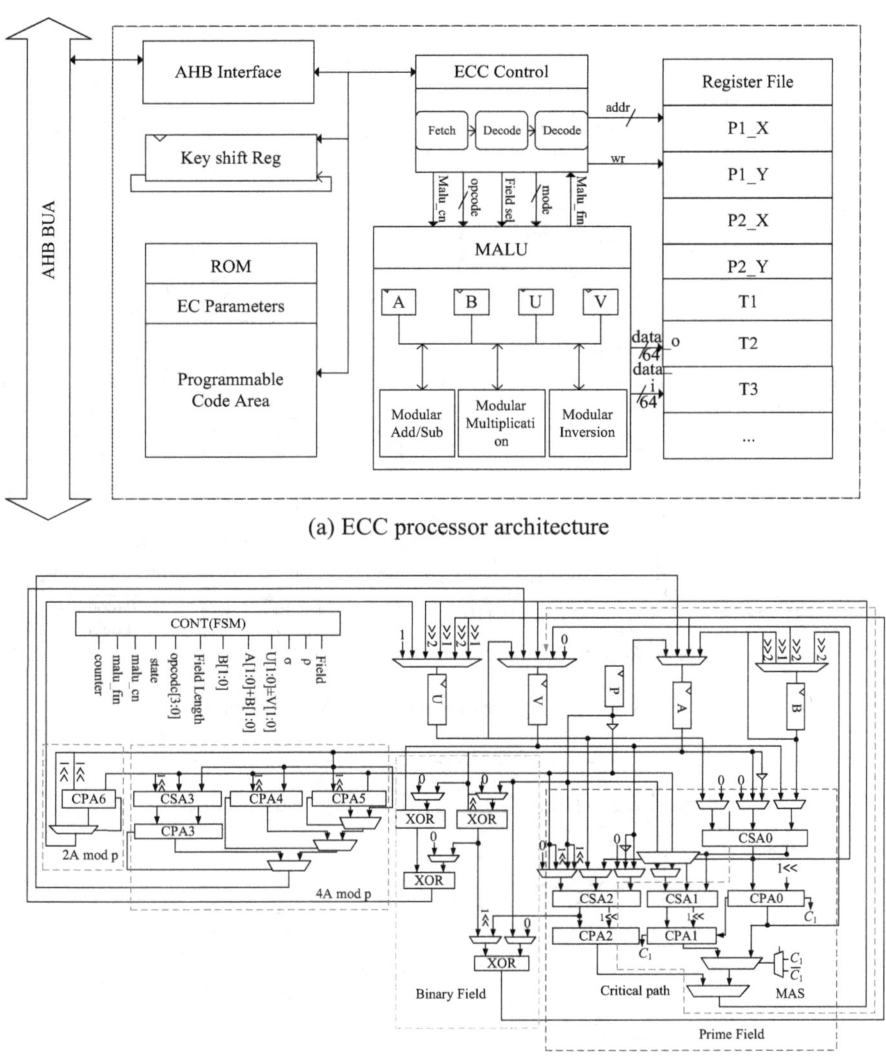

(a) ECC processor architecture

(b) MALU architecture

Fig. 1.13 ECC and its MALU architecture [36]

carry out a 163-bit ECC algorithm and 6.75 ms to perform a 571-bit ECC algorithm. When the ECC was implemented on Xilinx Virtex-4 FPGA, a time duration of 7.29 ms for 192 bits ECC and 49.6 ms for 521 bits ECC was required.

2. Low Power Consumption and High Area Efficiency Design

The key technologies for low power consumption and high area efficiency design include function module multiplexing and circuit-level design with low power consumption and low area overhead [37–39]. Zhang et al. design an AES

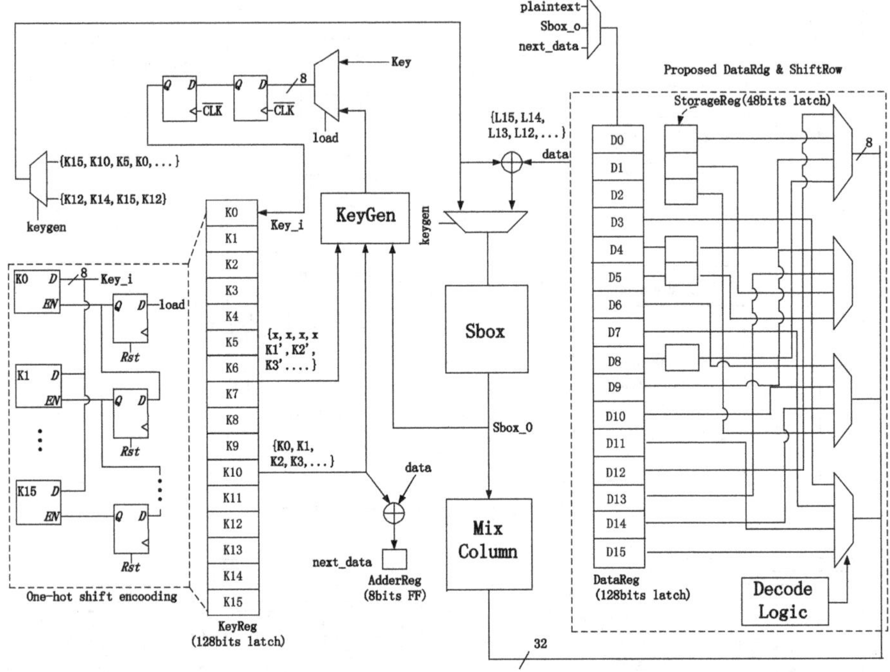

(a) Overall architecture of the AES with 8-bit datapaths

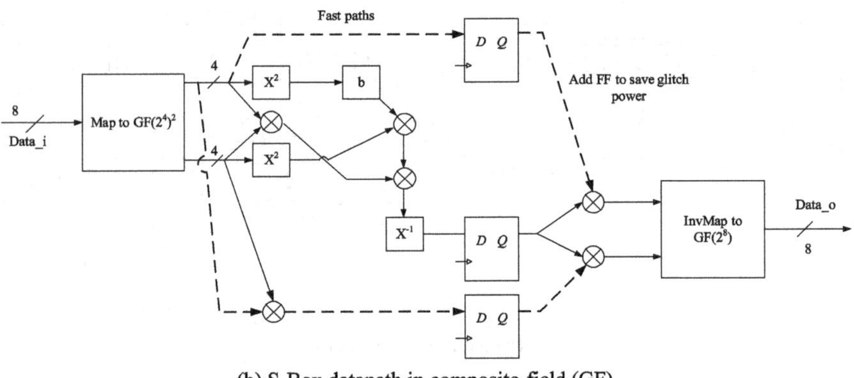

(b) S-Box datapath in composite-field (GF)

Fig. 1.14 AES architecture and S-box datapaths in composite field [37]

hardware accelerator with high energy efficiency [37]. Compared to conventional designs, the proposed architecture reduces the area by 25% and power consumption by 69%. This is achieved by eliminating ShiftRows stage in the round function, and using the retiming technology to replace the flip-flops in data and key storage with latches. The 8-bit datapath of the proposed AES architecture is shown in Fig. 1.14a. The design eliminates the ShiftRows in the AES architecture by directly loading the plaintext to the latches in the ShiftRows byte-order. The computation of S-Box is carried out in native $GF(2^4)^2$ composite field. As shown in Fig. 1.14b, a two-stage pipeline and glitch reduction technologies are used. The glitch reduction technique equalizes path delays by retiming the S-Box and adding flip-flops to the path. This architecture is implemented in 40 nm CMOS process with only 2228 equivalent gates. The energy efficiency is 446 Gb/s/W, and the throughput is 46.2 Mb/s 0.47 V.

Mathew et al. present a lightweight 8-bit nanoAES accelerator for a mobile system on chip (SoC) with an ultra-low power consumption [38]. The datapath of the proposed 8-bit nano AES architecture is shown in Fig. 1.15a. The ShiftRows operation in nanoAES is moved to the beginning of the round operation, and shift is implemented through a serial scan-chain. NanoAES uses only one 8-bit S-Box and adopts 4-bit logic for basic operations such as addition, squaring, and inverse operations. The S-box circuit is shown in Fig. 1.15b. To reduce the delay of the critical path, the mapping transform is moved from the critical AES loop with all operations are completed in $GF(2^4)^2$ field, and the polynomial optimization technology is used. The circuit area is reduced by 18%, and the critical path delay is reduced by 12%. As only one S-Box is available, nanoAES can process only one byte of data in each cycle. An 8-bit serial-accumulating MixColumns circuit is designed as shown in Fig. 1.15c. This circuit supports a maximum of 32-bit serial MixColumns operation. This architecture is implemented with 22 nm tri-gate/high-k/metal-gate CMOS process with total die area of 0.19 mm^2, among which 2200 μm^2 is occupied by the 1947-gate encryption accelerator and 2736 μm^2 is occupied by the 2090-gate decryption accelerator. Experimental results show that the accelerator functions well over a supply voltage range of 340 mV–1.1 V, and the peak energy efficiency is 289 Gb/s/W.

Henzen et al. design an area-efficient hardware architecture for the BLAKE algorithm (one of the SHA-3 second-round candidates), as shown in Fig. 1.16a [39]. To reduce the area, the round function G is implemented by ten iterations of a 32-bit adder. The module used for computation of the function G consists of two 32-bit exclusive OR operations, one rotation selector, and one adder (item ② in Fig. 1.16a). The selected status words are sorted and used in the computation of the chain value h (item ③ in Fig. 1.16a). The values stored in the intermediate register can come from a new chain value or from the computation result of ① in Fig. 1.16a. In addition, semi-custom 4 × 32-bit memories based on clock-gated latch arrays are designed, as shown in Fig. 1.16b. This memory is used to store salt value, and the proposed architecture works with five memories. Compared to a flip-flop-based memory, the overall memory reduction achieves 34%. The architecture is implemented with UMC 1P/6 M 0.18 μm process, and the total area of BLAKE-32 is 0.127 mm^2.

Table 1.1 compares the performance of different cryptographic processor architectures and lists the algorithms, key technologies, tape-out processes, operating frequencies, throughputs, power consumptions, and the number of equivalent gates supported by different processors. The number of equivalent gates indicates the circuit area evaluated based on the number of 2-input NAND equivalent gates, and frequency indicates the maximum operating frequency obtained from the critical path delay. Throughput is the throughput at the maximum operating frequency. Results given in [37, 40, 41] are obtained by synthesizing with Synopsys tool in the NanGate 15 nm standard unit library based on the same optimization conditions, and other data are from the original manuscripts.

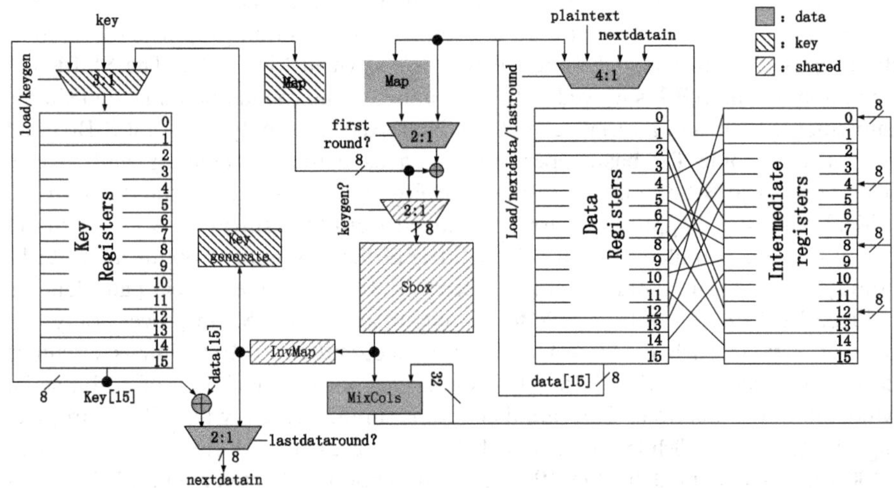

(a) 8-bit nanoAES datapath

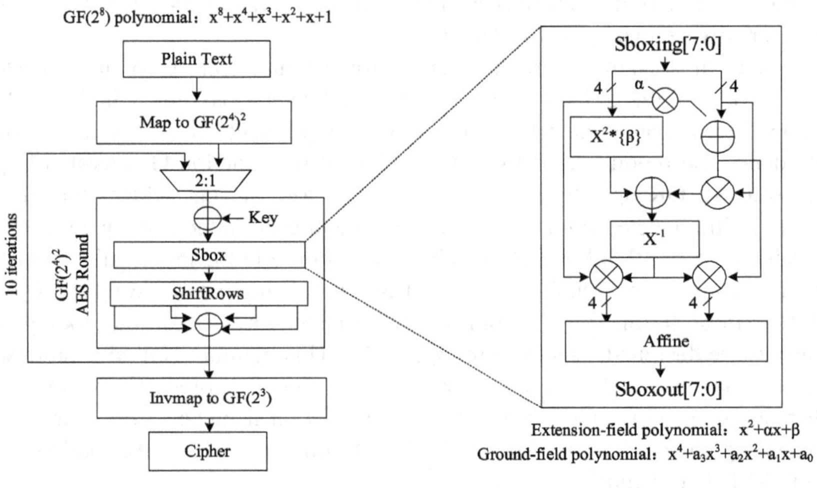

(b) S-Box circuit

Fig. 1.15 8-bit nanoAES architecture [38]

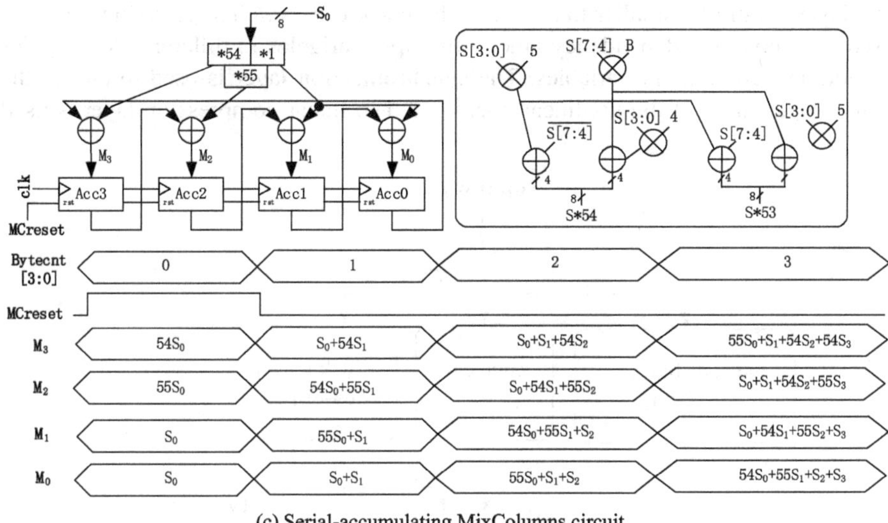

(c) Serial-accumulating MixColumns circuit

Fig. 1.15 (continued)

3. Physical Attack Resistance of ASIC Cryptographic Processors

As the emergence of physical attacks, the emphasis on the security of cryptographic algorithm is not only limited to math but also hardware implementation. Currently, leading chip assessment institutions regard physical attack resistance as a major metric to measure the security of chips. For example, among 31 security inspection standards for smart cards chips proposed by MasterCard, there are at least eight rules involving physical attacks. Next, physical attack resistance of ASIC cryptographic processors will be presented by taking some common physical attacks such as power analysis attack, electromagnetic attack, and fault attack.

Reparaz et al. present a countermeasure against power attack based on consolidated masking schemes [51]. This technique inherits all features of the threshold implementation (TI) [52] and uses the remasking technology in the ISW [53] to eliminate the dependency among multiple variables in different clock cycles. Therefore, this scheme can solve not only the glitch attack problem which is difficult to solve for traditional masking, but also the problem of mutual information attack and collision attack which is difficult to solve for the TI scheme. Figure 1.17 is the schematic diagram of the second-order CMS with 2-input AND gates. There are five and three shares in the left and right figures, respectively. The CMS scheme can be divided into five layers: nonlinear layer (N), linear layer (L), refreshing layer (R), synchronization layer (S), and linear compression layer (C). The nonlinear layer contains all the shared terms of the two input AND gates to ensure the correctness of the masking algorithm and all shared terms should be distributed uniformly. The linear layer is used to preserve the non-completeness of the TI algorithm and the number of shares which are used in the same XOR operation on this layer should be smaller than that of the mask orders d. The refreshing layer is used to remove the dependency among multiple variables in different clock cycles by adding new random variables. The synchronization layer is used to ensure the non-completeness of the nonlinear operation. The linear compression layer is used

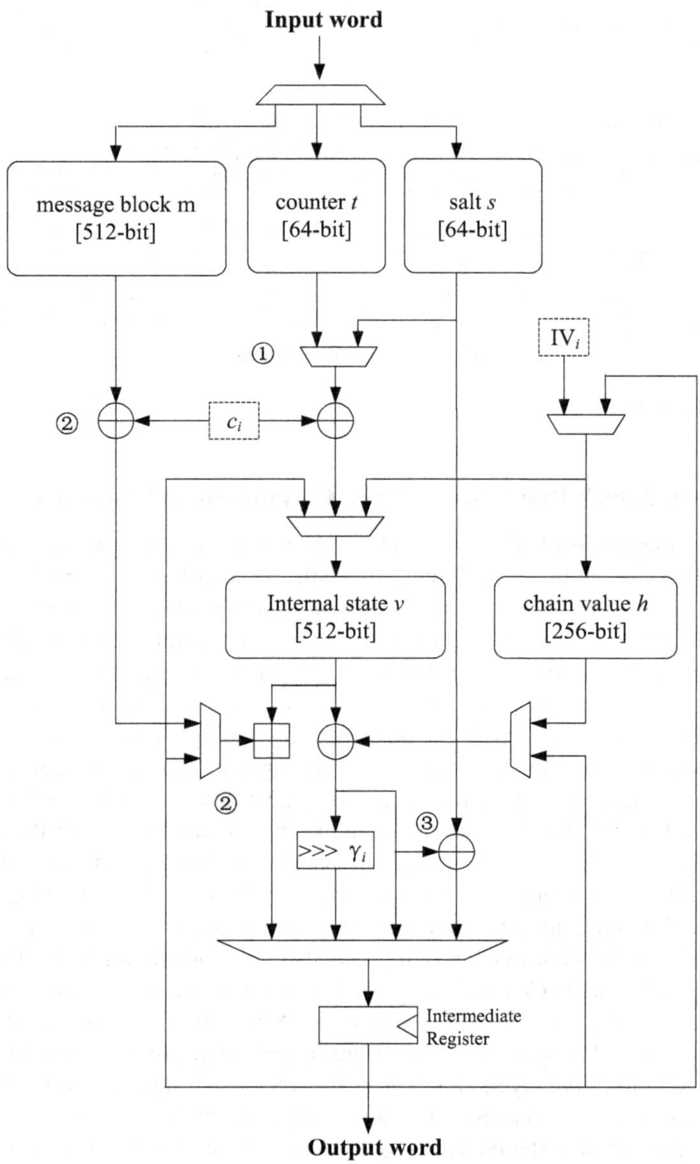

(a) BLAKE-32 hardware architecture

Fig. 1.16 BLAKE-32 hardware architecture [39]

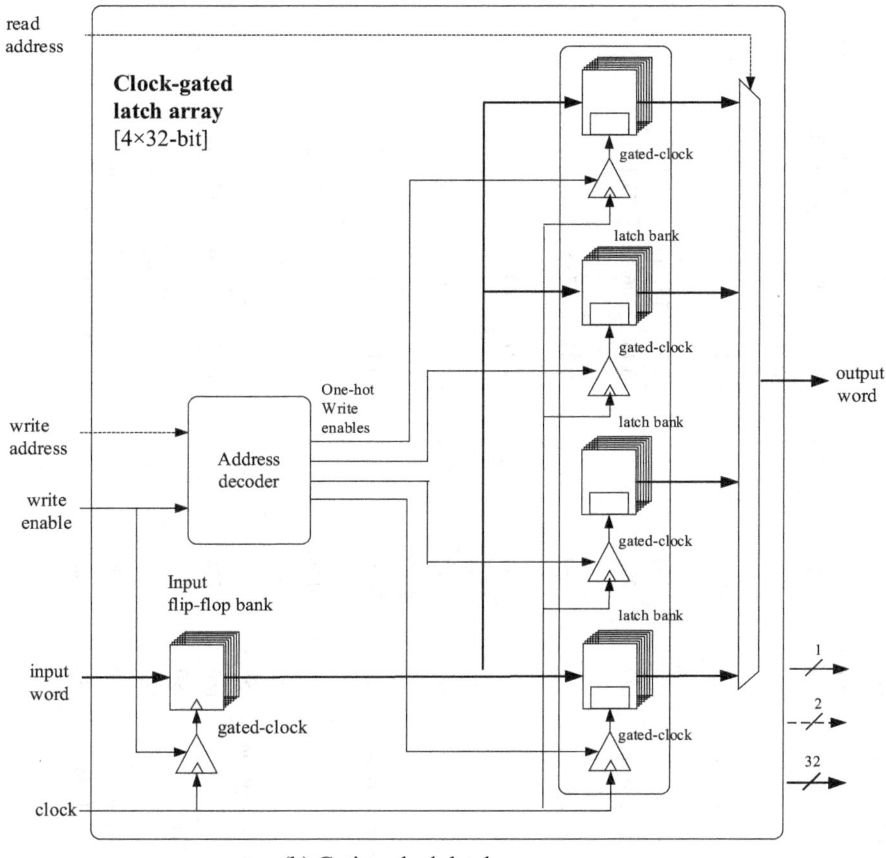

(b) Gating clock latch array

Fig. 1.16 (continued)

to reduce the number of output shares on the synchronization layer to be equal to those on the nonlinear layer. As proposed by De Cnudde et al., the CMS scheme is used in AES and uses $d + 1$ mask components [54]. The architecture of the second-order CMS mask of its S-Box is shown in Fig. 1.18. Compared to other masking techniques, the proposed method is of higher security but with smaller area. However, the amount of randomness increases and the complicated masking scheme for different algorithms requires custom design, which makes it difficult to implement masking.

In [55], a hiding countermeasure against power and electromagnetic attacks based on current equalizer is put forward. By inserting a three-state switching capacitor, this method enables the current of the chip after each power charging/discharging to be identical, instead of varying with the data during the actual circuit operation. As shown in Fig. 1.19, the power supply for the AES circuit can be divided into three stages. Firstly, connect the charging capacitor so that the power

Table 1.1 Performance comparison among different ASIC cryptographic processors

Design	Algorithm type	Key technologies	Process/ nm	Frequency/ MHz	Throughput/ Gbps	Power/ mW	Number of equivalent gates/kGE
Zhang [37], Liu [41]	AES	Tower field optimization	110	224	2.61	–	21.3
Lutz [40]	AES SERPENT	Reorganized round function that shares LUT	15	4800	61.44	–	25.692
Su [42]	AES	Four-stage pipeline	250	250	2.3	230.6	58.4
Hodjat [43]	AES	Separate data and control stream, hierarchy of control, block pipeline	180	295	3.43	86	73.2
Hamalainen [44]	AES	Parallel computing, module multiplexing	130	290	0.23	17.98	3.2
Lin [34]	AES	Two-stage pipeline	130	333	4.27	40.9	173
Liu [41]	AES	Module multiplexing	15	3014	35.07	–	15.758
Good [45]	AES	Shared SubBytes operator	130	12	3.24×10^{-5}	0.099	5.5
Mathew [46]	AES	(folded ShiftRows), (fused MixColumns)	45	2100	53	125	–
Mathew [38]	Lightweight AES	Module multiplexing, polynomial optimization	22	1100	0.42	13	1.947
Ueno [35]	AES	Retiming, two-stage pipeline	15	6118	71.19		17.232
Zhang [37]	AES	Operation-reordering, retiming	40	1300	0.48	4.39	2.228
Liu [36]	ECC	Interleaved multiplication, module multiplexing	55	–	–	–	189
Henzen [39]	SHA-2	Rescheduling technology	180	215	0.132	1.26	13.575
Lee [47]	ECC	Priority-oriented scheduling algorithm	90	238	–	–	313
Dao [48]	ECC	Clock gating	180	59	–	0.61	27.5
Guo [49]	SHA-3	Throughput-to-area ratio oriented design	130	200	1.51	5.18	21.67
Koo [50]	RSA	Radix-4 Montgomery's module multiplication algorithm	180	300	1.233	–	192

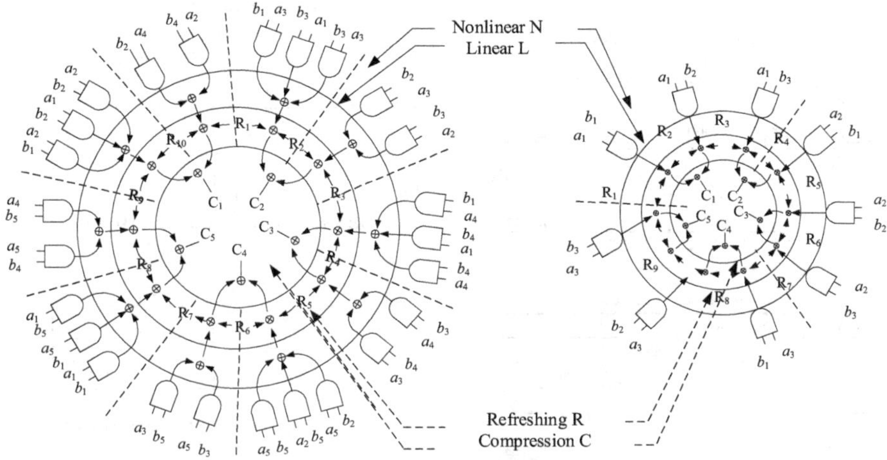

Fig. 1.17 Schematic diagram of the second-order CMS of two input AND gates [53]

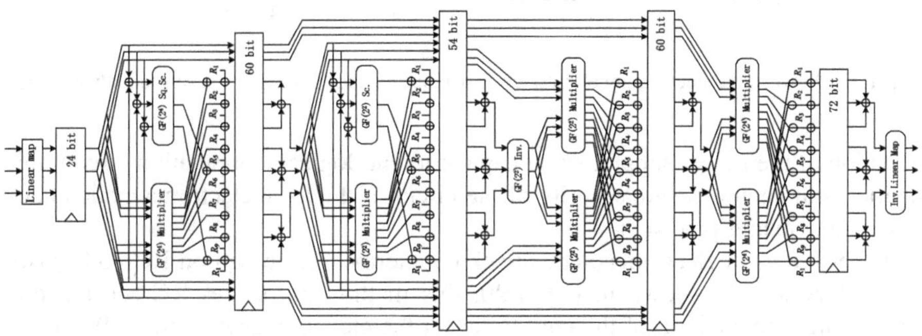

Fig. 1.18 Architecture of the second-order CMS mask of AES S-Box [53]

supply of the chip can charge the shunt capacitor through the charging capacitor; after the completion of charging, shut down the charging capacitor and start the capacitor connected with the AES circuit to supply power for the circuit; after the completion of charging, shut down the capacitor connected with the logical port and discharge the shunt capacitor to the specified value to prepare for the next charging. This method separates the power supply from the actual computation circuit by using a shunt capacitor, so that the capacitor directly supplies power for the encryption module. As the total power consumption of the capacitor during charging and discharging is fixed, this method ensures that the entire encryption module has a fixed power consumption during each encryption and will not leak any information. As a hiding countermeasure, this method, in addition to power attack resistance, can also provide certain resistance against the electromagnetic attacks from a large-size probe. As the probe can identify the electromagnetic

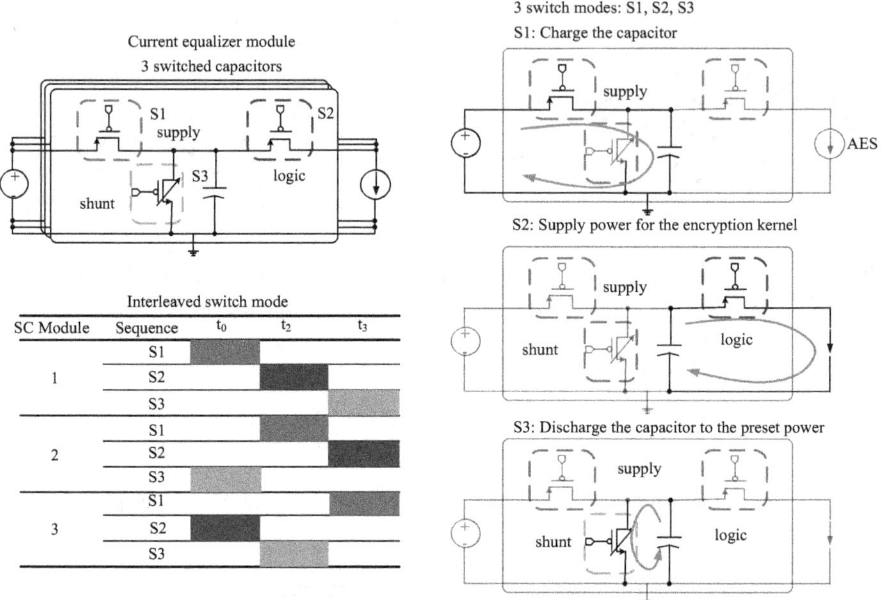

Fig. 1.19 Power and electromagnetic attack resistance method based on current equalizer [55]

radiation of the power supply, shunt capacitor, and logical computation circuit, this method still cannot ensure security of the circuit if the local electromagnetic attack has a high spatial precision.

In [56], a local electromagnetic attack countermeasure based on RC coil sensor is put forward. As shown in Fig. 1.20, this method introduces RC sensing coil around the active circuit of the chip to detect the electromagnetic probe. When the attacker places the electromagnetic probe near to the chip surface to make an electromagnetic attack, the induced current generated by the electromagnetic probe will generate interference to the original electromagnetic field of the chip. These interference signals will be captured by the RC coil and then cause an alarm or stop the ongoing encryption. Compared to traditional electromagnetic attack countermeasures, this method which is based on RC coil sensor will not be affected by the size and sensitivity of the electromagnetic probe, and thus can provide resistance against local electromagnetic attacks featuring precise sampling. This is implemented at the cost of the extra circuit area and power consumption introduced by the RC coil. As the power consumption of the RC module is much higher than a digital circuit module, it is required to shut down the RC coil during non-encryption period to reduce the power consumption. In addition to a high power consumption, this method has another limitation for the direction of the electromagnetic probe; that is, it can provide resistance for only electromagnetic attacks with probe parallel to the chip surface (in Fig. 1.20, the attack made when the plane of the probe coil is parallel to the plane of the chip). As an electromagnetic probe parallel to the chip

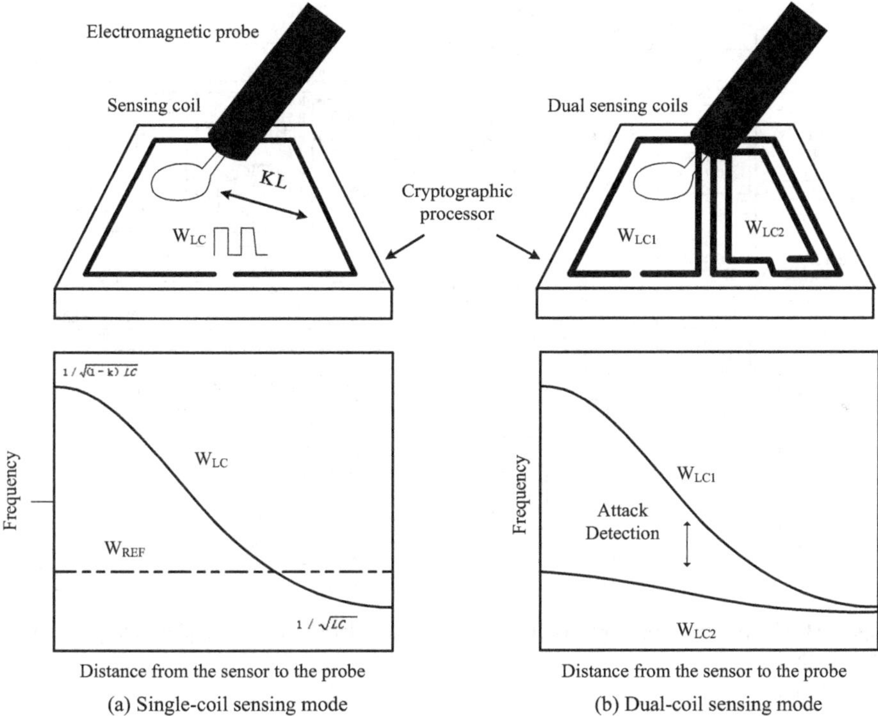

Fig. 1.20 Local electromagnetic attack countermeasure based on RC coil sensor [56]

plane samples mainly the electromagnetic field perpendicular to the chip surface, the interference to the original electromagnetic field from this electromagnetic probe is perpendicular to the chip surface and can be efficiently sampled by the RC coil. When the attacker adopts an electromagnetic probe which samples perpendicular to the chip surface, what the electromagnetic probe samples will be the electromagnetic information parallel to the chip surface. In this case, the interference to the original magnetic field from this electromagnetic probe concentrates only on the direction parallel to the chip surface and the generated interfering electromagnetic field is distributed parallel to the RC coil. Therefore, the RC coil can seldom sense the information from the interfering magnetic field and security vulnerability may exist.

In [57], a cooperative resistance method for fault attack and side-channel attack based on complementary redundant circuits is put forward. As shown in Fig. 1.21, adopt the hardware redundancy method for fault attack resistance in the round function module (RND_EXE) for normal encryption and the key expansion module (KEY_EXPANDER) for key generation respectively, that is, create two duplicates of the circuit and compare them to obtain the comparison result. When a difference appears in the result, that is, a fault is detected, the fault pattern will be spread to the cipher text, so that the attacker will be unable to get the cipher text output for fault

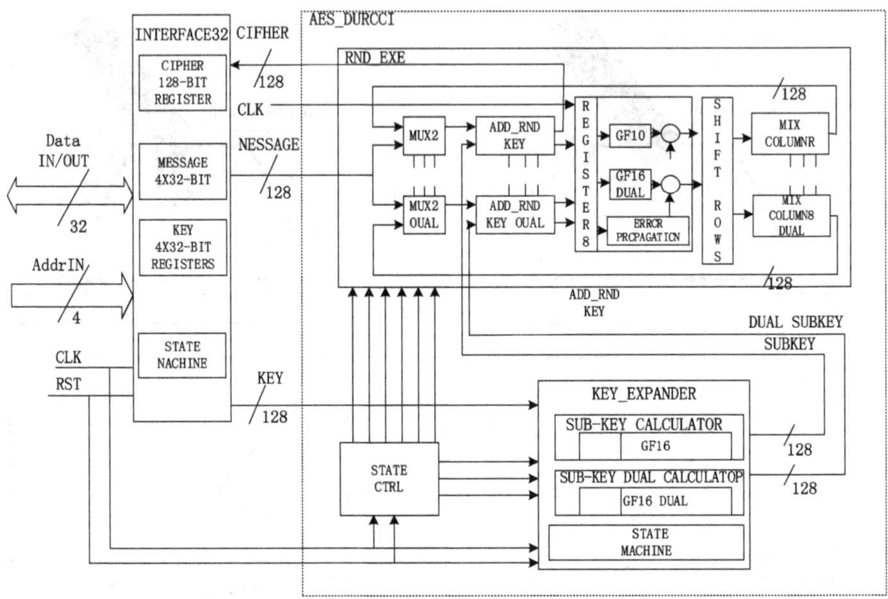

Fig. 1.21 Cooperative countermeasure for physical attacks based on complementary redundant circuits [57]

analysis. In addition to traditional redundancy and detection methods, this method also uses the complementary attack countermeasure for power attacks and electromagnetic attacks. By adopting complementary logic gates on the redundant path or main path, the power consumption of the whole circuit can be equalized. For example, for the XOR gate implementation on the main processing path, XNOR gate will be used for its corresponding implementation on the redundant path. This method combines the hardware redundancy fault attack countermeasure with the hiding power/electromagnetic attack countermeasure to reduce the overhead of cooperative attack resistance, but the limitation of the two original attack countermeasures still exists. On the one hand, as the complementary circuits exist separately, high-resolution electromagnetic attacks may separate the radiation of the original execution circuit from that of the complementary circuit; on the other hand, when the attacker injects faults into the two paths respectively by using double fault attacks, the attack countermeasure may become ineffective.

1.3.2 ISAP Cryptographic Processors

GPU, DSP, and GPP in ISAP are universal and are not specially designed for cryptographic algorithms. Therefore, this book will focus on application-specific instruction set processor, that is, ASIP cryptographic processor. On the premise that the commonality is kept, ASIP better matches the processing features of algorithms in a field, and thus can more efficiently implement the computation tasks in this field. As cryptographic algorithms have a large computation strength and a high computation complexity, and its processing process is usually simple to control, customized design can be conducted for the instructions, overall architecture, and function modules of ASIP according to the features of cipher processing to achieve faster data processing and lower processor power consumption. Next, we will introduce the research status and physical attack resistance of cryptographic processors of this kind by taking block ciphers, public-key ciphers, and hash functions as examples.

1. Architecture of ISAP Cryptographic Processors

In [58], a 32-bit reduced instruction set computer (RISC) extension set for the AES algorithm is put forward. Figure 1.22 shows the block diagram of their function unit (ISE FU). Input two 32-bit operands of the function unit and express them with op1 and op2. The operations of the function unit are configured by the

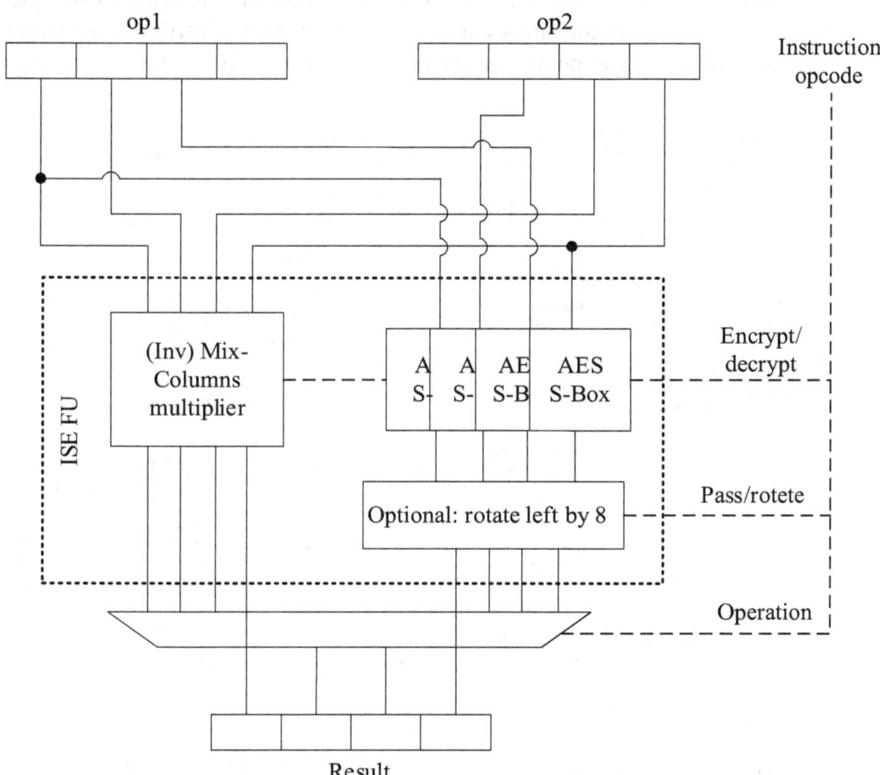

Fig. 1.22 Block diagram of the function unit of AES extended instruction set [58]

instruction opcode. The function unit can execute such operations as AES SubBytes, ShiftRows, MixColumns, and their inverse operations.

To improve the operation parallelism, this extended instruction set has designed Sbox4 and MixCol4 instructions which can operate four bytes at a time. The instruction architectures of these two extended instructions are shown in Fig. 1.23a, b. Sbox4 instruction can implement SubBytes for all the four bytes of the source register at a time and place them in the destination register, so as to conduct a selective rotation for the result. imm is used to select S-Box or inverse S-Box, and specify the rotation distance of the result. Thus, Sbox4 instruction can also give effective support for key expansion. MixCol4 instruction computes all the four result bytes of the MixColumns or inverse MixColumns operation according to the intermediate value. Sbox4 and MixCol4 extended instructions have a speed 4.86 times higher than pure software implementation, but its performance is not four times higher than the instructions which operate one byte (with a speed 1.74 times higher than that of pure software implementation) due to the bottleneck of the ShiftRows operation of the AES. To solve this problem, the author further improves the circuit and instructions, including the instructions Sbox4s, iSbox4s, Sbox4r, MixCol4s, and iMixCol4s shown in Fig. 1.23c, d. The improved instruction system uses two input registers. In this way, the instruction system can extract two bytes from both source registers and thus carry out implicit AES ShiftRows operation. Compared to software implementation, the improved instruction system has a speed

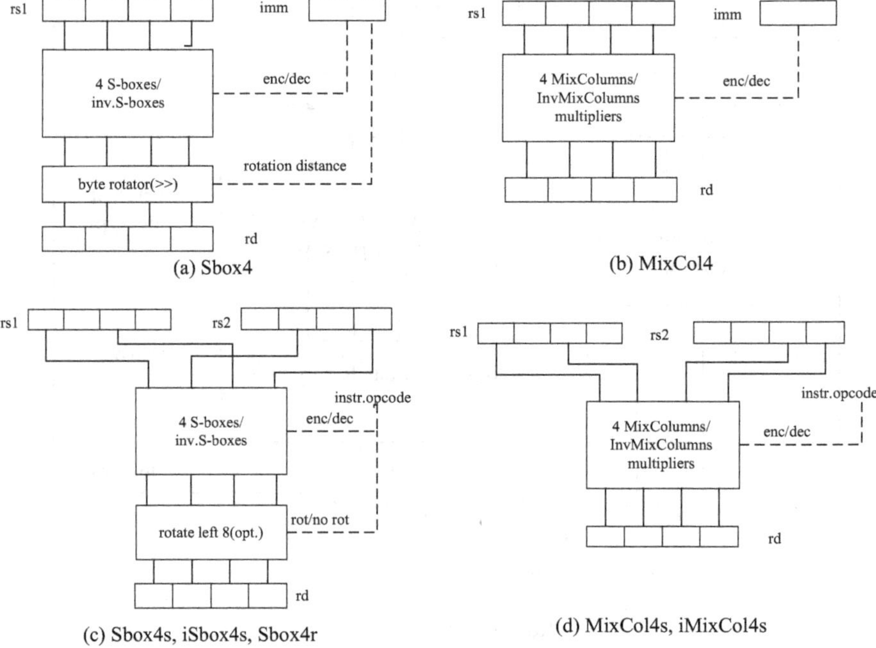

Fig. 1.23 Extended instructions [58]

7.47 times (or 8.35 times in case of loop unrolling) higher and a code quantity 81% lower than that of the former.

In [59], a hardware architecture of Koblitz curve ECC coprocessor based on 16-bit microcontroller (such as TI MSP430F241x or MSP430F261x) is put forward, as shown in Fig. 1.24. A coprocessor is composed of arithmetic logic unit (ALU), address-generation unit (ADDRESS), random access memory (RAM), and control unit (CU) consisting of hierarchical finite state machines. The ALU has 16 bits of datapaths and is composed of the circuit of a 16-bit integer adder/subtracter, a 16-bit binary multiplier, and two binary adders. The memory block is a single-port RAM shared by the ECC coprocessor and a 16-bit microcontroller. The adder circuit of the address block is an 8-bit adder which computes the physical address based on the read/write deviation and base address. The control unit is composed of a group of hierarchical FSMs and is used to generate the control signals required by the system. The author adopts the 130 nm CMOS process for hardware implementation of the architecture put forward. This architecture has an area of 4323 equivalent gates. It takes 98 ms to compute 283-bit point multiplication in case of a clock frequency of 16 MHz. The power consumption generated is 97.70 μW, and the corresponding energy is 9.56 μJ.

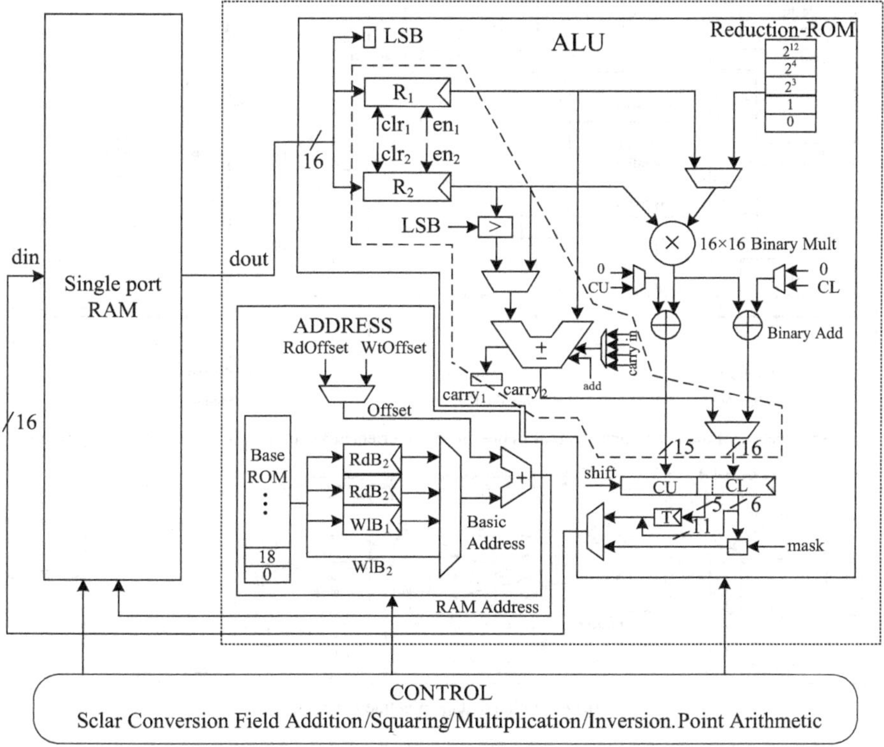

Fig. 1.24 Hardware architecture of the ECC coprocessor [59]

In [60], a heterogeneous multicore processor oriented to public-key cryptographic algorithms is put forward. This processor has the advantages of low delay and high throughput. This processor is composed of two clock domains with different functions, and its architecture is shown in Fig. 1.25. A high-frequency clock domain consists of four processing elements (PEs), and a low-frequency clock domain consists of 1 RISC core. These two parts interconnect with each other through FIFO, and the RISC generates microinstructions used to control PEs to execute computation function. The PEs in this processor are programmable. They can provide high-performance arithmetic computation like long-word-length modular multiplication and addition, has a five-stage pipeline structure, and can execute 292-bit long-word-length modular addition. The author adopts TSMC 65 nm CMOS process for hardware implementation of the architecture put forward. This architecture has a maximum frequency of 960 MHz and takes 0.087 ms to complete an encryption for a 1024-bit RSA.

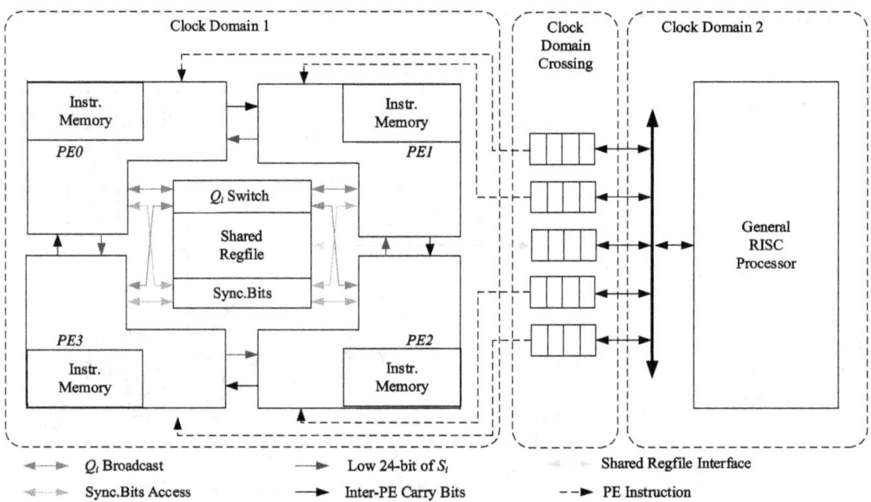

(a) Architecture of the heterogeneous multi-core processor oriented to public-key cryptographic algorithm

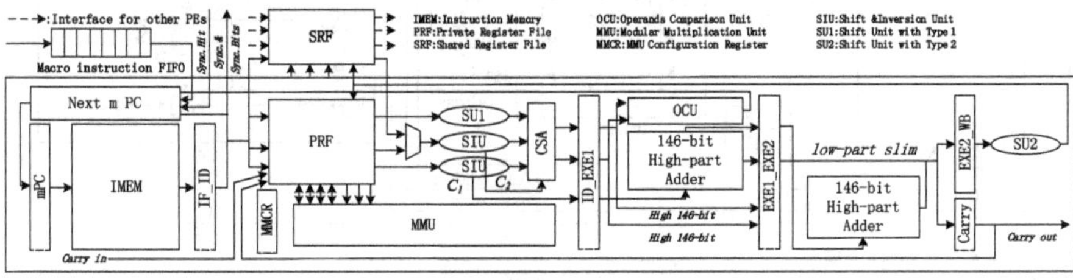

(b) PE with a 5-stage pipeline structure

Fig. 1.25 Architecture of the heterogeneous multicore processor oriented to public-key cryptographic algorithms and its PE [60]

In [61], an ARM extended instruction set applicable to 128-bit interfaces is developed for KECCAK. This extension set extends six instructions with the feature of single instruction stream multiple data stream (SIMD), which are shown in Fig. 1.26. To enable the extended instructions to be compatible with the currently popular instruction set processors and provide the maximum computing performance, the following methods are used. Divide the data-dependency graph (DDG) into multiple subgraphs to adapt to the customized instruction set; shorten the schedule length of the subgraphs as much as possible to reduce the logical complexity of the customized instruction set for parallel operation; divide the length of the status word of KECCAK so that the status word adapts to the storage length of the processor (e.g., for the SIMD instruction which is in format of two 128-bit input and one 128-bit output, it should be divided into multiple 128-bit blocks); use bit interleaving access instructions to adapt to the KECCAK algorithm. When the above-mentioned

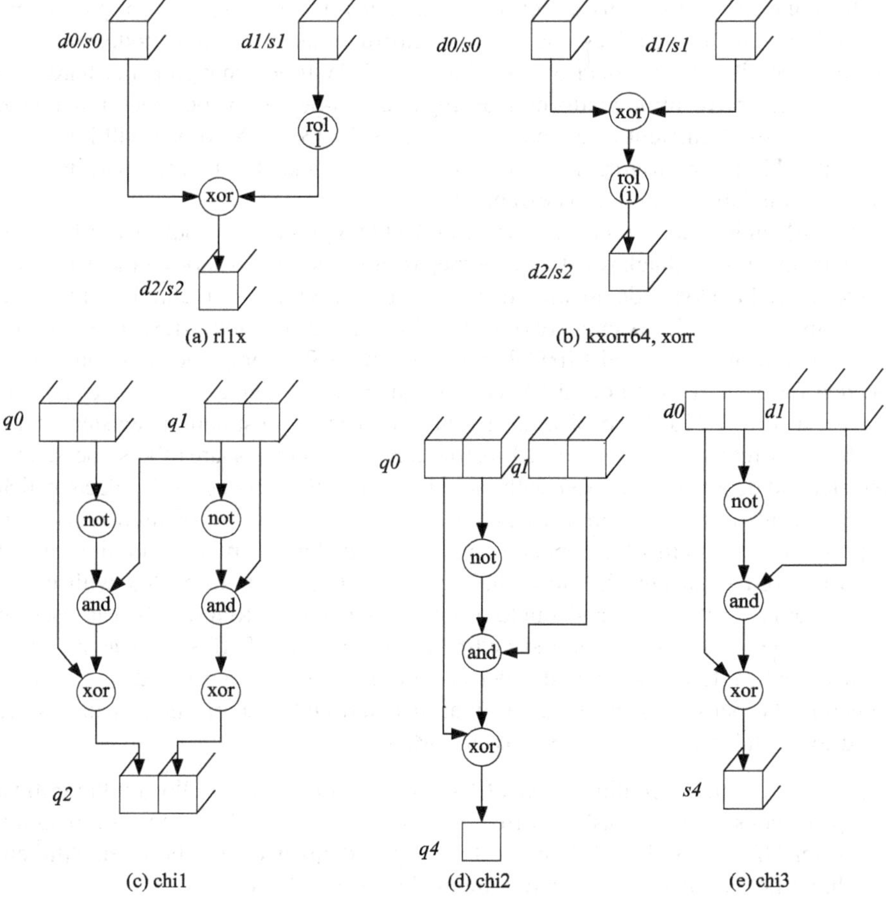

Fig. 1.26 Extended instruction set [61]

extended instructions are applied to ARM V7 based on NEON instructions, the instruction set system provides a much higher performance than the best software implementation based on NEON instructions available currently, specifically, 2.2 times higher than that of SHA-3, 2.6 times higher than that of RIVER KEYAK, 1.6 times higher than that of KETJESR, and 1.4 times higher than that of KETJEJ. In addition, this processor has an equivalent gate of 4658 in case of 90 nm process and its hardware complexity is much lower than the existing processor.

Table 1.2 compares the performance of different cryptographic processor architectures and lists the algorithms, key technologies, extended instruction sets, datapath bit width, operating frequency, execution cycle, operating platform, and execution time.

2. Physical Attack Resistance of ISAP Cryptographic Processors

Like other kinds of cryptographic processors, ISAP also faces threats from physical attacks. As the ISAP cryptographic processor adopts the instruction set architecture, its instructions are executed in series when it executes cryptographic algorithms, and it organizes and executes instructions in a fixed mode, and it is more susceptible to physical attacks than ASIC. Without changing the hardware architecture or execution platform, masking and non-deterministic execution are the commonly used efficient attack countermeasures for ISAP. Next, we will introduce how the ISAP cryptographic processor resists attacks by taking masking and non-deterministic execution as examples.

In [66], an instruction set used for 32-bit LEON3 processor is put forward, which implements power attack resistance by means of masking. To shorten the execution cycle and reduce the code quantity introduced by masking, this method is optimized in many aspects. It extends two instruction operations for AES: Execute the combination of S-Box and MixColumns; execute S-Box only (no MixColumns is needed for the last round of AES). The extended instruction architecture of AES is shown in Fig. 1.27a. It consists of a source register, a destination register, and a 12-bit imm register. This extended instruction architecture shares the same source register and destination register with the original LEON3 processor. It selects which extended instruction operation to execute by setting the value of imm register. It supports a maximum of 2^{12} operations. The extended instruction architecture of AES first-order masking is shown in Fig. 1.27b. It consists of S-Box with mask, MixColumns with mask, and random number generator (RNG). RNG is used to generate the mask values required by first-order masking of AES. This architecture is also applicable to the second-order and third-order masking of AES. To avoid affecting the performance of the original datapath and system, three methods are used to optimize masking for S-Box and MixColumns.

(1) Rearrange the execution order of S-Box and ShiftRows, reallocate the critical paths consisting of masked S-Box and MixColumns, and map the computation from $GF(2^8)$ to $GF\ (2^4)^2$, move the map and inverse map between different finite fields outside the round computation of the AES.

Table 1.2 Performance comparison among different ASIP cryptographic processors

Design	Algorithm type	Key technologies	Extended instruction set	Datapath bit width	Frequency/ MHz	Execution cycle	Operating platform	Time/ ms
Tillich [58]	AES	Real-time key expansion, word-oriented operation	Sbox4, Mixcol4, Sbox4s, iSbox4s, Sbox4r, MixCol4s, iMixCol4s	32	–	196	ARM	–
Soliman [62]	AES	Pipeline parallel architecture, increase the bit width of memory and number of ports	Immediate, register, key expansion, SetLen	128	444	–	UltraSPARC	–
Roy [59]	ECC	Koblitz curve scalar conversion	–	283	16	1566000	Xilinx	97.89
Han [60]	RSA'ECC	Montgomery algorithm optimization	AOP'AOP. c'CAOP, CAOP. C, MM256, MM521, MM1024, MM2048	292	960	–	Heterogeneous multicore architecture [69]	0.087
Rawat [61[1]	SHA-3	DFG division; subgraph scheduling optimization, division of the state word length of KECCAK, bit interleaving access instructions	rllx, kxorr64, xorr, chi1, chi2, chi3	128	–	–	ARM	–
GroB [63]	PRESENT	4-bit S-Box, S-Box batch operation, acceleration state, internal key permutation	Sbox, Perm, Swap, SboxRot, AddC, SubC	80	–	3406	Faraday[2]	–
Grabher [64]	DES, Serpent, AES, PRESENT, SHA-1	Bit slice implementation based on look-up table, implementation of permutation of any bit based on two instructions	CLUT'ULUT'GRP, SHIFT PAIR	AES-32	–	1662	Xilinx	–
O'Melia [65]	DES, 3DES, IDEA, AES	Loop unrolling, real-time key expansion, embedded CPA-based modular multiplication unit, embedded finite field modular multiplication unit	mmull 6, deskey, desf, desipl, desipr, desfpl, desfpr, desdir, aessb, aessbs, aessb4, aessb4s, gfmkld, gfmmul, mixcol4	32	117	–	LEON2 RISC processor	–

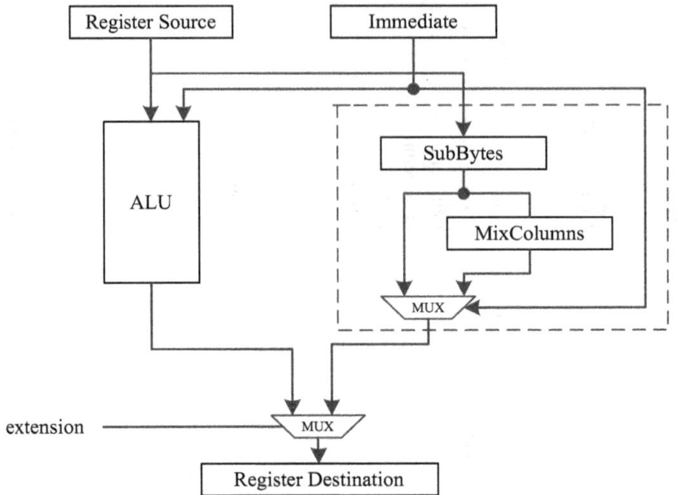

(a) Extended instruction architecture of AES

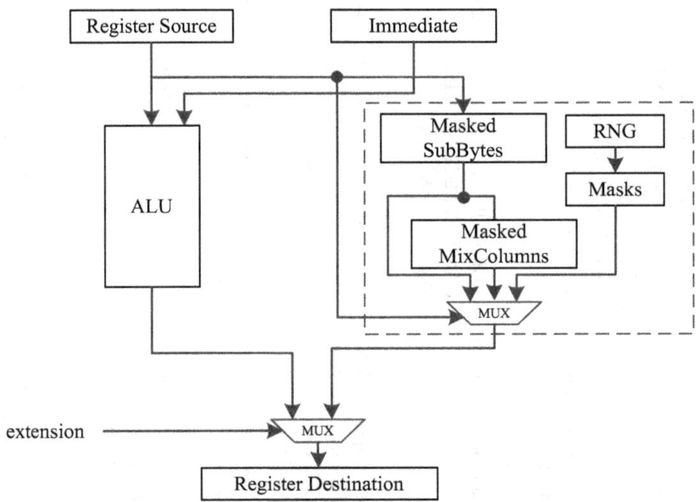

(b) Extended instruction architecture of AES first-order masking

Fig. 1.27 Extended instruction architectures of AES and its first-order masking [66]

(2) Design an optimal transformational matrix to reduce the critical path length of MixColumns masking.
(3) Reuse the S-Box of the first-order masking in higher-order masking to reduce the circuit area.

In terms of the execution cycle, the execution cycle of the first-order masking based on the original LEON3 processor is 18.4K, the execution cycle of third-order masking is 29.2K, and the execution cycle of the expanded ASIP keeps almost unchanged, is about 3.3K. In terms of code quantity, the code quantity of first-order masking of ASIP is reduced by 11% and that of the third-order masking is reduced by 12.82% relative to GPP reference design. Extended instructions used for masking need a special compiler and are also modified for the instruction architecture accordingly. Therefore, these instructions are not universal.

The non-deterministic execution works like this: Reassign the circuit power in the time domain so as to minimize the correlation between power consumption and operation instructions [67]. In [68], a side-channel attack countermeasure based on non-deterministic execution is put forward. This method inserts stall and flush into the pipeline operations through software, so as to ensure that pseudo-operations exist in each encryption; that is, each instruction is no longer executed at a fixed moment in a computation. After obtaining the leakage curve, the power and electromagnetic attacks need to align the curve. The operation of curve alignment is carried out to ensure that the correct key has the maximum correlation during the later correlation computation. After a pseudo-operation is inserted, the attacker will not be able to easily align the curve. This will finally affect the correlation of the correct key among all guessed keys. The probability of a pseudo-operation is determined by the extra controller (Ghost Hazards in Fig. 1.28). The controller determines the time when a pseudo-operation is inserted and the number of the pseudo-operations which have been inserted by comparing the random number generated by the true random number generator with the threshold. Another point that should be noted is that non-deterministic execution can also provide resistance against fault attacks which need precise injection, that is, make it more difficult for the attacker to successfully inject faults in the specified period. Non-deterministic execution is easy to design and implement, but the pseudo-operation it inserts may lower the throughput of the processor. The more pseudo-operations are inserted, the more the throughput drops. In addition, the attacker can reduce the attack resistance of non-deterministic execution by taking multiple samples. Therefore, a certain gap may exist between the actual attack resistance and the expected attack resistance when non-deterministic execution is used alone.

1.3.3 Limitation of Traditional Cryptographic Processors

An ASIC-based cryptographic processor directly processes data flow graphs of cryptographic algorithms through hardware, and thus it has an advantage over other processors in terms of energy efficiency. However, cryptographic algorithms are constantly evolving, and cipher standards and security protocols are also updating. This requires the hardware to adapt to these changes in time. For an ASIC-based cryptographic processor, as no functional modification can be made after it is manufactured, new hardware must be designed and produced to enable it to support a new algorithm or modify the existing algorithm, or adapt to a new cipher standard. Such factors as design investment, non-repetitive engineering cost carry significant limitations to these processors. That is, ASIC cannot meet the

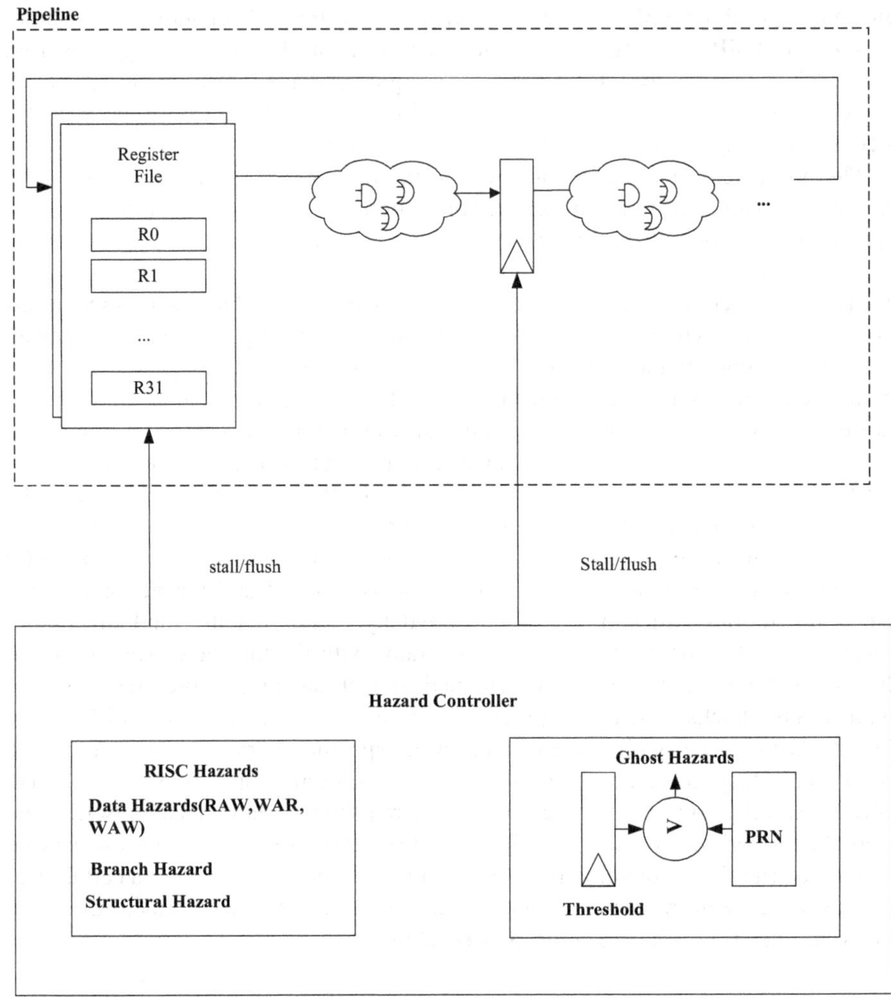

Fig. 1.28 Side-channel attack countermeasure based on non-deterministic execution [68]

requirements of a cryptographic processor for flexibility. In terms of security, an ASIC-based cryptographic processor is susceptible to attacks from reverse engineering and is not applicable to applications where the cryptographic algorithm itself needs be kept secret. An ISAP-based cryptographic processor can implement powerful instruction set and flexibly execute various encryption tasks. However, as its instruction structure cannot perfectly match the cryptographic algorithm, it cannot give full optimization for processing of cryptographic algorithm, and thus it can provide only a low processing speed and a low energy efficiency. In terms of security, an ISAP-based cryptographic processor works by executing instructions and is easily modeled. Compared to hardware implementation, it is more susceptible to threats from physical attacks. Therefore, it cannot meet the security requirements of the fields which need a high security level, including national defense, communication, banking, and finance. Sometimes, some cryptographic chips may combine ASICs with instruction set structures (such as GPP, ASIP, and DSP+ASIC) to design complex systems. No matter what a complex structure the chip uses, the inherent defects of the two implementation styles still exist as long as the chip is implemented by their combinations.

1.4 Reconfigurable Cryptographic Processors

1.4.1 Overview of Reconfigurable Computing

A reconfigurable cryptographic processor is a successful application of the reconfigurable computing technology in the security area. It mainly benefits from the perfect match of the algorithm features and reconfigurable computing features in this field. To better explain the design and working principle of a reconfigurable processor, we will give an overall introduction to the reconfigurable computing technology.

1. Development and Definition of Reconfigurable Computing

The concept of reconfigurable computing dates back to the 1960s, when Estrin [69] from the University of California Los Angeles pointed out that a computer can be composed of a main processor and a group of reconfigurable hardware, with the main processor in charge of the actions of the reconfigurable hardware and the reconfigurable hardware supporting tailoring and reorganizing based on the computation features of the task, so as to accelerate a specific task. In the above description, Estrin put forward the concept of reconfigurable hardware for the first time and pointed out the key feature of reconfigurable hardware; that is, its computing functions and computing architecture can be modified after it is manufactured. It is not unusual that a predictable idea receives a snub because it arrives before their time. Limited by the backward process of integrated circuit which cannot meet the hardware requirements of reconfigurable computing at that time, the concept of reconfigurable computing was ignored for a long time after it was put forward.

At the time when the reconfigurable computing was put forward, integrated circuit was in its initial step. In 1958, engineer Kilby from Texas Instruments in America developed integrated circuit and successfully integrated three electronic components into a silicon chip. Integrated circuits at that time was of a very small scale and was often called small and medium-sized integrated circuit, which allowed integrating only dozens of or hundreds of electronic components into a monocrystalline silicon chip with an area of several square millimeters. After that, integrated circuits mainly developed toward larger-scale ones. On November 15th in 1971, Hoff from INTEL successfully developed the first microprocessor 4004, which contained 2,300 transistors, used a 4-bit system, had a clock frequency of 108 kHz, and can execute 60,000 instructions per second. The integrated circuit at this time was called large-scale integrated circuit and allowed integrating hundreds of components into a silicon chip. After 1920s to 1980s, very large-scale integrated circuit (VLSI) appeared, which had developed to allow integrating tens of thousands or even millions of components into a silicon chip. The period from 1960s to 1980s sees the growth of both integrated circuits and chips of general-purpose processors. As the microprocessor developed from the earliest one 4004 processor to Intel Pentium 4 processor in 1995, the number of transistors on a microprocessor chip increases from 2,300 to 42,000,000, and the computation power of the chip is improved from 60,000 times per second to 10,000,000 times per second or even 100,000,000 times per second. Microprocessor is the absolute leading role in this period. As only very limited transistors can be integrated into a chip, scientists even had no enough hardware capability to enhance the microprocessor's functions, let alone had spare energy to develop reconfigurable computing which claimed much higher hardware capability.

Since 1980s to 1990s, reconfigurable computing has been given much attention by the international academia and has become a hot area of research. In 1999, Dehon and Wawrzynek from Reconfigurable Technology Research Center in University of California, Berkeley, put forward a definition of reconfigurable computing which was generally accepted at the ACM Design Automation International Conference [70]; that is, each computing form can be recognized as reconfigurable computing if it has the following two features.

(1) The functional units can be reconfigured after the chip is manufactured; that is, the computing function can still be modified according to the application task even after the silicon implementation, which is different from traditional ASIC.
(2) Spatial mapping from an algorithm to a computing engine can be implemented to a great extent, which is different from traditional ISAP.

The academia has carried out various and comprehensive research on reconfigurable computing. Early research focused on fine-grained (for the concept and definition, refer to the subsequent sections) reconfigurable structure and such structures as Ramming machine, PAM machine [71], GARP [72] were put forward. Later research focused on coarse-grained reconfigurable structures, and such structures as MATRIX [73], RAW [74], MorphoSys [75], PipeRench [76], ADRES [77], DySER [78] were put forward. Meanwhile, mixed-grained (i.e., combination of coarse-grained and fine-grained) reconfigurable computing structure has also become an important research direction. Some representative structures are Morpheus [79], TRIPS [80], and TIA [81].

In 1980s or 1990s, the integration degree of a chip was able to meet the requirements of reconfigurable computing, and reconfigurable chips gradually became popular in the industry. Early-stage hardware implementations of reconfigurable computing include complex programmable logic device (CPLD) and field-programmable gate array (FPGA). These two hardware implementations have similar structures and are both composed of a great number of programmable logic devices whose configurations are static and will not be changed during runtime. Comparatively, CPLD is more suitable for implementing algorithms and large combinational logics, while FPGA has a higher applicability and flexibility. Just as Freeman, the founder of Xilinx and inventor of FPGA, expected, both flexibility and customization are attractive for many applications if they are implemented properly. Maybe FPGA can only be used in prototype design originally, but it may replace customized chips in a broader sense in future. From XC2064 developed by Xilinx in 1985, which is the first to use 2 μm process in the world and works as an verification platform for chip design, to Virtex-7 also developed by Xilinx, which uses 28 nm process and works as efficient acceleration hardware for computation-intensive, data-intensive, and communication-intensive applications, FPGA products have greatly developed in both hardware performance and application scope. Now, FPGA has also become an important cornerstone in the information age, just like general-purpose processor.

After 1990s, the industry began to focus on coarse-grained and mixed-grained reconfigurable computing structures. Take FPGA as an example, an early FPGA is a fine-grained reconfigurable structure. However, it brings a great waste of hardware resources and requires too much configuration information, and the functional modules implemented by using it cannot meet the performance requirements. As a result, FPGA is also gradually integrated with coarse-grained processing units, including multiplier, IO module, Block RAM, CPU, and clock management units. In addition to improving the FPGA structure, the industry also designs brand-new coarse-grained reconfigurable processors, such as XPP [82] from PACT XPP Technologies, DRP [83] from NEC, and PicoArray [84] from PicoChip (purchased by Mindspeed Technologies).

2. Why to Study Reconfigurable Computing

Why reconfigurable computing has been popular in both the academia and industry for a long time and developed constantly in recent dozens of years is because it combines the advantages of the two traditional computing structures (i.e., ASIC and ISAP). It avoids the disadvantages of ASIC and ISAP by using their complementary structures and makes a balance between the power consumption and flexibility. Figure 1.29 shows the distribution of different computing forms in performance and flexibility. ASIC and ISAP are two extreme computing forms. The former customizes hardware based on the software and thus has the highest performance. The latter customizes (or compiles) the software based on the hardware and thus has the highest flexibility. However, both of them have obvious disadvantages. ASIC lacks flexibility while ISAP has limited performance and energy efficiency. Therefore, both of them also have seen unprecedented challenges in recent years. Reconfigurable computing will become an important development direction in computing chip area in the future because it has both high flexibility and high energy efficiency.

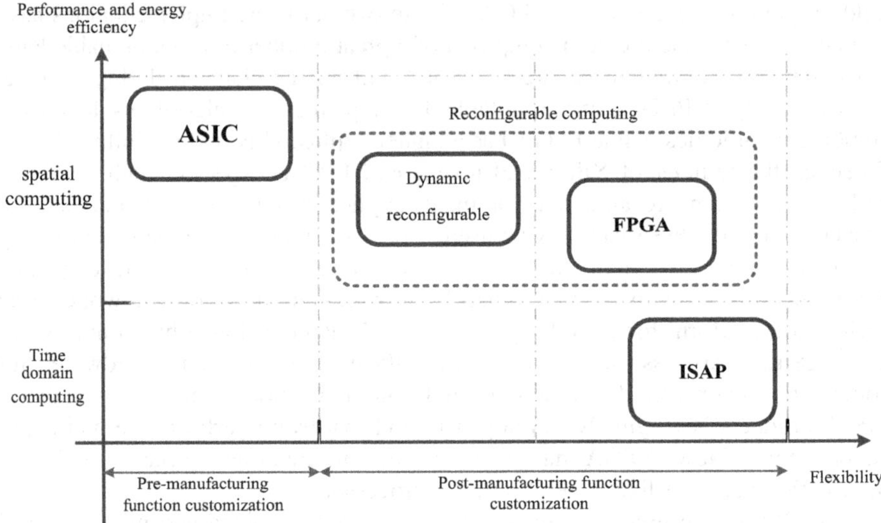

Fig. 1.29 Comparison of different computing forms in performance and flexibility

(1) Flexibility

ASIC implements the application functions by means of hardware function units and fixed interconnections. To provide applications with the hardware structure which implements the optimal performance, area, and power consumption, it requires the hardware engineers to spend a lot of time in design and optimization. It scarcely provides any hardware flexibility. On the one hand, ASIC cannot adapt to other applications and even cannot adapt to the updated version of the application where it is used. It has very poor structure scalability, and basically each function update needs redesigning the chip. On the other hand, the labor cost and time cost of ASIC design become higher and higher, which causes a longer time-to-market. Meanwhile, each design mistake may generate a high non-recurring engineering (NRE) cost (Fig. 1.30), which limits ASIC to application fields where sufficient demand exists for chips. Therefore, insufficient flexibility will limit ASIC to narrower and narrower application scope. Flexibility will be a very important ability for computing chips in future.

(2) Energy efficiency

ISAP usually indicates a computer of von Neumann architecture. It is mainly composed of arithmetic logic unit, memory unit, control unit, and IO interface. As the application software will compile the ISAP-based instruction set to assembly instructions which can be executed by the processor, ISAP has a high flexibility and can meet any application requirements (a computing field which can be described by a Turing machine or finite state machine). However, ISAP has a very low performance and energy efficiency, which is mainly for the following two reasons.

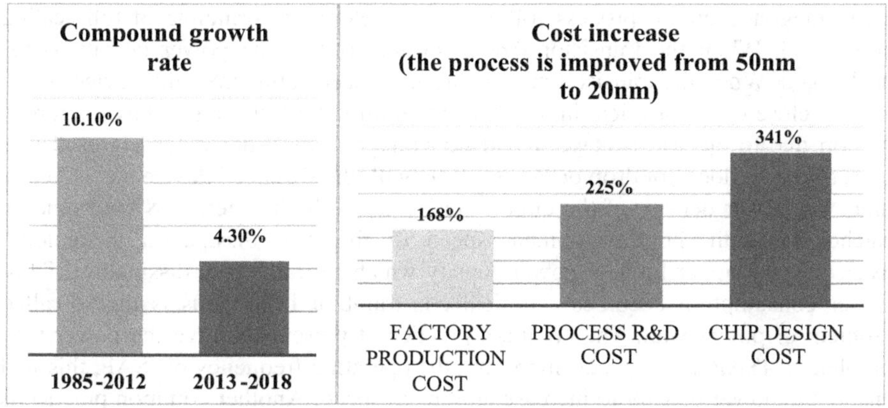

Fig. 1.30 Decreased capital growth rate in contrast with increased cost of chip design and tapping out in the integrated circuit industry (*Source* Morgan Stanley, PWC)

Firstly, von Neumann architecture computes based on the time division multiplexing principle (time domain computation) and cannot well support spatial parallelism. Secondly, von Neumann architecture takes a lot of time and energy in such stages as instruction fetching, decoding, register access, execution, and data write-back, while only execution is a true computational process, and a lot of time and energy is spent in data preparation, instruction preparation, and instruction analysis processes (these processes do not exist in ASIC). As a result, von Neumann architecture provides a high flexibility at the cost of performance and power consumption. In addition, ISAP makes many improvements in spatial parallelism based on von Neumann architecture to improve the performance, including using such technologies as superscalar, multi-issue, and out-of-order execution. The performance is improved at the cost of area and power consumption, and the energy efficiency is actually lowered.

Energy efficiency is though the letdown of ISAP, but it never becomes a key constraint of ISAP design. This is because the power density of the chip is stable when the technological level of integrated circuit is low. As shown in Table 1.3,

Table 1.3 Comparison between full scaling down and fixed voltage scaling down

Parameters	Full scaling down	Fixed voltage scaling down
Transistor density	S^2	S^2
Operating frequency	S	S
Load capacitor	$1/S$	$1/S$
Power voltage	$1/S$	1
Power consumption density	1	S^2
Transistor utilization	1	$1/S^2$
Transistor power consumption	$1/S^2$	1

early integrated circuit process follows the development principle of full scaling down [85]. When the transistor size is shrunk to 1/S, the power density keeps unchanged. When the transistor size is shrunk to deep submicron, the device size is already close to the physical limit. Thus, the traditional full scaling down is forced to stop due to electric leakage and fixed voltage scaling down [86] is used (the power voltage does not drop or drops a little with the shrink of device size). At this time, the power density of the circuit increases at a high speed of S^2 and quickly catches up with the development speed of the heat dissipation technology, exceeding the upper limit of power density which restrains heat dissipation. Thus, power consumption becomes a critical constraint of ISAP; this is the so-called problem of power wall. The common practice at present to solve the power wall problem of ISAP is no longer improving the operating frequency of ISAP; this will enable the power density to increase at a linear speed. Another common practice is to improve the spatial parallel computing capability of ISAP and then the energy efficiency by using the multicore technology. Finally, using a small core which has a low performance and power consumption at appropriate time can also reduce the power consumption of the system. To obtain the performance gains brought by the new process, future computing chip must improve the energy efficiency through innovative design of the architecture.

To sum up, advantages of reconfigurable computing come from two intrinsic features: The functional units can be reconfigured after the chip is manufactured; that is, the hardware can adapt to the function of the application software through configuration modification after being manufactured, and provide sufficient flexibility; the chip supports both spatial parallel computing (data-driven data stream execution model) and spatial algorithm mapping, and provides a high energy efficiency.

3. General Architecture of Reconfigurable Computing

The reconfigurable computing structure combines the advantages of both ASIC and ISAP. Next, we will discuss the forming of general reconfigurable computing architecture by taking the ISAP structure of von Neumann architecture as the starting point. Figure 1.31a is the schematic diagram of von Neumann architecture [87], which consists of arithmetic logic unit, control unit, memory unit, input and output. This architecture looks simple but has a far-reaching influence. It is the basis of all computer structures nowadays. All computer structures, no matter it is Harvard architecture (where the instructions and data are stored separately), pipeline, multithread, multi-issue, out-of-order execution, very long instruction word, dual-core, multicore, or many-core processor, are within the scope of this architecture. Most ASICs also follow the principle of von Neumann architecture. Figure 1.31b is the schematic diagram of ASIC. You will find that it is actually a transformational and tailed von Neumann architecture if you observe it carefully. Similar to Harvard architecture, an ASIC memory is divided into two independent parts: One part stores the data and the other part stores control signals. There are two main differences between ASIC and ISAP: reinforcement of datapath and

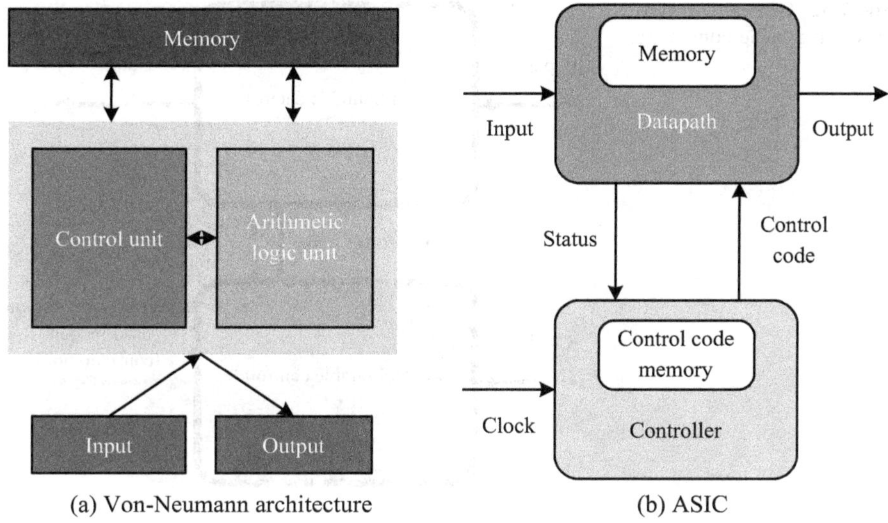

Fig. 1.31 Block diagram showing the basic principles of ISAP and ASIC

weakening of control unit. For the former, the datapath of ASIC is reinforced from few ALUs of ISAP to become the hardware implementation structure of spatial mapping of the target application, including many fixed arithmetic logic resources (different from flexible ALUs), storage units, and interconnection resources. For the latter, a control unit of ISAP obtains instructions from the memory to carry out complex processes including decoding, operand fetching, and execution, while a control unit of ASIC is usually a finite state machine (FSM) which outputs control codes based on the state signals reported by the datapath and controls only the critical system state of the datapath. These two differences, that is, the powerful spatial parallel computing power brought by the spatial structure and the efficient control process of the data stream-driven structure, are the key reasons that explain why ASIC has a higher performance and energy efficiency. The problem of ASIC is that it goes to the other extreme. It completely gives up hardware flexibility. The reconfigurable computing structure is the trade-off between ISAP and ASIC. On the one hand, it improves the computing power of datapath while ensuring the function flexibility of datapath. On the other hand, it tailors the functions of the control unit while keeping the control of the control unit over datapath.

The general structure of reconfigurable computing is shown in Fig. 1.32, mainly consists of two parts: reconfigurable datapath (RCD) and reconfigurable controller (RCC). The function of the RCD is to process the input data stream by means of configurable spatial concurrent computing, and the function of the RCC is to manage the configuration, switching, and scheduling of the reconfigurable datapath. Table 1.4 lists the hardware implementation respectively corresponding to the RCD and RCC in some typical reconfigurable computing structures.

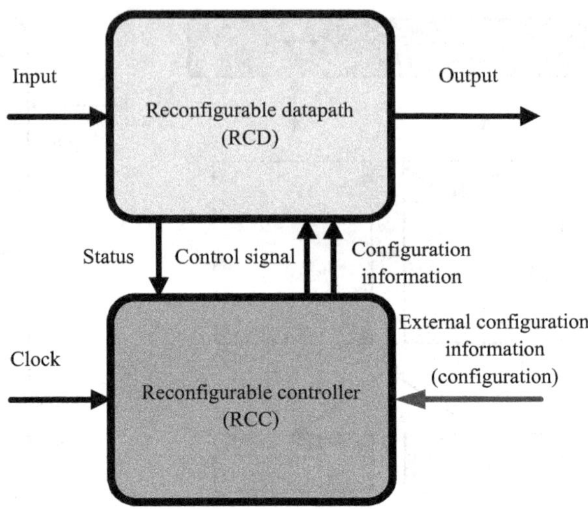

Fig. 1.32 General structure of reconfigurable computing

Table 1.4 Some typical reconfigurable computing structures

Reconfigurable computing structure	RCC	RCD	
		Computing resources	Interconnection resources
FPGA	Finite state machine logic	Look-up table logic block	Segment switching network
ADRES	VLIW processor	Coarse-grained reconfigurable cell array	Two-dimensional MESH interconnection
DYSER	OpenSPARC	Coarse-grained reconfigurable cell array	Two-dimensional MESH interconnection
DRP	Sequence generator	Coarse-grained reconfigurable cell array	Two-dimensional MESH interconnection
REMUS	ARM processor	Coarse-grained reconfigurable cell array	One-dimensional systolic interconnection

(1) Reconfigurable datapath

The structure of reconfigurable datapath usually consists of processing element array (PEA), memory, data interface, and configuration interface, as shown in Fig. 1.33. The configuration interface obtains control signals and configuration information from the reconfigurable controller and outputs the system state signals of datapath meanwhile. Then, the configuration interface parses the configuration information and configures the functions and task execution sequence of the processing element array. After being configured, the processing element array begins to be driven and executed by data streams, like ASIC. The input data are obtained

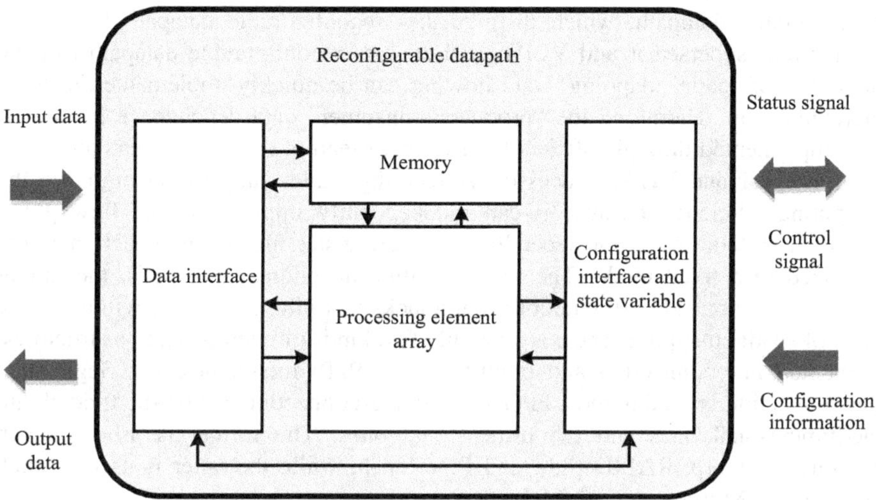

Fig. 1.33 General structure of reconfigurable datapath

from the data interface, the intermediate data are cached to the memory module, and the computation result of data streams is output from the data interface to the external device.

The processing unit array is composed of many PEs and configurable interconnection structures, and a processing unit is usually composed of ALU and register, as shown in Fig. 1.34. Interconnection is an important feature of

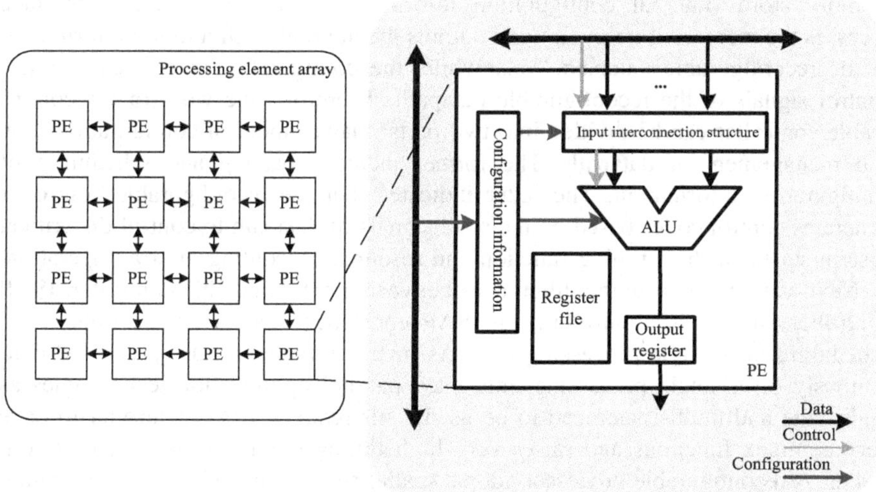

Fig. 1.34 General structure of processing unit array

reconfigurable datapath, which distinguishes reconfigurable datapath from such structures as superscalar and VLIW, and enables reconfigurable datapath to have the ability of spatial mapping. Data flowing can be quickly implemented through interconnection. Therefore, the "producer–consumer" data dependence relation is now implemented through efficient hardwires, instead of the register read and write of superscalar and VLIW processor. A reconfigurable datapath can improve the computing efficiency because it can independently implement data flow graph (DFG) of suitable size, not depending on the register file. Interconnection is not organized in a fixed mode. The more flexible the interconnection is, the higher hardware cost it needs. An interconnection lacks flexibility, however, will affect the effect of spatial mapping. There are roughly two kinds of interconnection structures: segmented interconnection and point-to-point (P2P) interconnection. Segmented interconnection is used in most FPGAs. P2P interconnection can be subdivided into one-dimensional ones and two-dimensional ones. The former is used in such structures as Garp, RAPID [88], and PipeRench, while the latter is used in such structures as Morphosys and ADRES.

Compared to the datapath of ISAP, processing unit array is a typical reconfigurable spatial computing structure and it can implement spatial mapping of algorithms and spatial parallel computing. Compared to the datapath of ASIC, processing unit array ensures that all resources of spatial computing are flexible and configurable.

(2) Reconfigurable controller

The structure of a reconfigurable controller is shown in Fig. 1.35, consisting of configuration interface, configuration management unit, and memory. The configuration management unit receives and parses external configuration information to generate top-layer control signals and internal configuration information. The memory stores internal configuration information. The configuration interface accesses the memory if necessary and outputs the internal configuration information to the reconfigurable datapath. Meanwhile, the configuration interface generates control signals of the reconfigurable datapath. Therefore, the work of a reconfigurable controller can be divided into two parts: management of configuration flow and management of datapath. The former includes parsing and scheduling of configuration information. The latter indicates that the reconfigurable controller generates control codes based on the state signals of datapath to control the critical system states, such as timing function and resource scheduling in array operation.

Next, we will analyze the differences between a reconfigurable controller, ISAP controller, and ASIC controller. An ISAP controller focuses on the timing and scheduling of a single processing unit. As instruction streams are processed continuously on a single processing unit, such parallel optimization technologies as multistage and multi-issue need to be used. This requires ISAP controller to have very complex functions and raises very high timing requirements for ISAP controller. A reconfigurable processor adopts spatial computing and uses configuration information instead of instructions. A reconfigurable controller handles resource

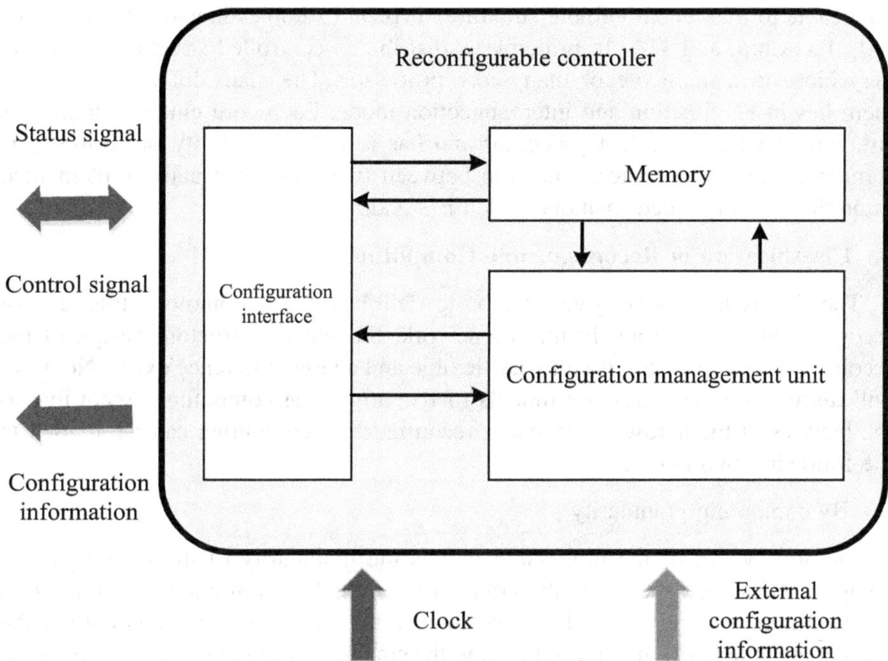

Fig. 1.35 General structure of reconfigurable controller

scheduling of processing unit array and focuses on functional control on array layer. Therefore, for a reconfigurable controller, the space/time utilization of the processing unit array is more important than the scheduling of a single processing unit. Due to fixed datapath structure, an ASIC controller has even lower functional requirements and it does not need to process problems about instructions or configuration information.

A reconfigurable controller conducts resource scheduling for processing unit array mainly in two modes: centralized control and distributed control. A reconfigurable controller in centralized control mode generates a configuration word sequence for the entire reconfigurable datapath. A reconfigurable controller in distributed control mode generates independent configuration word sequence for each PE. In a reconfigurable structure in centralized control mode, the PEs are executed passively and cannot control their own configuration flows. Therefore, this structure is also called passive-controlled PEA, with the entire reconfigurable datapath corresponding to a thread. Typical examples include MATRIX, ADRES, TRIPs, DYSER, and REMUS [89]. On the contrary, in a reconfigurable structure in distributed control mode, the PEs are executed actively and each distributed controller in the PE can control its own configuration flow independently. Therefore, this reconfigurable structure is also called active-controlled PEA, with each PE corresponding to a thread. Technologies related to thread-level parallelism (TLP) is applicable to this reconfigurable structure. Typical examples include RAW, AsAP [90], PicoChip, and TIA. In principle, a distributed-controlled structure is already very close to a multicore or many-core processor. The main difference between them lies in PE function and interconnection mode. Each core unit in a multicore structure is an independent processor and has a higher flexibility but a low performance because the interconnection between two cores is usually implemented through a bus or shared memory which has a slow speed.

4. Classification of Reconfigurable Computing

The discussion above gives the basic definitions and common structures of reconfigurable computing. In this framework, the specific structure design of the reconfigurable computing is still very flexible and a large difference exists. Next, we will discuss different structure models of reconfigurable computing. According to the features of the hardware structure, reconfigurable computing can be divided in the following two manners:

(1) By computing granularity

The data width of the processing unit is the granularity of the reconfigurable datapath. Generally, the larger the granularity is, the less configuration information the reconfigurable processor has, the quicker the functions are reconfigured, the smaller area the hardware circuit has, and the smaller power consumption the circuit has. On the other hand, the smaller the granularity is, a higher adaptability and programmability the reconfigurable processor has.

By granularity, reconfigurable processors can be divided into find-grained processors, coarse-grained processors, medium-sized-grained processors, and mixed-grained processors. The academia calls the granularity not exceeding 4 bits fine-grained, and the granularity equal to or greater than 4 bits coarse-grained. For example, the processing units of some FPGA are four-input look-up tables (LUTs) of single-bit width, which are fine-grained reconfigurable processors. A reconfigurable processor whose processing unit data has a bit width of 8 bits, 16 bits, or above is regarded as a coarse-grained reconfigurable processor. A granularity of 4 bits is also called medium-sized-grained [91], which, however, is not common. If a reconfigurable processor has processing units which are of more than one granularity, it can be called a mixed-grained reconfigurable processor. What is worth mentioning is that mixed-grained and coarse-grained are not defined so strictly, and their definitions are sometimes interchanged. For example, a processing unit array which has both 8-bit processing units and 16-bit processing units belongs to a mixed-grained processing unit array, but it is occasionally called coarse-grained processing unit array.

(2) By reconfiguration mode

Reconfiguration mode is an important feature of a reconfigurable processor, and it can be described from two dimensions: time and space. Time feature of reconfiguration indicates whether function reconfiguration can be carried out during the period when the processor works normally, which depends on whether function reconfiguration can be completed within a certain period to ensure continuous and real-time work for the reconfigurable processor. Space feature of reconfiguration indicates whether the reconfiguration of the reconfigurable processor is regionally and locally and whether it is possible to reconfigure the functions of only parts of processing units, which mainly depends on the flexibility of the system structure of the reconfigurable processor.

According to time feature, reconfiguration of reconfigurable processors can be divided into static reconfiguration and dynamic reconfiguration. Static reconfiguration indicates the function reconfiguration which can be carried out only before the computation of the datapath of the reconfigurable processor but cannot be carried out for datapath during computation due to a too high time cost. The most typical static reconfigurable processor is FPGA. An FPGA usually loads configuration bit streams from the off-chip memory during system power-on for function reconfiguration. As the reconfiguration of FPGA usually takes dozens to hundreds of milliseconds or even several seconds, it often interrupts the normal work of FPGA. FPGA can resume its work only after the function reconfiguration has completed. Contrary to static reconfiguration, dynamic reconfiguration indicates the function reconfiguration which can be carried out for the reconfigurable processor during computation of the datapath and thus has a low time cost. The reconfiguration of a typical dynamic reconfiguration processor usually takes several to dozens of nanoseconds, which seldom affects the normal work of the processor. Coarse-grained reconfigurable array (CGRA) is the most common dynamic reconfiguration structure. Function reconfiguration of CGRA takes place between different computing tasks during CGRA's work. When one specific computing task is finished, the new configuration information can be quickly loaded for function reconfiguration. As CGRA usually has a small amount of configuration information, its reconfiguration usually lasts only several to hundreds of clock cycles. After the function reconfiguration is completed, CGRA can continue to execute the computing tasks of the new configuration. From the perspective of the application layer, the time of function reconfiguration of the computing tasks accounts for only a small proportion of the total application time, and the computing tasks are basically continuously executed. Therefore, dynamic reconfiguration is also called runtime reconfiguration.

According to the spatial feature of reconfiguration, reconfigurable processors can be divided into partially reconfigurable processors and completely reconfigurable processors. The datapath of a reconfigurable processor can be divided into multiple areas in space, and each area can be independently reconfigured without affecting the state of other areas. This feature is called partial reconfiguration. A reconfigurable processor which does not support partial configuration is a completely reconfigurable processor. As the amount of configuration information decreases during partial reconfiguration, the feature of partial reconfiguration can shorten the configuration time. This is common on a reconfigurable processor which needs a long configuration time. For example, some commercial FPGAs introduce system designs supporting partial reconfiguration to implement dynamic reconfiguration. One method for partial reconfiguration is a difference-based reconfiguration technology [25]; that is, only the different part between two neighboring configuration information is changed during function reconfiguration. Partial reconfiguration can reduce the configuration information, but it also has a big problem; that is, it can obtain a big benefit only where there is a small difference between two neighboring modules. This obviously brings considerable limitation to its actual application. Another method for partial reconfiguration is partitioning the hardware structure, so that each partition can complete relatively independent functions and the function reconfiguration on partial structure will not affect normal work of other parts. PlanAhead in Xilinx is a typical example [92], and the module-based design technology put forward by Altera also uses similar principle. Such operations as module division and layout planning in these methods need a great amount of manual intervention to complete. Thus, these methods are also severely limited in actual application.

What should be noted is that reconfigurable processors are not often divided by the dimension of time or by the dimension of space independently. It is also common that reconfigurable processors are divided into processors of dynamic partial reconfiguration, processors of static partial reconfiguration, processors of dynamic complete reconfiguration, and processors of static complete reconfiguration by the combination of time and space. Dynamic partial reconfiguration can improve the hardware utilization of datapath and thus improve the energy efficiency of the entire processor. Static partial reconfiguration is mainly used to shorten the configuration time and minimize the influence of reconfiguration on the operation of the reconfigurable processor.

Similarly, reconfigurable processors are not divided by granularity or by reconfiguration method independently. It is difficult for a fine-grained reconfigurable processor to implement dynamic reconfiguration because it takes a long time to reconfigure. Even though it adopts the partial reconfiguration technology, it is still difficulty to reduce the reconfiguration time in orders of magnitude. Therefore, the reconfigurable processors we see are usually reconfigurable processors of fine-grained static (partial/complete) reconfiguration (e.g., FPGA) and reconfigurable processor of coarse-grained dynamic (partial/complete) reconfiguration.

5. Application Prospect of Reconfigurable Computing

The key advantages of reconfigurable computing lie in its flexible spatial parallel computing. It adopts data-driven dataflow execution models, which provide high energy efficiency than the sequential execution model of ISAP processors. It implements explicit data dependence through spatial mapping and uses distributed interconnect communication instead of centralized register file or memory and thus has a high internal communication bandwidth and performance. It adopts datapath based on spatial computing structure to implement algorithm parallelism at different levels (instruction/pipeline/loop/thread) and thus has a higher algorithm implementation efficiency. Therefore, reconfigurable computing is perfect for compute-intensive and data-intensive applications. For application in these fields, a very high algorithm parallelism can be explored, and the spatial parallel computing capability of reconfigurable computing can be exploited and utilized to the greatest extent. As the reconfigurable technologies develop fast and more and more serious problems are challenging ASIC and ISAP, reconfigurable computing will gain wider and wider application. The following lists some application fields where reconfigurable computing has achieved great success or is very promising, cipher, image processing, digital signal processing, deep learning, and data center. Reconfigurable computing in the cipher field will be introduced in the next chapter.

(1) Image processing

Applications in image processing field are mainly to process images and videos and extract effective advanced information from the images and videos. Such as, target detection and tracking in video surveillance, defect detection in production line, facial recognition, and scenario classification, etc. The features of image processing are intensive data and stream processing. Each image or each video contains many pixels. Therefore, an image processing application always needs to perform such processes as filtering, convolution, feature extraction, target recognition, image matching on a great amount of image data. Such applications have a very high data parallelism. When implemented on ISAP, these applications have a low efficiency and an obvious limitation in storage bandwidth, and real-time processing is almost impossible. In addition, these algorithms process pixels in line with the principle of localization and neighborhood, and the computation is very regular, and thus are very suitable to reconfigurable computing. Applications in the image processing field even gave the research of early-stage reconfigurable computing a tremendous boost. Reconfigurable computing structures which have achieved success in image processing field include ADRES, XPP-III, and REMUS.

(2) Digital signal processing

Applications in the digital signal processing field include digital recording, storage, recovery of audio and video signals, wireless communication, and medical imaging diagnosis. The common feature of these applications is that they all contain such processes as time domain signal processing, filtering, transformation, convolution, modulation and demodulation, channel multiplexing, and signal generation. Many researches show that reconfigurable computing is perfect for the digital signal processing field, because a digital signal processing application itself has considerable parallelisms, including the parallelism of coarse-grained operations and fine-grained operations as well as the parallelism of a great amount of data. The form of reconfigurable computing can provide not only a great number of mixed-grained processing units supporting spatial parallel computing, but also a high memory access bandwidth and IO bandwidth. Therefore, a reconfigurable computing structure used to process digital signals has a performance higher than that of ISAP and DSP even it works at an operating frequency below 1/10 of theirs. FPGA has been successfully applied in the digital signal processing field and has already been able to substitute some low-capacity ASICs.

(3) Deep learning

Deep learning is a machine learning field, and it originates from the research on the artificial neural network. Its main objective is to simulate human brain's neural network for analysis and learning, so as to implement artificial intelligence. Deep learning is a hot research field emerging in recent years and will bring profound influence on the future. Deep learning forms abstract high-level features by combining low-level features of the samples, so as to discover the distributed feature representation of data. A reconfigurable computing structure is also suitable for deep learning applications. Take the convolutional neural network (CNN) as an example. CNN is a directed acyclic graph composed of many computing layers, with each computing layer responsible for extracting the feature of a higher level from the input data. Processing of a great amount of input data raises a high requirement for the computing power of the computing layers, and intra-computing-layer communication and inter-computing-layer communication bring a great pressure to system bandwidth. A reconfigurable computing structure, however, just has the spatial computing feature which can provide powerful spatial parallel computing power and a high internal communication bandwidth, and thus become a popular accelerator for deep learning. Some successful cases at present include Eyeriss [93] and Thinker [94] structures.

(4) Data center

The main body of a data center is server cluster, and a data center is built to provide Internet services and other various data services. Data center applications require a high computing capability, flexibility, energy efficiency, and a low cost. It is difficult for a common server chip to satisfy all these requirements, especially the power consumption requirement which has become a critical factor limiting the performance of the data center. Using a dedicated server chip can solve the problem of energy efficiency, which, however, is not a good solution because a data center itself needs a unified hardware platform which is easy to manage. Quick update of application requirements of Internet services deprives non-programmable servers of almost survival space. Therefore, reconfigurable computing which has both flexibility and energy efficiency advantages will represent the future development trend of data center. Microsoft has introduced FPGA into its data center applications [95, 96] and integrated 6×8 FGPG arrays with two-dimensional ring interconnection into the server cabinet to greatly improve the processing performance of webpage search applications.

1.4.2 Reconfigurable Cryptographic Processors

Cryptographic algorithms are both compute-intensive and data-intensive, and a reconfigurable processor is suitable for processing such applications because it can achieve a better balance between the flexibility and energy efficiency. With the fast development of cryptographic technologies, increasingly higher requirements are raised for the performance, flexibility, and security of cryptographic processors, and the research on reconfigurable cryptographic processors has also gained extensive attention.

1. Reconfigurable Cryptographic Processor Architecture

COBRA [97] is a reconfigurable processor which supports various symmetric cryptographic algorithms jointly put forward by University of Massachusetts Lowell and Ruhr University of Bochum in 2005. This processor has 4×4 arrays, with each 32-bit datapath composed of four reconfigurable crypto-elements (RCEs). It can form a 128-bit datapath by interconnecting four 32-bit datapaths together to process both 64-bit and 128-bit data. Through interconnection, it can support two commonly used block cipher implementation structures: Feistel network and substitution permutation network (SPN). There are two kinds of RCEs in COBRA: single-progression RCE without multiplier and dual-progression RCE with a multiplier. There are no big differences between these two RCEs. The following part will only introduce RCEs without multiplier. The processing units of RCE provide the following functions:

A: Bit operation: XOR, AND, or OR;
B: Modular addition and subtraction in 2^8, 2^{16}, or 2^{32} domain;
C: Look-up table: 8-to-8 bit mapping or 4-to-4 bit mapping;
D: Modular multiplication in 2^{16} or 2^{32} domain, square in 2^{32} domain;
E: Left shift, right shift, or rotation left shift;
F: Multiplication in $GF(2^8)$ domain.

The overall architecture of COBRA and the architecture of RCE without multiplier are shown in Fig. 1.36.

To design this cryptographic processor architecture and optimize it, we will make deep analysis of over 40 commonly used symmetric cryptographic algorithms. In terms of flexibility, COBRA is designed for the symmetric cryptographic algorithm. Its reconfigurable computing units contain all operators required by the block cryptographic algorithm in addition to modular inversion operation, and thus can support most current block cryptographic algorithms and the new cryptographic algorithms containing these operators. The flexibility analysis result of COBRA is shown in Table 1.5. However, this also brings a lot of redundant hardware resources when COBRA is used to implement different cryptographic algorithms and greatly increase the required configuration information. As a result, more time is required for configuration and the performance and area efficiency of system is limited. Mapping and implementation is conducted for over 40 symmetric cryptographic algorithms based on this architecture, and the performance results of algorithms are partly shown in Table 1.6.

Celator [98] is a reconfigurable architecture for block cryptographic algorithm and hash function developed by Aix-Marseille University. As a coprocessor among cryptographic processors, this architecture is composed of a main controller, a PE

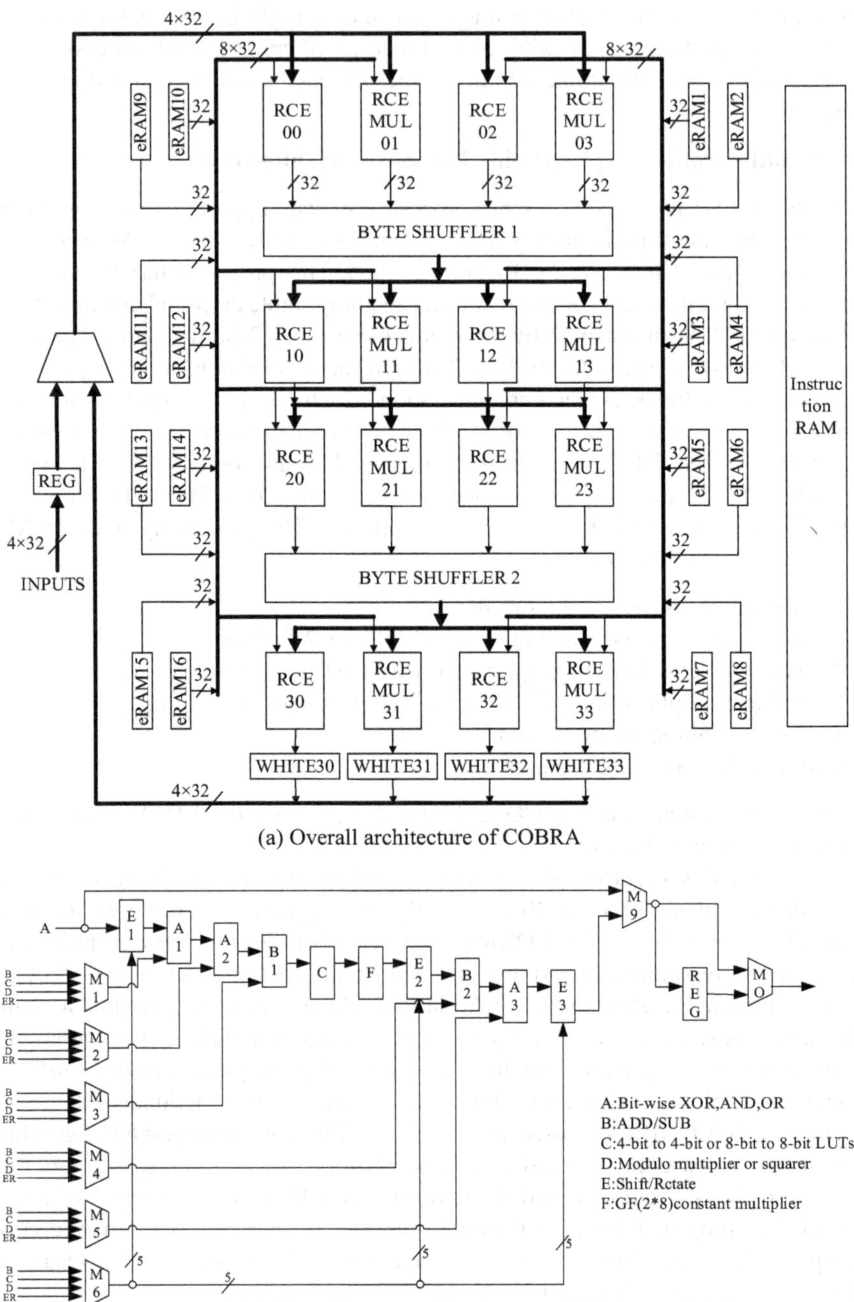

(a) Overall architecture of COBRA

(b) RCE architecture without any multiplier

Fig. 1.36 COBRA architecture and RCE architecture [97]

Table 1.5 Flexibility analysis of reconfigurable cryptographic processor

CGRA	Algorithm																		
	Block cipher								Hash function				Public-key cryptography						
	AES	DES	IDEA	RC6	Serpent	Twofish	Camellia	SEED	SM4	MD4	MD5	SHA-1	SHA-2	RSA-512	RSA-1024	RSA-2048	ECC-163	ECC-239	ECC-571
COBRA [97]	✓	✓	✓	✓	✓	✓	✓	–	–	–	–	–	–	–	–	–	–	–	–
Celator [98]	✓	✓	–	–	–	–	–	–	–	–	–	–	✓	–	–	–	–	–	–
Cryptoraptor [99]	✓	✓	✓	✓	✓	✓	✓	✓	–	✓	✓	✓	✓	–	–	–	–	–	–
HPURC [100]	–	–	–	–	–	–	–	–	–	–	–	–	–	✓	✓	–	✓	✓	✓
CGRAC [101]	✓	✓	✓	✓	✓	✓	✓	–	✓	–	–	–	✓	✓	✓	–	✓	✓	✓
SRCP [102]	✓	✓	✓	✓	–	–	–	–	✓	–	–	–	–	–	–	–	–	–	–
REPROC [103]	–	–	–	–	–	–	–	–	–	–	–	–	–	–	–	–	–	–	–

Table 1.6 Performance analysis of reconfigurable cryptographic processor

CGRA	Algorithm															
	Block cipher (Mb/s@MH$_2$)							Hash function (Mb/s@MHz)			Public-key cipher (ms@MH$_2$)					
	AES	DES	IDEA	RC6	Serpent	Twofish	SM4	MD5	SHA-1	SHA-2	RSA-512	RSA-1024	RSA-2048	ECC-163	ECC-239	ECC-571
COBRA [97]	1451.3 @102	–	–	3902. 4@61	2306.3 @54	–	–	–	–	–	–	–	–	–	–	–
Celator [98]	46@ 190	26@ 190	–	–	–	–	–	–	–	36 @190	–	–	–	–	–	–
Cryptoraptor [99]	6553.6 @1000	2734.1 @1000	–	–	–	3276. 8 @1000	–	4075. 5 @1000	4372. 5@1000	1638.4 @1000	–	–	–	–	–	–
HPURC [100]	–	–	–	–	–	–	–	–	–	–	0.98 @415	3.84 @415	–	44 @415	–	4.9 @415
CGRAC [101]	17755. 5 @200	11717. 9 @200	4466.1 @200	4763. 6 @200	2317.4 @200	3020.3 @200	–	12143.8 @200	9684. 9 @200	3051. 8 @200	2. 63 @200	10.50 @200	41. 96 @200	0.21 @200	0. 44 @200	2. 50 @200
SRCP [102]	304 @100	78 @100	100 @100	86. 5 @100	–	–	–	–	–	–	–	–	–	–	–	–
REPROC [103]	2212 @400	2785 @400	–	–	–	–	6451 @400	–	–	–	–	–	–	–	–	–

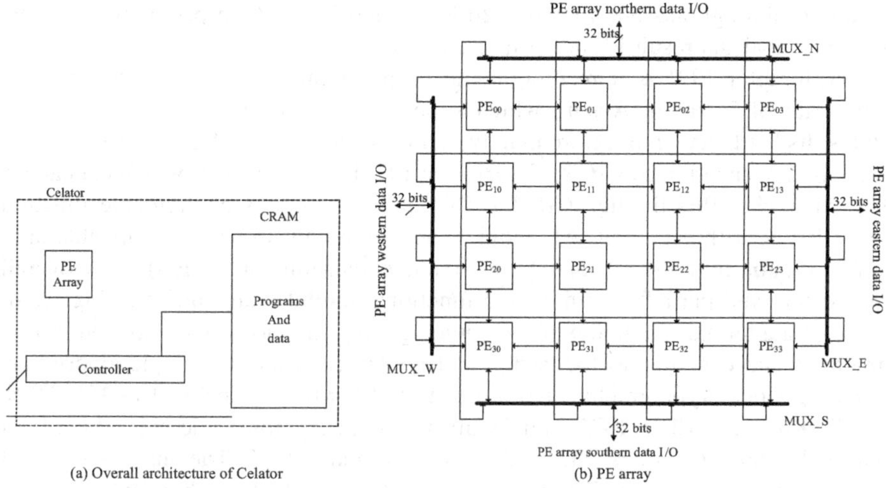

(a) Overall architecture of Celator (b) PE array

Fig. 1.37 Overall architecture of Celator and PE array [98]

array, and a Celator RAM (CRAM). The overall architecture and PE array of Celator are shown in Fig. 1.37. The PE array is designed based on the systolic structure, and it adopts a two-dimensional interconnection structure to implement data transmission between computing units. The 4 × 4 two-dimensional PE array is composed of 8-bit processing units. The finite state machine is used to control the data flow and arithmetic operation between computing units. A dual-port RAM is used as the CRAM to store data and instructions. The 4 × 4 two-dimensional PE array provides four 32-bit data I/O interfaces and four multiplexers (MUX_N, MUX_E, MUX_W, and MUX_S) which can implement data selection in four directions (east, south, west, and north). Each PE contains one arithmetic logic unit, two 8-bit registers (register A and register B), four 8-bit data input interfaces (one interface per direction), one 8-bit data output interface which is connected with the adjacent data input interface, four multiplexers (two of them are used to select the input or output data from four directions and two of them are used to select the data in registers A and B).

This architecture is designed for three mainstream cryptographic algorithms (AES, DES, and SHA-512) and thus has a large limitation in flexibility. The analysis result is shown in Table 1.5. On design, Celator improves the utilization of hardware resources. However, an obvious transmission bottleneck exists in its data flow and configuration flow, which limits Celator's final implementation efficiency. In addition, this architecture provides a low computing efficiency for permutation operation which is common in cryptographic algorithms. In ECB mode, Celator takes 514 clock cycles to execute AES-128 and provides a maximum throughput of 47 Mb/s; in CBC mode, it takes 524 clock cycles to execute AES-128, 476 clock cycles to execute DES, and 2,720 clock cycles to execute SHA-512, and the maximum throughputs are 46 Mb/s, 26 Mb/s, and 36 Mb/s respectively. The performance analysis result is shown in Table 1.6.

Cryptoraptor [99] is a reconfigurable cryptographic processor developed by University of Texas at Austin, which supports hundreds of cryptographic algorithms. Its hardware architecture mainly includes state engine (SE), register file, PE array, and connection row (CR). Every four PEs form a PE line, which exchanges data with other PEs through CR. Each PE has five bypass configurable function units, that is, arithmetic unit (AU), logical operation unit (LOU), look-up table unit, shifter–rotator unit (SRU), and permutation–expansion unit (PEU). The overall architecture and partially configurable functional modules are shown in Fig. 1.38.

Cryptoraptor has high flexible computing unit functions and interconnection structures and can support the design and implementation of multiple algorithms, including block cryptographic algorithms AES, Blowfish, Camellia, Cast128, DES, GOST, Kasumi, SEED, RC5, and Twofish, stream cryptographic algorithms RC4 and Phelix, hash functions MD4, MD5, SHA-1, and SHA-2. The analysis result of its flexibility is shown in Table 1.5. It provides a high throughput when implementing block cryptographic algorithm. It can meet the requirements of various cryptographic algorithms for the maximum concurrent memory access and storage, and has the highest performance and area efficiency so far. It adapts to as many cryptographic algorithms as possible to ensure the flexibility. This, however, brings the problem of over-designed computing units. The problem of excessive hardware redundancy may exist for some block cryptographic algorithms when they are mapped to Cryptoraptor. This brings large limitation to the area efficiency of the processor Cryptoraptor. Table 1.6 lists the performance of Cryptoraptor when different cryptographic algorithms are used.

In [100], a high-performance universal reconfigurable cryptography processor (HPURC) is put forward. The reconfigurable datapaths of this processor are suitable for computing any prime and irreducible polynomial in the prime field and binary extension field. Therefore, this architecture can be used to compute public-key cipher RSA and elliptic curve cryptography. A reconfigurable cell array (RCA), as shown in Fig. 1.39a, can implement various cipher operations, including Montgomery modular multiplication, square, and multiplication and division in a finite field. An RCA contains 1,024 reconfigurable cells (RCs) and supports runtime configuration so as to execute specific operations. Each RC contains a universal reconfigurable executional unit (REU) used to execute arithmetic operation and some multiplexed local registers (MLRs) used to select specific operands.

Some universities in China have made researches on the structure of reconfigurable cryptographic processors and have gained some research achievements. The PLA Information Engineering University has made in-depth research into reconfigurable cryptographic processors and put forward a reconfigurable cryptographic processor model CGRAC [101], which supports block cryptographic algorithm, hash function, and public-key algorithm. CGRAC is mainly composed of interface module, global control module (GCM), directly memory access (DMA) control module, data buffer module (DBM), configuration and instruction module (CIM), and reconfigurable array (RA) module. RA is a core computing device of CGRAC,

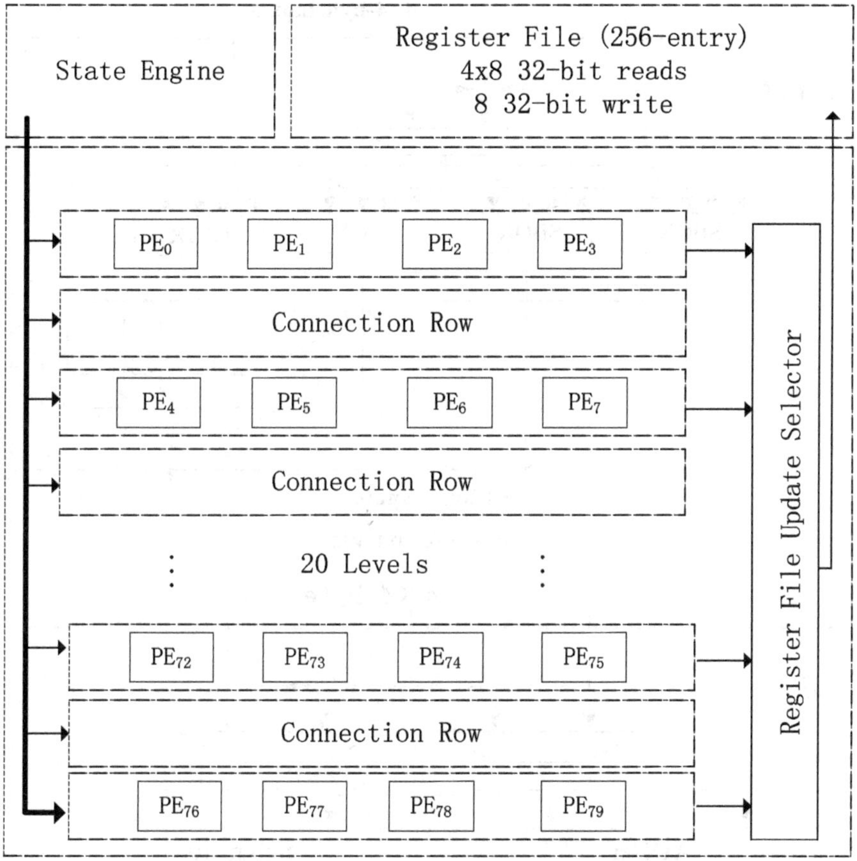

(a) Overall architecture of cryptoraptor

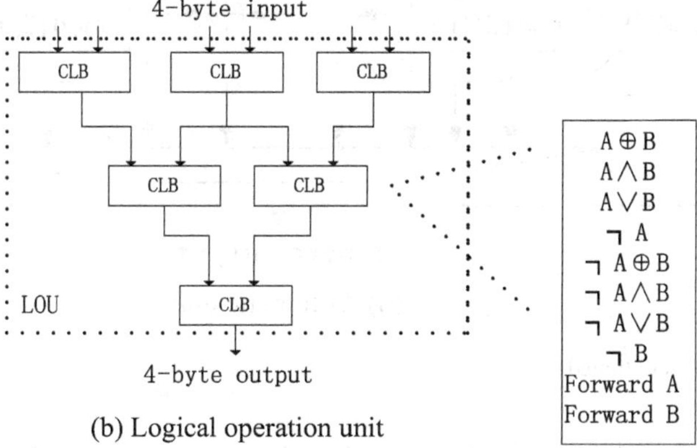

(b) Logical operation unit

Fig. 1.38 Overall architecture and some configurable function modules of Cryptoraptor [99]

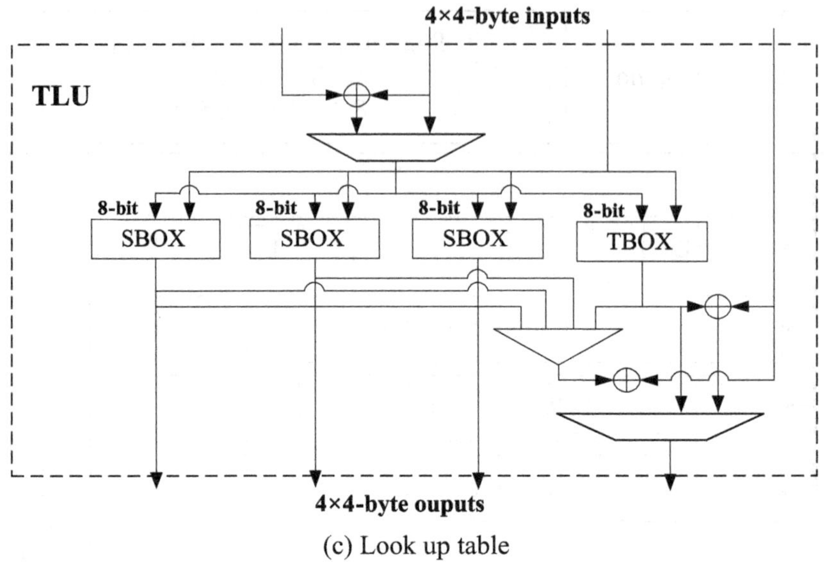

(c) Look up table

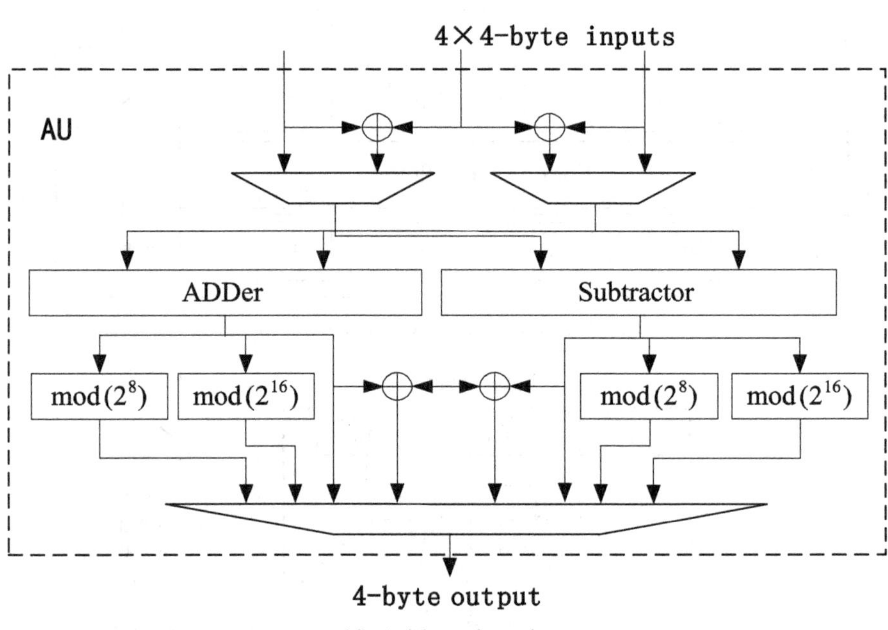

(d) Arithmetic unit

Fig. 1.38 (continued)

and it is mainly composed of reconfigurable function unit (RFU), external bus, and reconfigurable interconnect network (RTN). By reconfiguring RFU and RTN, RA can implement various functions to match different algorithms. RFU is mainly

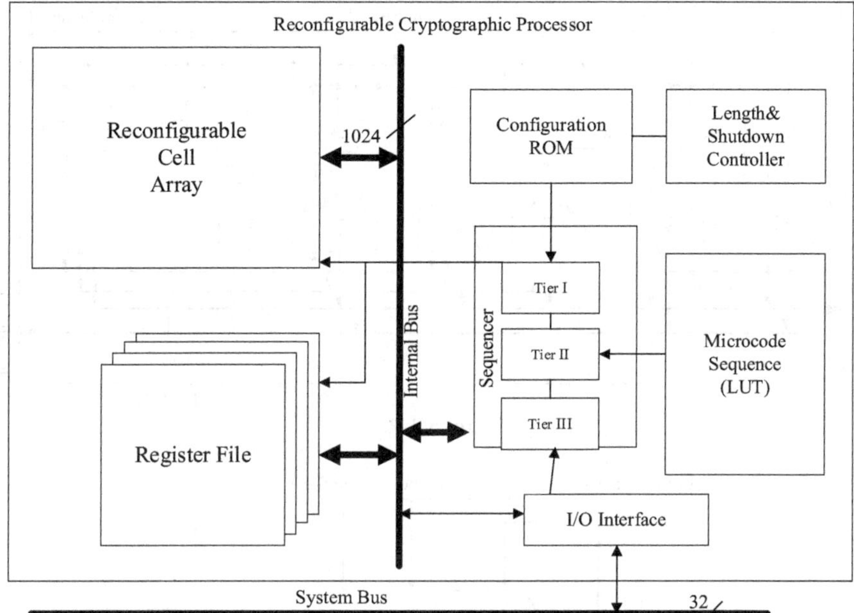

(a) Overall architecture of HPURC

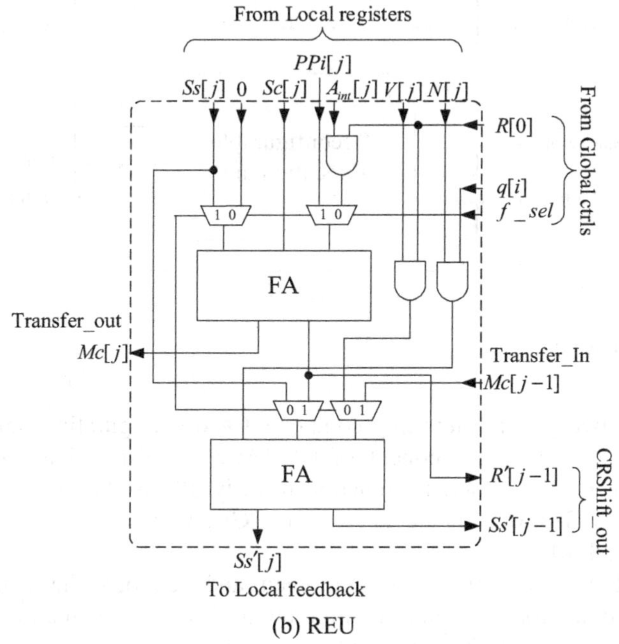

(b) REU

Fig. 1.39 Overall architecture and its RCs of HPURC [100]

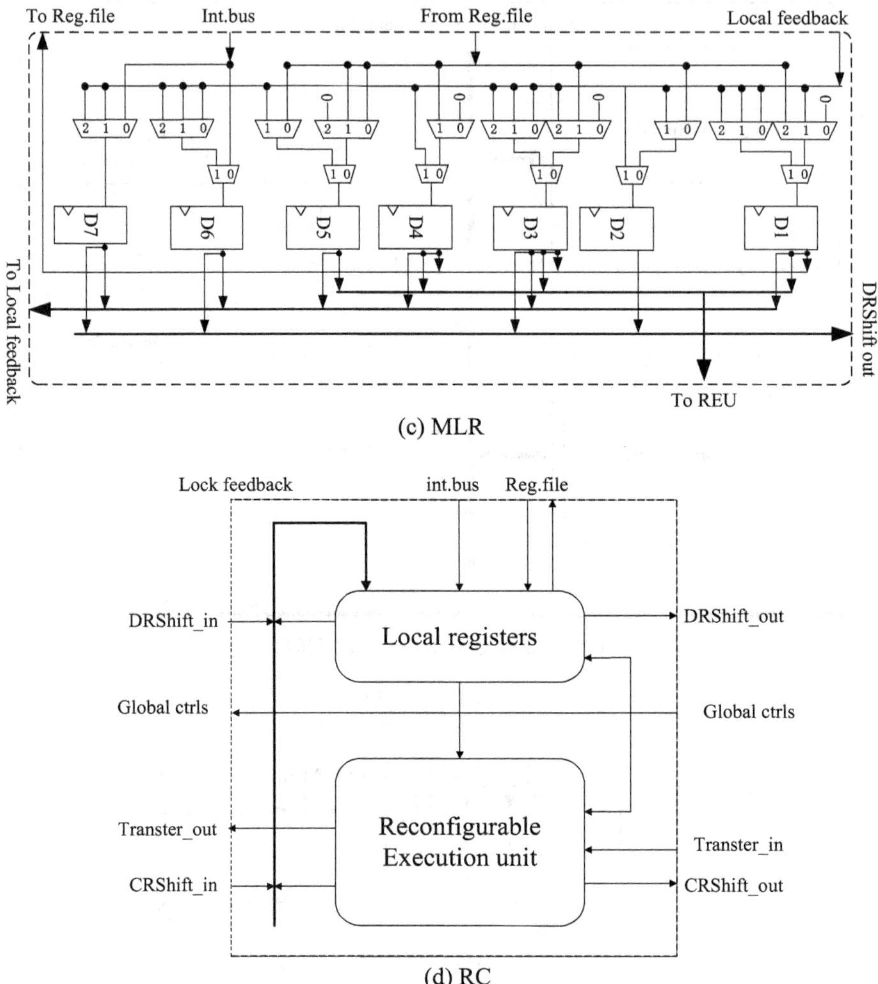

(c) MLR

(d) RC

Fig. 1.39 (continued)

composed of two parts: computing component and configuration buffer. It is not only a basic operation component of CGRAC, but also a basic configuration component of CGRAC, and its function directly affects the performance of the entire structure. The overall architecture of CGRAC and its RAs and RFUs are shown in Fig. 1.40.

The CGRAC structure is coarse-grained and supports the combination of dynamic configuration and static configuration. The implementation of interconnection structure inherits the interconnection features of linear array, two-dimensional grid, and crossbar. It supports horizontally parallel processing and vertical pipeline processing, and it can flexibly change the horizontal parallelism

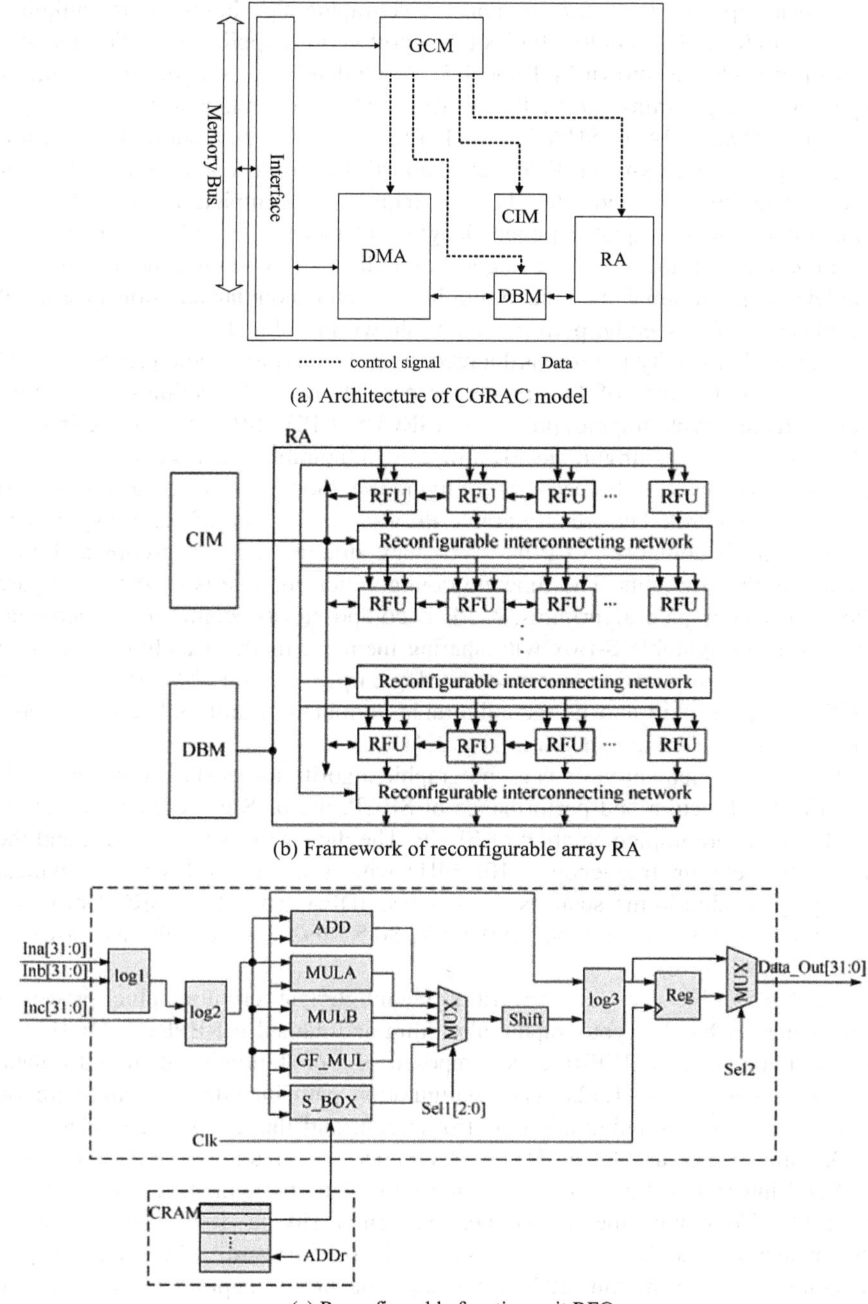

(a) Architecture of CGRAC model

(b) Framework of reconfigurable array RA

(c) Reconfigurable function unit RFO

Fig. 1.40 Overall architecture and its RA and RFU of CGRAC [101]

and vertical pipeline depth for different cryptographic algorithms via reconfiguration, so as to form different topologies to support various applications. The analysis result of flexibility is shown in Table 1.5. On CGRAC, such algorithms as block cryptographic algorithms (DES, IDEA, RC6, AES, Serpent, and Twofish), hash functions (MD5, SHA-1, SHA-2, and RIPEDMD-160), and public-key cryptographic algorithms (RSA and ECC) are mapped. For RSA, one execution time of the complete algorithm indicates the time required to complete a modular exponentiation which has equal exponent length and modulus length. For ECC, one execution time of the complete algorithm indicates the total time required to complete all polynomial modular multiplication and modular addition in a point multiplication. The specific performance is shown in Table 1.6.

Southeast University put forward a reconfigurable cryptographic processor SOC [102]. The architecture of this processor consists of such modules as security reconfigurable cryptographic processor (SRCP), CPU, SRAM, and peripheral. SRCP consists of reconfigurable PE array, reconfigurable registers, function configuration module, countermeasure configuration module, interconnect bus, and buffer output control. The reconfigurable PE array is the core of the cryptographic processor and is composed of PEs, which have different register groups and configurable arithmetic units. The function configuration module is used to configure different cryptographic algorithms. A PE is composed of reconfigurable permutation unit, reconfigurable S-Box with sharing memory, modular addition operation unit, modular multiplication unit, and shift logic operation unit. The architecture of SRCP system on chip and its reconfigurable permutation unit and reconfigurable storage shared S-Box are shown in Fig. 1.41.

SRCP can implement various cryptographic algorithms, as shown in Table 1.5. To verify the function and performance of SRCP, the 0.18 μm CMOS process is used for hardware implementation of SRCP. The die size is $4.1 \times 4\ \text{mm}^2$, and the maximum operating frequency is 100 MHz with a supply of 1.8 V. For typical cryptographic algorithms such as AES, DES, IDEA, and RC6, SRCP provides throughputs 304 Mb/s, 78 Mb/s, 100 Mb/s, 86.5 Mb/s, respectively, as shown in Table 1.6.

Tsinghua University put forward a reconfigurable cryptographic processor architecture for block cryptographic algorithm and named it REPROC [103]. The overall architecture of REPROC is composed of a configuration path and a datapath, as shown in Fig. 1.42a. The configuration path includes the configuration storage unit and its configuration control circuit, and the datapath includes input FIFO, output FIFO, and RCA. The RCA architecture includes reconfigurable cells (RC) and interconnection network composed of 2-to-1 multiplexers, as shown in Fig. 1.42b. To reduce the interconnection complexity, a design method called interconnection tree between rows (ICTR) is put forward, where each RC is interconnected with the nine RCs in the adjacent lines to support bit-wise permutation and irregular rotation. To reduce the configuration and accelerate dynamic configuration, a design method called hierarchical context organization (HCO) is put forward, where the configuration is divided into top context, group context, and core context. For different encryption algorithms, the total configuration can be

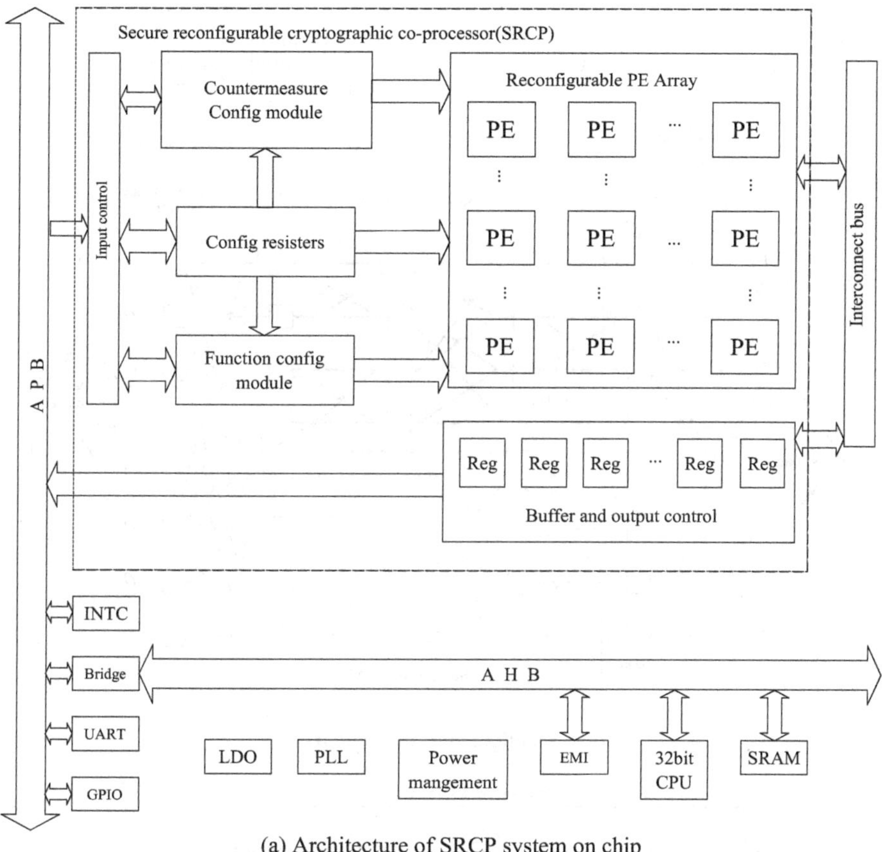

(a) Architecture of SRCP system on chip

Fig. 1.41 Architecture of SRCP system on chip [102]

reduced by 82.8–93.6%. This architecture is applicable to multiple encryption algorithms, such as AES, DES, SHACAL-1, SMS4, and ZUC (as shown in Table 1.5). The 65 nm CMOS process is used for the hardware implementation of REPROC. The chip has a die size of 51.36 mm^2, a maximum operating frequency of 400 MHz, an area efficiency of 0.99 Gb/s/mm^2, and an energy efficiency of 87.6 Gb/s/W. The throughputs for different typical cryptographic algorithms are shown in Table 1.6.

The above reconfigurable cryptographic processors are specially designed for multiple cryptographic algorithms and can meet certain energy efficiency metric without compromising the flexibility. Some researchers even try to directly map the cryptographic algorithm to the universal reconfigurable cryptography processor FPGA. FPGA has a very high flexibility and can implement almost any digital logic regardless of its capacity. FPGA is a fine-grained (the configuration granularity is usually less than 4 bits) reconfigurable processor, and its functions can be changed

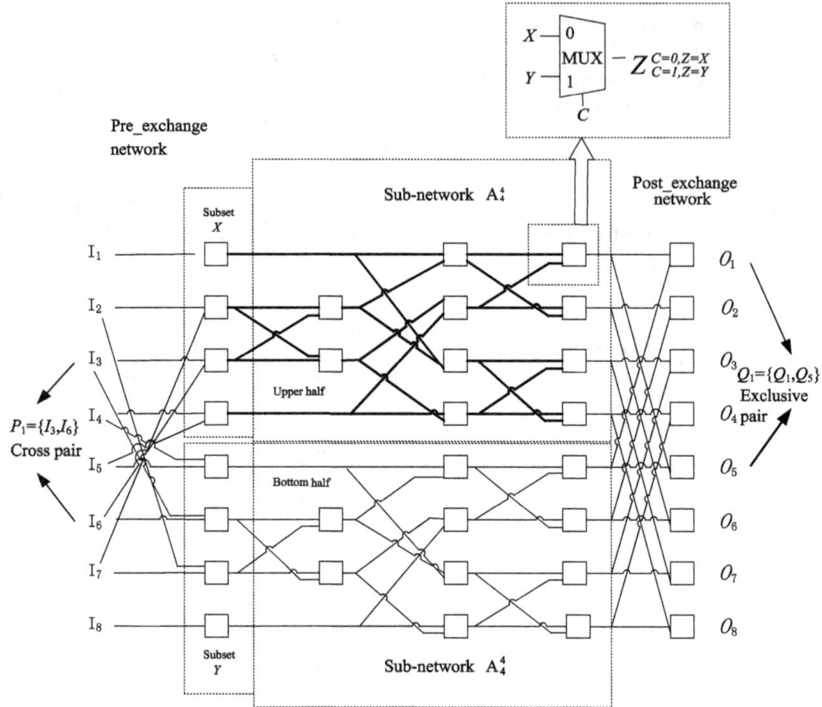

(b) Reconfigurable permutation unit

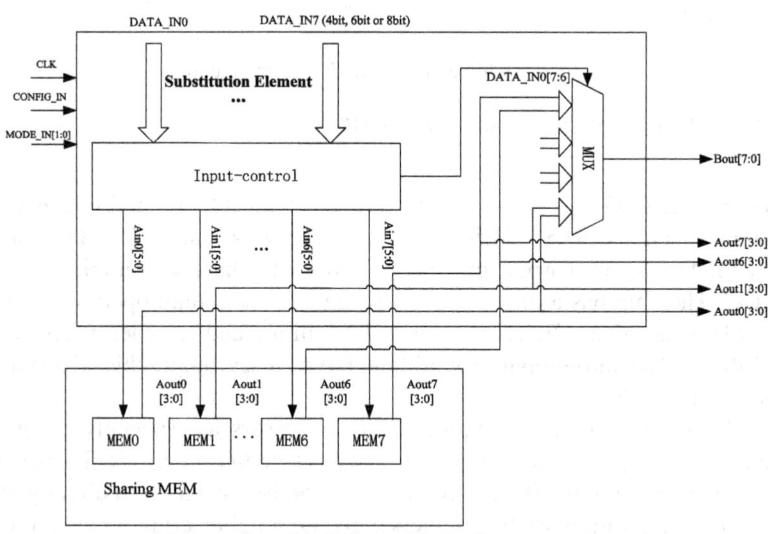

(c) Reconfigurable S-Box with sharing memory

Fig. 1.41 (continued)

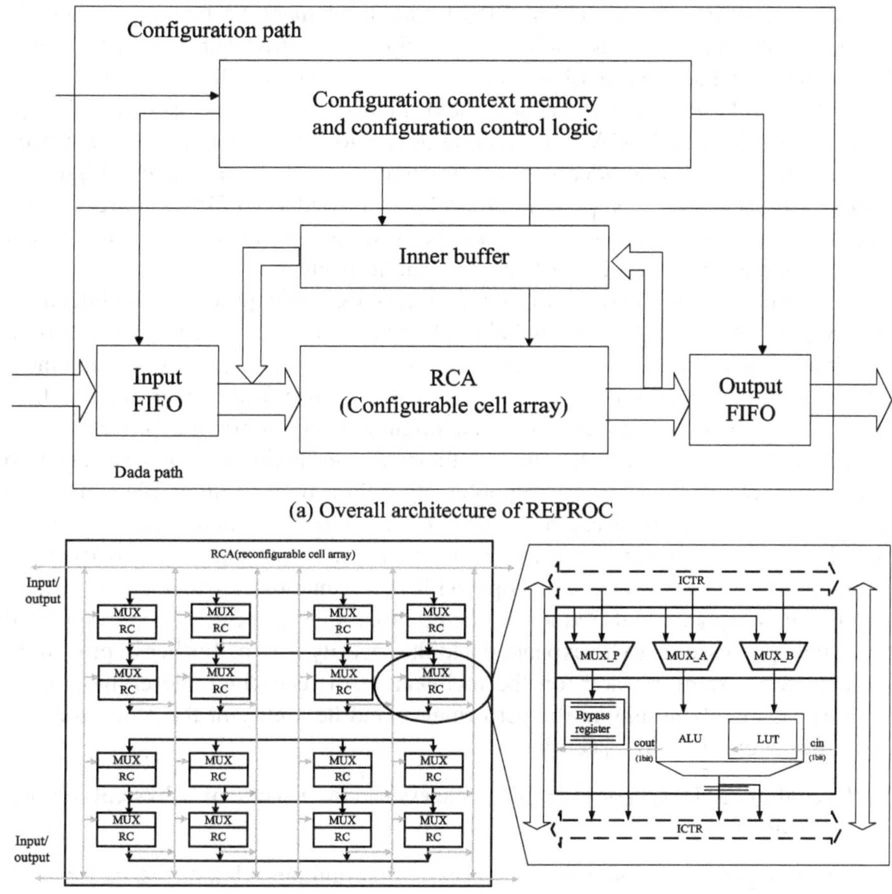

(a) Overall architecture of REPROC

(b) Architecture of reconfigurable array

Fig. 1.42 Architecture of REPROC system [103]

via configuration. It implements programmable interconnection by means of switching unit and connecting unit and stores configuration on chip by means of cross-coupling transistors (like SRAM unit). In theory, it can implement any algorithms. In [104], a complete pipeline architecture is used to implement SHACAL algorithm, which can provide a maximum throughput of 17 Gb/s. In [105], two AES architectures with different performance are put forward, which can provide a maximum throughput of 25 Gb/s. By means of iteration structure, these two architectures can reduce the hardware complexity and provide a throughput of 2.2 Mb/s when only 124 slices in Xilinx Spartan-II XC2S15 are used. In [106], DSP and BRAM modules on FPGA are used to implement the AES algorithm, so as to reduce the usage of FF and LUT. In [107], the reconfiguration feature of FPGA is used to implement iteration and full expansion design of symmetric keys and ciphers. It can be seen that FPGA can implement various cryptographic algorithms and meet various application requirements. However, FPGA has some disadvantages which are hard to surmount and make it difficult to gain wide application in the information security field. Firstly, from the perspective of cryptographic application, FPGA is a common device to some extent and has not been customized for encryption/decryption computation. It does not have a high efficiency in processing cryptographic algorithms. Secondly, FPGA is more susceptible to threats from physical attacks because its hardware architecture and configuration are organized in a fixed and public manner.

Therefore, though many reconfigurable cryptographic processor architectures have been put forward at home and abroad, there is still a gap between the current situation and wide application of reconfigurable cryptographic processors, and there are many critical scientific problems to be solved. Specifically, there is a lack of mathematical models for designing reconfigurable cryptographic processor and unified methods to evaluate the energy efficiency and flexibility; there has been no mature mapping method for cryptographic algorithms on reconfigurable computing array; there is a lack of research on methods for scheduling and managing reconfigurable hardware resources; only frequent partial reconfiguration and dynamic reconfiguration technologies can cope with the scenarios where there is a high configuration capacity and frequency; the cryptographic algorithm scope supported by a single processor is to be expanded and the security is to be further improved. In addition, the current research on the architecture of reconfigurable cryptographic processor is mainly aiming at the performance and flexibility of the processor, and pays litter attention to the security.

2. **Reconfigurable Cryptographic Processors in Terms of Physical Attack Resistance**

A reconfigurable cryptographic processor has unique advantages in terms of security. Compared to the software implementation of ISAP, it resists physical attacks easier and receives no threat from reverse engineering due to its "blank chip" feature. By fully developing the partial dynamic reconfiguration feature of reconfigurable cryptographic processor and the array computing mode, it is possible to implement physical attack resistance with less cost on performance, area, and power consumption, and presents a new solution to the endless new attacks. However, though a reconfigurable cryptographic processor has a great potential in resisting physical attacks, the current research is still mainly limited to how to introduce traditional physical attack resistance measures, mechanisms or methods into reconfigurable cryptographic processors, and the special advantages of a reconfigurable cryptographic processor on hardware architecture and computing form have not yet been exploited and utilized. In [108], a countermeasure based on space redundancy is implemented on a reconfigurable architecture, and the normal computation execution path and fault detection redundancy execution path are respectively mapped to different parts of the computing array for fault detection and fault attack resistance. In [21], power attack countermeasures based on hiding

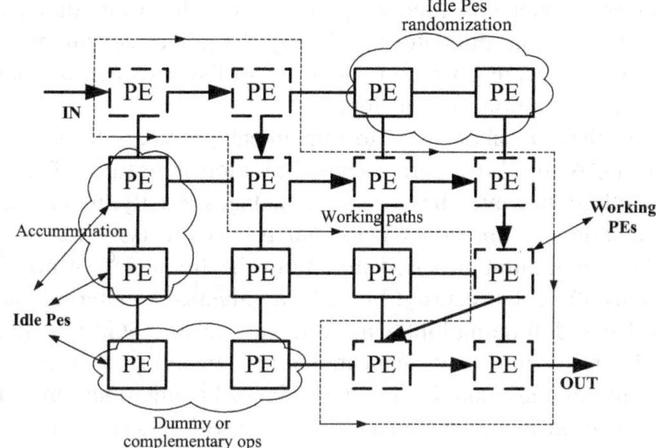

(a) Implement physical attack resistance by means of idle PEs

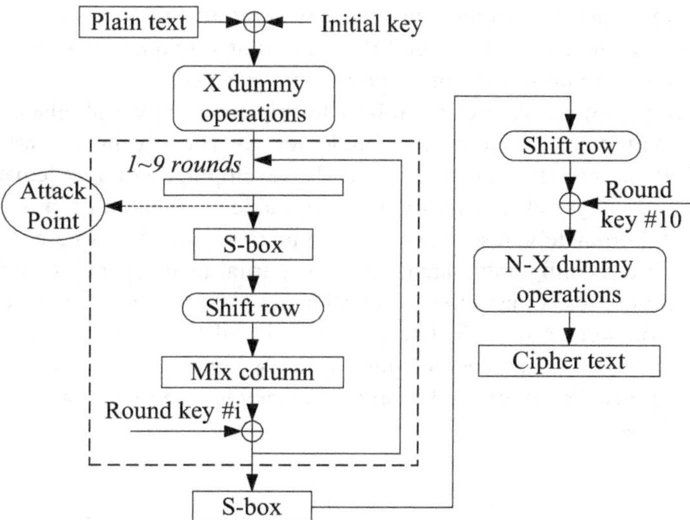

(b) Implement physical attack resistance by means of pseudo-operations

Fig. 1.43 Power attack resistance measures based on hiding technology [108]

technology (Fig. 1.43) are adopted. The original idle units in the reconfigurable array are configured as power generating body (e.g., a power generating body used to generate the power complementary to the original power consumption curve), or dummy operations are added to disturb the execution timing, so as to reduce the signal-to-noise ratio of sensitive signals and thus resist power attack. In [109], countermeasures against physical attack are put forward for reconfigurable computing architecture for cryptographic algorithms, including using random-source-driven clock signals to control the reconfigurable computing resources so as to generate random power noise, inserting dummy operations randomly to increase the difficulty in aligning the power consumption curve, and using idle PEs to execute operations opposite to that executed by active PEs for hiding power consumption information.

Some researches are about how to implement physical attack resistance technologies on FPGA. In [110], partial reconfiguration is used for fault attack resistance. This method, however, has two disadvantages. Firstly, it is not universal and requires extracting the function blocks and routes precisely for reconfiguration. Secondly, the reconfiguration speed may limit the frequency of position randomization and thus affect the security. In [111], a universal countermeasure applicable to an SRAM-based time-multiplexing soft gate array (SGA) is studied, which transforms flexibility into a mechanism for implementing universal countermeasures. Compared to traditional FPGA, this universal countermeasure can implement power attack resistance and fault attack resistance with less cost on time, speed, and power consumption. In [112], a self-evolution hardware system supporting fault tolerance and fault correction is put forward. However, as it is difficult for FPGA to implement real partial reconfiguration and dynamic reconfiguration, these researches on physical attack resistance on FPGA are limited to a large extent and cannot meet the security requirements of cryptographic processors.

Objectively, current researches on how to implement physical attack resistance through reconfigurable computing technologies are actually in its start-up stage; many critical technologies are still constantly developing, and a systematic design method against physical attacks for reconfigurable cryptographic processors has been formed. Fortunately, researches have begun to explore how to use such special advantages of a reconfigurable architecture as partial reconfiguration and dynamic reconfiguration to resist many new physical attacks which cannot be coped with by the current countermeasure effectively. It is believed that in near future, a reconfigurable cryptographic processor which can cope with various physical attacks with high security, flexibility, and energy efficiency is sure to gain wide application in various fields.

References

1. Stallings W (2006) Cryptography and network security: principles and practice. Pearson Education, Delhi
2. Chen L (2014) Research on and application of IoT-based information security transmission system. Master's thesis of North China University of Technology, Beijing
3. Yanting R (2015) Utilization of information about side channel attack for cipher chips and research on comprehensive defense technologies. Master's thesis of Tsinghua University, Beijing
4. Koç ÇK (2009) Cryptographic engineering. Springer, New York
5. Kent S, Corp B, Atkinson R. Security architecture for the internet protocol [EB/OL]. https://tools.ietf.org/html/rfc2401.html. Accessed 01 Nov 1998
6. Elgamal T, Hickman KEB. Secure socket layer application program apparatus and method. U.S. Patent 5, 657, 390, 1997-8-12

7. Rueppel RA (1986) Analysis and design of stream ciphers. Springer, New York, pp 177–190
8. Gura N, Patel A, Wander A et al (2004) Comparing elliptic curve cryptography and RSA on 8-bit CPUs. In: Proceedings of the 2004 international workshop on cryptographic hardware and embedded systems-CHES Cambridge, MA, USA, 11–13 August 2004, pp 119–132
9. Society IC. IEEE Std 802.3-2008. LAN/MAN Standards Committee, 2008
10. Hiertz GR, Denteneer D, Stibor L et al (2010) The IEEE 802.11 universe. Commun Mag IEEE 48(1):62–70
11. Nishida Y, Kawai K, Koike K (2010) A 2 Gb/s network processor with a 24 mW IPsec offload for residential gateways. In: 2010 IEEE international solid-state circuits conference digest of technical papers (ISSCC), pp 280–281
12. Prasithsangaree P, Krishnamurthy P (2004) Analysis of energy consumption of RC4 and AES algorithms in wireless LANs. In: Global telecommunications conference. IEEE, pp 1445–1449
13. Burd TD, Brodersen RW (1995) Energy efficient CMOS microprocessor design. In: Hawaii international conference on system sciences, p 288
14. Mangard SOEPT (2008) Power analysis attacks: revealing the secrets of smart cards. Springer Science & Business Media, Dordrecht
15. Anderson R, Kuhn M (1996) Tamper resistance: a cautionary note. In: Conference on proceedings of the second Usenix workshop on electronic commerce, p 1
16. Sergei S (2011) Physical attacks on tamper resistance: progress and lessons. In: Proceedings of 2nd ARO special workshop on hardware assurance
17. Skorobogatov PS (2005) Semi-invasive attacks: a new approach to hardware security analysis. University of Cambridge doctor dissertation, Cambridgeshire
18. Bar-El H, Choukri H, Naccache D et al (2006) The sorcerer's apprentice guide to fault attacks. Proc IEEE 94(2):370–382
19. Kocher P, Jaffe J, Jun B (1999) Differential power analysis. In: International cryptology conference on advances in cryptology, pp 388–397
20. Gandolfi K, Mourtel C, Olivier F (2001) Electromagnetic analysis: concrete results. In: International workshop on cryptographic hardware and embedded systems, pp 251–261
21. Shan W, Shi L, Fu X et al (2014) A side-channel analysis resistant reconfigurable cryptographic coprocessor supporting multiple block cipher algorithms. In: Design automation conference, pp 1–6
22. Genkin D, Shamir A, Tromer E (2014) RSA key extraction via low-bandwidth acoustic cryptanalysis. In: International cryptology conference, pp 444–461
23. Genkin D, Pipman I, Tromer E (2015) Get your hands off my laptop: physical side-channel key-extraction attacks on PCs. J Cryptogr Eng 5(2):95–112
24. Briais S, Cioranesco JM, Danger JL et al (2012) Random active shield. In: The workshop on fault diagnosis and tolerance in cryptography, pp 103–113
25. Karaklaji D, Schmidt JM, Verbauwhede I (2013) Hardware designer's guide to fault attacks. IEEE Trans Very Large Scale Integr Syst 21(12):2295–2306
26. Joye M, Manet P, Rigaud JB (2007) Strengthening hardware AES implementations against fault attacks. IET Inf Secur 1(3):106–110
27. Herbst C, Oswald E, Mangard S (2006) An AES smart card implementation resistant to power analysis attacks. In: International conference on applied cryptography and network security, pp 239–252
28. Tiri K, Verbauwhede I (2004) A logic level design methodology for a secure DPA resistant ASIC or FPGA implementation. In: Design, automation and test in Europe conference and exhibition, proceedings, p 10246
29. Schramm K, Paar C (2006) Higher order masking of the AES. In: Cryptographers' track at the RSA conference, pp 208–225
30. Wang B, Liu L, Deng C et al (2016) Against double fault attacks: injection effort model, space and time randomization based countermeasures for reconfigurable array architecture. IEEE Trans Inf Forensics Secur 11(6):1151–1164

31. Ghalaty NF, Yuce B, Taha M et al (2014) Differential fault intensity analysis. In: 2014 workshop on fault diagnosis and tolerance in cryptography (FDTC), pp 49–58
32. Beroulle V, Candelier P, Castro SD et al (2014) Laser-induced fault effects in security-dedicated circuits. In: IFIP/IEEE international conference on very large scale integration-system on a chip, pp 220–240
33. Genkin D, Pachmanov L, Pipman I et al (2015) Stealing keys from PCs using a radio: cheap electromagnetic attacks on windowed exponentiation. In: International workshop on cryptographic hardware and embedded systems, pp 207–228
34. Lin SY, Huang CT (2007) A high-throughput low-power AES cipher for network applications. In: Design automation conference, Asia and South Pacific, pp 595–600
35. Ueno R, Morioka S, Homma N et al (2016) A high throughput/gate AES hardware architecture by compressing encryption and decryption datapaths. In: International conference on cryptographic hardware and embedded systems, pp 538–558
36. Liu Z, Liu D, Zou X (2017) An efficient and flexible hardware implementation of the dual-field elliptic curve cryptographic processor. IEEE Trans Industr Electron 64(3): 2353–2362
37. Zhang Y, Yang K, Saligane M et al (2016) A compact 446 Gbps/W AES accelerator for mobile SoC and IoT in 40 nm. In: 2016 IEEE symposium on VLSI circuits (VLSI-circuits), pp 1–2
38. Mathew S, Satpathy S, Suresh V et al (2015) 340 mV–1.1 V, 289 Gbps/W, 2090-gate nanoAES hardware accelerator with area-optimized encrypt/decrypt GF (2 4) 2 polynomials in 22 nm tri-gate CMOS. IEEE J Solid-State Circuits 50(4):1048–1058
39. Henzen L, Aumasson JP, Meier W et al (2011) VLSI characterization of the cryptographic hash function BLAKE. IEEE Trans Very Large Scale Integr Syst 19(10):1746–1754
40. Lutz AK, Treichler J, Gürkaynak FK et al (2002) 2Gbit/s hardware ealizations of RIJNDAEL and SERPENT: a comparative analysis. Lect Notes Comput Sci 2523:144–158
41. Liu PC, Chang HC, Lee CY (2009) A 1.69 Gb/s area-efficient AES crypto core with compact on-the-fly key expansion unit. In: Proceedings of ESSCIRC, pp 404–407
42. Su CP, Lin TF, Huang CT et al (2003) A high-throughput low-cost AES processor. Commun Mag IEEE 41(12):86–91
43. Hodjat A, Schaumont P, Verbauwhede I (2004) Architectural design features of a programmable high throughput AES coprocessor. In: Proceedings of the international conference on information technology: coding and computing, pp 498–502
44. Hamalainen P, Alho T, Hannikainen M et al (2006) Design and implementation of low-area and low-power AES encryption hardware core. In: Euromicro conference on digital system design: architectures, methods and tools, DSD 2006, pp 577–583
45. Good T, Benaissa M (2010) 692-nW advanced encryption standard (AES) on a 0.13-μmCMOS. IEEE Trans Very Large Scale Integr Syst 18(12):1753–1757
46. Mathew S, Sheikh F, Agarwal A et al (2010) 53 Gbps native GF(24) 2 composite-field AES-encrypt/decrypt accelerator for content-protection in 45 nm high-performance microprocessors. In: 2010 IEEE symposium on VLSI circuits (VLSIC). IEEE, pp 169–170
47. Lee JW, Chung SC, Chang HC et al (2013) Efficient power-analysis-resistant dual-field elliptic curve cryptographic processor using heterogeneous dual-processing-element architecture. IEEE Trans Very Large Scale Integr Syst 22(1):49–61
48. Dao VL, Nguyen VT, Hoang VP (2016) Low power ECC implementation on ASIC. In: International conference on advances in information and communication technology, pp 332–339
49. Guo X, Srivastav M, Huang S et al (2012) ASIC implementations of five SHA-3 finalists. In: Design, automation and test in Europe conference and exhibition, pp 1006–1011
50. Koo B, Lee D, Ryu G et al (2006) High-speed RSA crypto-processor with radix-4 modular multiplication and Chinese remainder theorem. Lect Notes Comput Sci 81–93
51. Reparaz O, Bilgin B, Nikova S et al (2015) Consolidating masking schemes. Lect Notes Comput Sci 9215:764–783

52. Nikova S, Rechberger C, Rijmen V (2006) Threshold implementations against side-channel attacks and glitches. In: International conference on information and communications security, pp 529–545
53. Ishai Y, Sahai A, Wagner D (2003) Private circuits: securing hardware against probing attacks. Lect Notes Comput Sci 2729:463–481
54. De Cnudde T, Reparaz O, Bilgin B et al (2016) Masking AES with d+1 shares in hardware. In: ACM workshop on theory of implementation security, p 43
55. Tokunaga C, Blaauw D (2009) Secure AES engine with a local switched-capacitor current equalizer. In: IEEE international conference on solid-state circuits conference-digest of technical papers, 2009, ISSCC 2009, pp 64–65, 65a
56. Miura N, Fujimoto D, Tanaka D et al (2014) A local EM-analysis attack resistant cryptographic engine with fully-digital oscillator-based tamper-access sensor. In: 2014 symposium on VLSI circuits digest of technical papers, pp 1–2
57. Doulcier-Verdier M, Dutertre JM, Fournier J et al (2011) A side-channel and fault-attack resistant AES circuit working on duplicated complemented values. In: IEEE international solid-state circuits conference
58. Tillich SGJ (2006) Instruction set extensions for efficient AES implementation on 32-bit processors. In: International workshop on cryptographic hardware and embedded systems, pp 270–284
59. Roy S, Järvinen K, Verbauwhede I (2015) Lightweight coprocessor for Koblitz curves: 283-bit ECC including scalar conversion with only 4300 gates. In: International workshop on cryptographic hardware and embedded systems, pp 102–122
60. Han J, Dou R, Zeng L et al (2015) A heterogeneous multicore crypto-processor with flexible long-word-length computation. IEEE Trans Circuits Syst I Regul Pap 62(5):1372–1381
61. Rawat HK (2016) Vector instruction set extensions for efficient and reliable computation of keccak. Virginia Polytechnic Institute and State University master dissertation, Blacksburg
62. Soliman MI, Abozaid GY (2011) FPGA implementation and performance evaluation of a high throughput crypto coprocessor. J Parallel Distrib Comput 8(71):1075–1084
63. Hannes PT (2013) On using instruction-set extensions for minimizing the hardware-implementation costs of symmetric-key algorithms on a low-resource microcontroller. In: International conference on radio frequency identification: security and privacy issues, pp 149–164
64. Grabher P, Großschädl J, Dan P (2008) Light-weight instruction set extensions for bit-sliced cryptography. In: Proceedings of the international workshop on cryptographic hardware and embedded systems—CHES 2008, pp 331–345
65. O'Melia S, Elbirt AJ (2010) Enhancing the performance of symmetric-key cryptography via instruction set extensions. IEEE Trans Very Large Scale Integr Syst 18(11):1505–1518
66. Wang Y, Ha Y (2014) A performance and area efficient ASIP for higher-order DPA-resistant AES. IEEE J Emerg Sel Top Circuits Syst 4(2):190–202
67. May D, Muller HL, Smart NP (2001) Non-deterministic processors. In: Proceedings of the information security and privacy, Australasian conference, pp 115–129
68. Bruguier F, Benoit P, Torres L et al (2016) Cost-effective design strategies for securing embedded processors. IEEE Trans Emerg Top Comput 4(1):60–72
69. Estrin G (1960) Organization of computer systems-the fixed plus variable structure computer. In: Western joint IRE-AIEE-ACM computer conference, pp 33–40
70. DeHon A, Wawrzynek J (2002) Reconfigurable computing: what, why, and implications for design automation. In: Proceedings of the design automation conference, pp 610–615
71. Dehon A (2000) The density advantage of configurable computing. Computer 33(4):41–49
72. Hauser JR, Wawrzynek J (1997) Garp: a MIPS processor with a reconfigurable coprocessor. In: Proceedings of the IEEE symposium on field-programmable custom computing machines, pp 12–21
73. DeHon (2002) MATRIX: a reconfigurable computing architecture with configurable instruction distribution and deployable resources. In: Proceedings of the IEEE symposium on FPGAs for custom computing machines, pp 157–166

74. Taylor MB, Kim J, Miller J et al (2002) The raw microprocessor: a computational fabric for software circuits and general-purpose programs. Micro IEEE 22(2):25–35
75. Singh H, Lee MH, Lu G et al (2000) MorphoSys: an integrated reconfigurable system for data-parallel and computation-intensive applications. IEEE Trans Comput 49(5):465–481
76. Goldstein SC, Schmit H, Budiu M et al (2000) PipeRench: a reconfigurable architecture and compiler. Computer 33(4):70–77
77. Mei B, Vernalde S, Verkest D et al (2003) ADRES: an architecture with tightly coupled VLIW processor and coarse-grained reconfigurable matrix. In: Proceedings of the international conference on field programmable logic and application, pp 61–70
78. Govindaraju V, Ho CH, Nowatzki T et al (2012) DySER: unifying functionality and parallelism specialization for energy-efficient computing. IEEE Micro 32(5):38–51
79. Thoma F, Kuhnle M, Bonnot P et al (2007) MORPHEUS: heterogeneous reconfigurable computing. In: International conference on field programmable logic and applications, pp 409–414
80. Sankaralingam K, Nagarajan R, Liu H et al (2003) Exploiting ILP, TLP, and DLP with the polymorphous TRIPS architecture. Micro IEEE 23(6):46–51
81. Parashar A, Pellauer M, Adler M et al (2013) Triggered instructions: a control paradigm for spatially-programmed architectures. ACM Sigarch Comput Archit News 41(3):142–153
82. Becker J, Vorbach M (2004) Coarse-grain reconfigurable XPP devices for adaptive high-end mobile video-processing. In: Proceedings of the IEEE international SOC conference, pp 165, 166
83. Suzuki M, Hasegawa Y, Yamada Y et al (2005) Stream applications on the dynamically reconfigurable processor. In: Proceedings of the IEEE international conference on field-programmable technology, pp 137–144
84. Duller A, Towner D, panesar G et al (2005) Picoarray technology: the tool's story. In: Design, automation and test in Europe, pp 106–111
85. Dennard RH, Gaensslen FH, Rideout VL et al (2007) Design of ion-implanted MOSFET's with very small physical dimensions. IEEE J Solid-State Circuits 9(5):256–268
86. Bohr M (2007) A 30 year retrospective on Dennard's MOSFET scaling paper. IEEE Solid-State Circuits Soc Newslett 12(1):11–13
87. Shaojun Wei, Leibo Liu, Shouyi Yin (2014) Reconfigurable computing. Science Press, Beijing
88. Ebeling C, Cronquist DC, Franklin P (1996) RaPiD-reconfigurable pipelined datapath. In: International workshop on field-programmable logic, smart applications, new paradigms and compilers, pp 126–135
89. Zhu M, Liu L, Yin S et al (2010) A reconfigurable multi-processor SoC for media applications. In: IEEE international symposium on circuits and systems, pp 2011–2014
90. Yu Z, Meeuwsen MJ, Apperson RW et al (2008) AsAP: an asynchronous array of simple processors. IEEE J Solid-State Circuits 43(3):695–705
91. Tessier R, Burleson W (2001) Reconfigurable computing for digital signal processing: a survey. J Signal Process Syst 28(1):7–27
92. Sarker MAL, Lee MH (2012) Synthesis of VHDL code for FPGA design flow using Xilinx PlanAhead tool. In: International conference on education and E-learning innovations, pp 1–5
93. Chen YH, Krishna T, Emer JS et al (2016) Eyeriss: an energy-efficient reconfigurable accelerator for deep convolutional neural networks. IEEE J Solid-State Circuits (99):1–12
94. Tu F, Yin S, Ouyang P et al (2017) Deep convolutional neural network architecture with reconfigurable computation patterns. IEEE Trans Very Large Scale Integr Syst 25(8): 2220–2233
95. Putnam A, Caulfield AM, Chung ES et al (2016) A reconfigurable fabric for accelerating large-scale datacenter services. Commun ACM 59(11):114–122
96. Ouyang J, Lin S, Qi W et al (2016) SDA: software-defined accelerator for large-scale DNN systems. In: Hot chips 26 symposium, pp 1–23
97. Elbirt AJ, Paar C (2005) An instruction-level distributed processor for symmetric-key cryptography. IEEE Trans Parallel Distrib Syst 16(5):468–480

98. Fronte D, Perez A, Payrat E (2008) Celator: a multi-algorithm cryptographic co-processor. In: International conference on reconfigurable computing and FPGAs, pp 438–443
99. Sayilar G, Chiou D (2014) Cryptoraptor: high throughput reconfigurable cryptographic processor. In: IEEE/ACM international conference on computer-aided design, pp 154–161
100. Chen JH, Shieh MD, Lin WC (2010) A high-performance unified-field reconfigurable cryptographic processor. IEEE Trans Very Large Scale Integr Syst 18(8):1145–1158
101. Yuliang W (2010) Research and design on coarse-grained reconfigurable structure oriented to cipher algorithms. Master's thesis of The PLA Information Engineering University, Zhengzhou
102. Shan W, Fu X, Xu Z (2015) A secure reconfigurable crypto IC with countermeasures against SPA, DPA, and EMA. IEEE Trans Comput-Aided Des Integr Circuits Syst 34(7):1201–1205
103. Wang B, Liu LB (2015) REPROC: a dynamically reconfigurable architecture for symmetric cryptography. In: Proceedings of the 2015 ACM/SIGDA international symposium on field-programmable gate arrays. ACM, p 269
104. Mcloone M, Mccanny JV (2003) Very high speed 17 Gbps SHACAL encryption architecture. Lect Notes Comput Sci 2778:111–120
105. Good T, Benaissa M (2005) AES on FPGA from the fastest to the smallest. Lect Notes Comput Sci 3659:427–440
106. Drimer S, Güneysu T, Paar C (2010) DSPs, BRAMs, and a pinch of logic: extended recipes for AES on FPGAs. ACM Trans Reconfig Technol Syst 3(1):3
107. Gaspar L, Fischer V, Bossuet L et al (2012) Secure extension of FPGA general purpose processors for symmetric key cryptography with partial reconfiguration capabilities. ACM Trans Reconfig Technol Syst 5(3):16
108. Gogniat G, Wolf T, Burleson W et al (2008) Reconfigurable hardware for high-security/high-performance embedded systems: the SAFES perspective. IEEE Trans Very Large Scale Integr Syst 16(2):144–155
109. Güneysu T, Moradi A (2011) Generic side-channel countermeasures for reconfigurable devices. In: Cryptographic hardware and embedded systems, pp 33–48
110. Mentens N, Gierlichs B, Verbauwhede I (2008) Power and fault analysis resistance in hardware through dynamic reconfiguration. In: The international workshop on cryptographic hardware and embedded systems, pp 346–362
111. Beat R, Grabher P, Page D et al (2012) On reconfigurable fabrics and generic side-channel countermeasures. In: International conference on field programmable logic and applications, pp 663–666
112. Salvador R, Otero A, Mora J et al (2011) Fault tolerance analysis and self-healing strategy of autonomous, evolvable hardware systems. In: 2011 international conference on reconfigurable computing and FPGAs (ReConFig), pp 164–169

Chapter 2
Analysis of the Reconfiguration Feature of Cryptographic Algorithms

This book focuses on the reconfigurable feature of cryptographic algorithms and analyzes the feasibility of implementing cryptographic algorithms with reconfigurable computing technologies, so as to provide a basis for the architecture design of the reconfigurable cryptographic processor. To study the reconfigurable cryptographic processor, a full understanding of cryptographic algorithms, the implementation object of the reconfigurable cryptographic processor, is a must. Based on the key factors of reconfigurable computing technologies, this book analyzes the features of cryptographic algorithms in terms of the execution process, algorithm structure, data width, computing granularity, core operations, parallelism, data dependency, common logic of algorithms computation, etc. This provides a basis for the architecture design of a reconfigurable cryptographic processor, including operator extraction, reconfigurable logic unit function, computing granularity, and scale of reconfigurable arrays. As each cryptographic algorithm has its unique features, this book will analyze the reconfigurable features of the block cipher, hash function, and public-key cipher separately. There are numerous types of symmetric cipher, and the information system has the most urgent demand for the flexibility of symmetric ciphers. Therefore, the next section will focus on symmetric ciphers.

2.1 Review and Classification of Cryptographic Algorithms

Cyptography has a very long history. As early as ancient times, people began to use ciphers to transfer secret messages so as to prevent secret leakage. In modern wars and foreign affairs, ciphers are widely used for battle commanding and information transferring. In addition, ciphers are also widely used in modern economic and social activities. The encryption technologies always evolve with the development of communication technologies and computing capabilities, developed. Before

L. Liu et al., *Reconfigurable Cryptographic Processor*,
https://doi.org/10.1007/978-981-10-8899-5_2

telegraph was invented, there was very limited communication traffic, and manual encryption and decryption can meet the requirements for information transfer. However, it becomes extremely difficult after the invention of telegraph technology as there is a great increase of communication traffic. Inspired by the idea of encoding from Morse telegraphy, people began to encrypt information by transforming English letters. In the late 1800s, radio communication was invented and there was an explosive growth in the communication traffic. This further stimulated the research on more efficient methods for encryption and decryption, and encryption technology entered the era of mechanical cipher. The period before *Communication Theory of Secrecy Systems* [1] was published in 1949 was called the era of classic ciphers. During this period, encryption was mainly conducted manually or mechanically, and the main encryption objects are words composed of letters.

After the electronic computer was invented, the capability of code breaking was enhanced greatly and all encryption technologies in the past can no longer ensure security. As such, cryptology was developed rapidly and gradually that became a branch of science. The era of modern cryptography began. Ciphers are applied not only in the military, but are also widely used in various social and economic activities such as foreign affairs, government affairs, finance, e-commerce, tax affairs, mobile communication, and Internet. The commonly used second-generation ID cards, bank cards, and mobile phones all use cipher technologies. During this period, encryption was mainly conducted by using computers or electronic equipment and the encryption objects are binary bits composed of 0 and 1.

This chapter classifies cipher algorithms into three categories, that is, the symmetric cryptographic algorithm, the hash algorithm, and the public-key cipher algorithm [2], based on different usages of the key. These three categories of cipher algorithms are presented separately, and symmetric cryptographic algorithm can be subdivided into stream cipher and block cipher.

1. Symmetric Cryptographic Algorithm

Symmetric cryptographic algorithms use the same key for encryption and decryption, and the encryption algorithm and the decryption algorithm are mutually reverse. The sender and the receiver must share the same key when using the symmetric cryptographic algorithm to transfer a message. As the security of the symmetric cryptographic algorithm depends on the key, the key must be secret. Otherwise, communication will no longer be secure and the information may be leaked. The key can be generated either by a secure third party and then be distributed to the sender and receiver through a secure channel, or by the sender and then be transferred to the receiver through a secure channel.

Figure 2.1 shows a basic symmetric cryptosystem model. The message sender encrypts the plaintext using the key obtained from the key source, and the ciphertext is then transferred to the receiver. The receiver, after receiving the ciphertext, decrypts it using the key obtained from the key source to produce the original

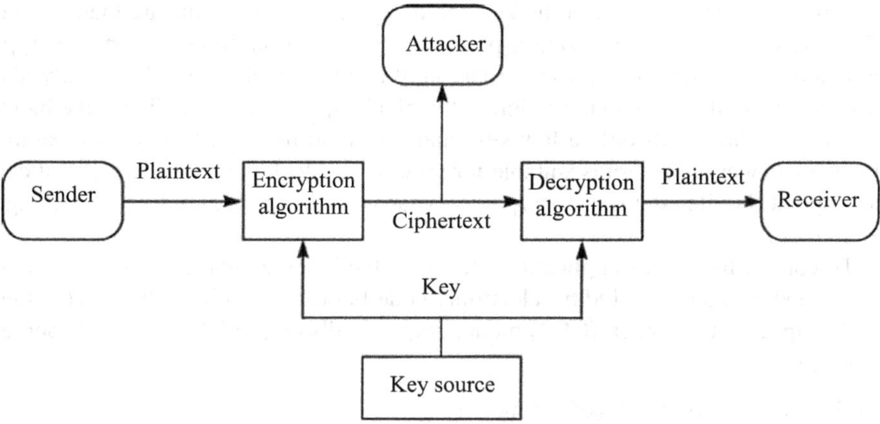

Fig. 2.1 Symmetric cryptosystem model

plaintext. In the above example, plaintext refers to the message to be sent and ciphertext refers to the encrypted message. The keys used by the sender and receiver are the same. The encryption algorithm is a method or process to transform the plaintext P into ciphertext C under the control of key (K), expressed as $C = E\,(K, P)$. The decryption algorithm is a method or process to transform the ciphertext C into plaintext P under the control of key (K), expressed as $P = D\,(K, C)$.

Block cipher algorithms operate on groups of bits of the plaintext, and the bit group is called a block. For a block cipher with b bits in each block, both its plain text space and its ciphertext space contain 2^b different elements. After both the encryption algorithm and keys are determined, a concrete encryption algorithm equals a substitution table from the plaintext space to the ciphertext space. The most basic application model of the block cipher algorithm is shown in Fig. 2.2 [3]. Each time, data of b bits are extracted from the plaintext and encrypted under the control of the key K, and then, the encrypted data of b bits are output. After a group of b-bit data is successfully encrypted, another group of b-bit data will be encrypted successively. The decryption process is similar. Each time, a group of b-bit data is

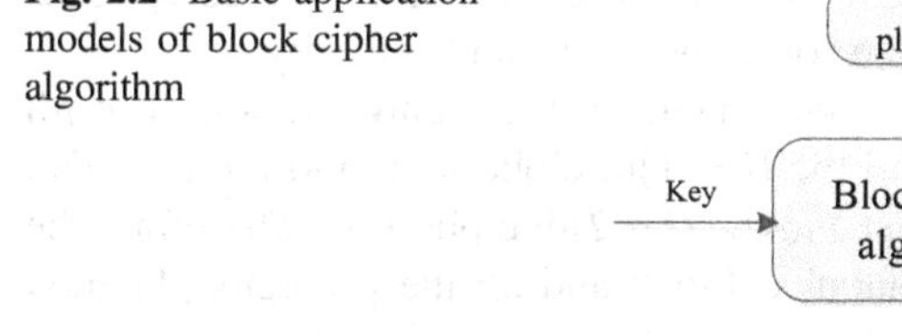

Fig. 2.2 Basic application models of block cipher algorithm

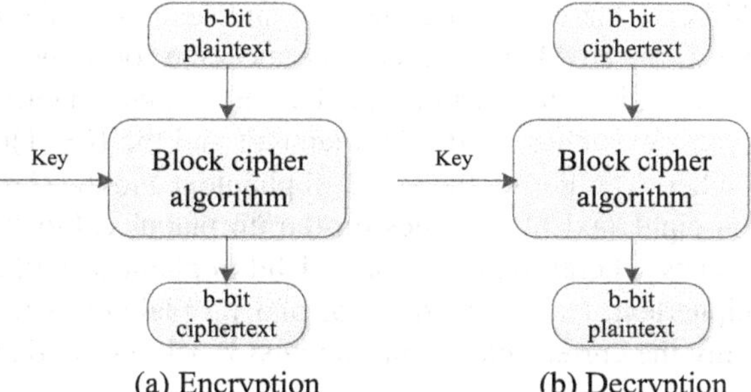

decrypted under the control of the key and decrypted plaintext data are then output. The difference between encryption process and decryption process is that decryption algorithm is the reverse transformation of encryption algorithm. Generally, the same key is used for each encryption in the block cipher algorithm. Thus, the block cipher algorithm needs only a few keys and is easy to manage. Due to this feature, the block cipher algorithm is suitable for processing block data. For example, it can be applied in file storage encryption, database encryption, and cloud storage encryption.

To cope with various application scenarios, the block cipher algorithm works in many modes, mainly including electronic code book mode, cipher block chaining mode, ciphertext feedback (CFB) mode, output feedback (OFB) mode, and counter mode [3].

(1) Electronic code book (ECB) mode

In this mode, the plaintext blocks are encrypted separately, and the blocks use the same key. There is no feedback and interference among blocks, and therefore, encryption can be carried out in parallel. ECB mode is the most basic working mode of the block cipher algorithm. As the blocks are encrypted separately in this mode, the plaintext blocks have the same ciphertext after being encrypted with the same key. The disadvantages of this mode include that the statistical and structural features are easy to be exposed, that the correct plaintext cannot be recovered once one bit in the ciphertext is lost, and that substitution attacks cannot be resisted in this mode.

(2) Cipher block chaining (CBC) mode

In this mode, every block of plain text is XORed with the ciphertext of the previous block before encryption. Then, exclusive OR operation is carried out between the ciphertext obtained after encryption and the plaintext of the next block, which affects the input of the encryption function of the next block. For the first block, the plaintext is XORed with an initial vector (IV) which is used as the ciphertext of Block 0. In CBC, when different IVs are used, distinct ciphertexts can be produced even though the same key is used to encrypt the same plaintexts. The same ciphertext can be produced only when the IV, key, and plaintext are all the same. The message authentication code (MAC) generated in the CBC mode can be used for message authentication. The steps are as follows: The sender attaches the MAC to the end of the message and sends it to the receiver, and the receiver uses the received MAC to check the message for authenticity and integrity.

The chaining attribute of the CBC mode enables self-synchronization. The ith ciphertext relies on the ith plaintext and the $(i-1)$th ciphertext, and the $(i-1)$th ciphertext relies on the $(i-1)$th plaintext and the $(i-2)$th ciphertext. Therefore, the ith ciphertext block relies on the ith plaintext block and all the previous plaintext blocks. Therefore, the error of 1-bit in plaintext will cause error to all the following ciphertext. However, the error of 1-bit plaintext will cause error to the plaintext of only the current block and the next block during decryption and will not affect the decryption of the ciphertext block after the next. Therefore, the CBC mode supports quick recovery in case of error of the ciphertext or loss of the entire block and has the nature of self-synchronization.

(3) Cipher feedback (CFB) mode

In this mode, the block cipher is used as a stream cipher by generating key streams. Each time, one part of the output of the block cipher is XORed with the plaintext of the same length. The operation result is shifted and fed back to the input of the next block. In the CFB mode, the initial vector IV is also needed and will be used as the input of the first block. In this mode, the first block will be divided into many segments. For example, a 128-bit block can be divided into 16 or even 128 segments. In this way, a 8-bit or a 1-bit segment is output for each operation. The one-bit output each time is essentially equivalent to a stream cipher output by bit. This method, however, provides a very low efficiency and thus is seldom adopted. The decryption in the CFB mode is similar to the encryption in this mode. The only difference between them is that during decryption, the exclusive OR operation is carried out between the received ciphertext and the output of the encryption function.

(4) Output feedback (OFB) mode

This mode is similar to the CFB mode. The only difference between these two modes is that in the CFB mode, ciphertext (the result of the exclusive OR operation between the output of the encryption function and the plaintext) is fed back to the shift register and is then used as the input of the encryption function, whereas in the OFB mode, the output of the encryption function is directly fed back to the shift register. Thus, in the OFB mode, the key stream is completely independent of the ciphertext and is only related to the initial vector IV and the key K.

(5) Counter (CTR) mode

The CRT mode is similar to CFB and OFB modes, and the difference is that the input of the encryption function of the CTR mode no longer comes from the output feedback of the encryption function of the previous block, but from an independent counter. In this way, CTR is not a feedback structure and no chaining relation exists between the blocks.

The stream cipher algorithm usually encrypts a single bit or byte in the plaintext, and its application mode is shown in Fig. 2.3 [3]. The key point of the stream cipher algorithm is generating random key streams. In the stream cipher algorithm, the plaintext is a bit stream composed of 0 s and 1 s, and the generated key stream is also a bit stream composed of 0s and 1s. During encryption, exclusive OR operation is carried out between the plaintext bit stream and the key stream by bit, and a ciphertext stream is obtained. During decryption, exclusive OR operation is performed again between the ciphertext stream and the key stream, and then, the plaintext stream is recovered. These plaintext streams, key streams, and ciphertext streams are of equal length. The features of stream ciphers enable them to be

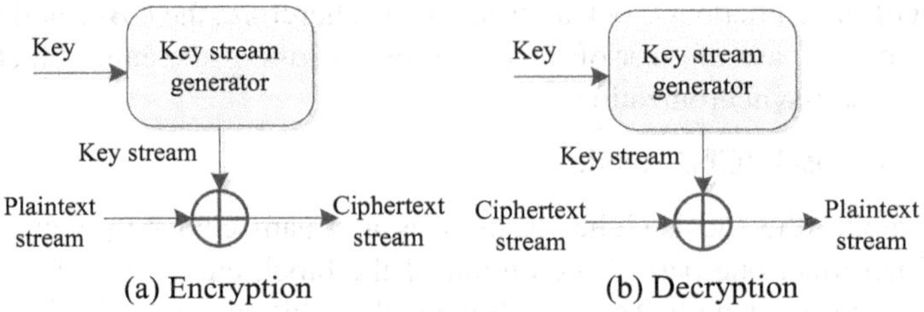

Fig. 2.3 Application model of stream cipher algorithm

specially suitable for transmission encryption, such as network transmission encryption and channel encryption.

2. Hash Algorithm

The hash algorithm is also called digest algorithm. This algorithm can transform input messages of any length to a fixed-length bit string. Its model is shown in Fig. 2.4 [3]. The fixed-length output of the hash algorithm is called the hash value or message digest of the message and is expressed as $h(m)$. In theory, there are chances that different messages have the same hash value. In reality, it is difficult, however, when a hash value $h(m)$ is given, to find a message whose hash value is also $h(m)$, and it is also difficult, when a message is given, to find another message which has the same hash value. In addition, it is also very difficult to find two messages that have the same hash value. This is a necessary condition for ensuring the security of the hash algorithm. The hash algorithm cannot be used to restore the original message from the hash value, but it can be used to verify the received messages.

Hash algorithms can be subdivided into hash algorithms with and without keys. The hash algorithm without any keys can generate a modifying detection code (MDC) and determine whether a message has been modified by attaching the MDC to the message. The hash algorithm with keys can generate and attach a MAC to

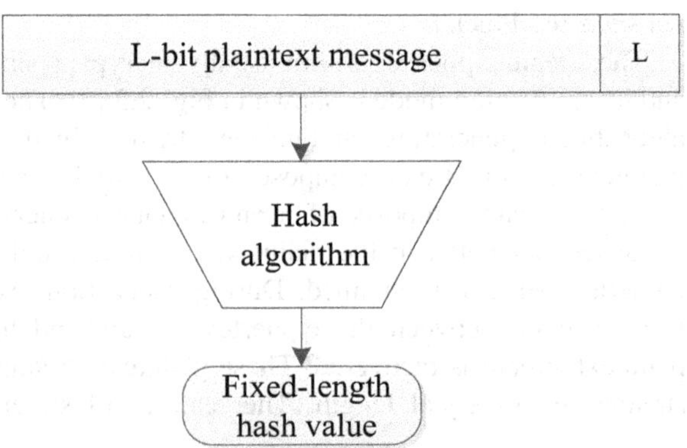

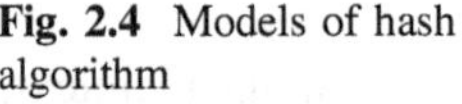
Fig. 2.4 Models of hash algorithm

check the authenticity and integrity of a message. The MDC or MAC of a message is usually sent to the receiver by attaching it to the end of a message and transferred to the receiver with the message. Using MDC or MAC, the receiver can authenticate and check the integrity of a message.

The hash algorithm is widely used in various security applications or network protocols. For example, it can be used to check data integrity for message authentication and to generate a one-way password file for identity authentication and digital signature, to construct hash algorithm-based message authentication codes, or to construct a deterministic random bit generator. Figure 2.5 shows a simple application of hash algorithm in typical message authentication application. The sender generates a group of hash values for the message to be sent using a hash algorithm. Then, the hash values and the message are sent together. The receiver uses the same algorithm to calculate the hash value and compares the calculated hash value with the received one. If they do not match, it is likely that the message and/or hash values have been modified. In addition, hash algorithm is also an important part of many security authentication protocols to realize efficient, reliable, and secure digital signature and authentication. For example, in digital signature, the message is not signed directly but is hashed to produce a short digest that is then signed because of inconsistent length or a large amount of computation of the public key.

3. Public-Key Cipher Algorithm

Before emergence of the public-key algorithm, symmetric cryptographic algorithms were used for traditional encryption. However, it is required to keep a secure channel to share the key, and this will generate a high cost. Only such institutions as government or a big bank can afford the cost, and thus, the application scope of symmetric algorithms was quite limited. The concept of public key was proposed by Diffie and Hellman in 1976 [4]. The emergence of a public key is a major revolution of the encryption technology and also a milestone of the development

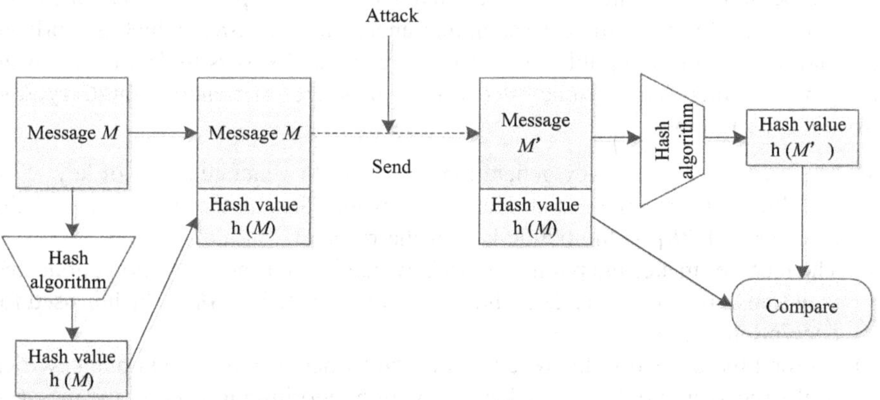

Fig. 2.5 Message authentication

history of cryptology. A public key does not need a secure channel to transfer keys and thus reduces the cost of the encryption system.

The public-key cipher algorithm is also called asymmetric cipher or dual-key algorithm as different keys are used for encryption and decryption. Public-key cipher algorithms does not require the two parties involved in communication to share one secret key, but uses different keys for encryption and decryption. One key is secret and called private key, and the other key is openly distributed and is called public key. Every user has a pair of keys, one public key and one private key. The core of public-key cipher algorithm is one-way trapdoor function; that is, it is easy to compute the function from one direction but difficult to compute in the opposite direction. It is easy to deduce the public key from the private key, but it is difficult to deduce the private key from the public key and recover the plaintext from the ciphertext and public key. The main advantage of the public-key cipher algorithm is that without secure channel information still can be exchanged securely. The sender and receiver no longer need to share the same key through a secure channel, only the public key is involved during transmission, and the private key is neither transferred nor shared.

The encryption of the public-key cipher algorithm is expressed as $C = E(K_e, P)$, and the decryption is expressed as $P = D(K_d, C)$, where K_e and K_d represent the encryption key and the decryption key, respectively. Unlike the symmetric cryptographic algorithm whose encryption key and decryption key are the same, the encryption key and the decryption key of the public-key cipher algorithm are different. The public-key cipher algorithm has two basic application models. One model is the encryption model where the public key is used as the encryption key and the private key is used as the decryption key. The other model is signature model where the private key is used as the encryption key and the public key is used as the decryption key.

Figure 2.6 shows the encryption model of the public-key algorithm, and the principle is that public key is used for encryption, while private key is for decryption. The key PK_B used for encryption can be open to the public and does not need to be transferred through a secret channel. The encrypted information can be decrypted only by using the corresponding encryption key SK_B, which is a private key and not open to the public. As only the receiver B knows his/her own private key, anyone other than B cannot decrypt the message. The encryption/decryption process is as follows.

(1) The receiver uses the key generation algorithm to generate a pair of keys SK_B and PK_B for encryption and decryption, where SK_B is the private key of the receiver and PK_B is the public key of the receiver.
(2) The receiver makes the public key PK_B which is used for encryption public and then transfers it to the sender, while keeps the private key SK_B which is used for decryption secret.
(3) To send message M to the receiver, the sender needs to use the public key PK_B of the receiver and the public key encryption algorithm to encrypt the message

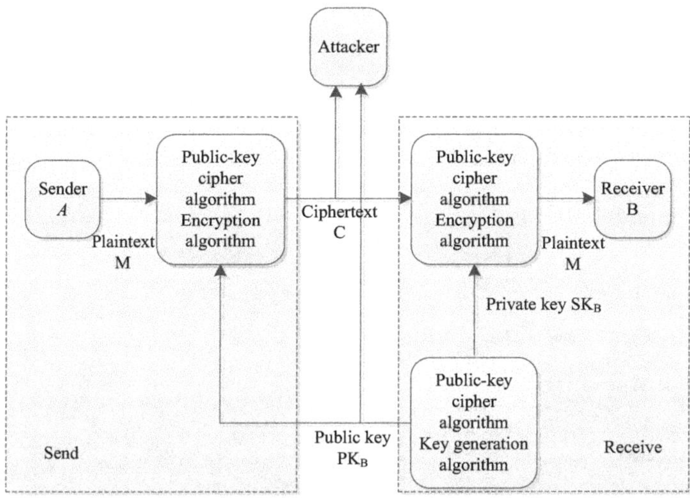

Fig. 2.6 Encryption model of the public-key cipher algorithm

M to obtain the ciphertext C ($C = E(PK_B, M)$) and then transfer the ciphertext C to the receiver.

(4) After receiving the ciphertext C, the receiver uses his/her own private key SK_B and the public key decryption algorithm to decrypt the ciphertext C to obtain the plain text M ($M = E(SK_B, C)$).

Figure 2.7 shows the signature model of the public key cipher algorithm which uses the private key for encryption and public key for decryption. The sender A uses his/her own private key SK_A to encrypt the plaintext. The receiver B, however, uses the public key PK_A of the sender A to decrypt the ciphertext and verify whether the message is surely from the sender A and has not been modified. As only the sender A has the private key SK_A, others cannot pretend to be the sender or modify the message M if they cannot get access to the private key of the sender. Thus, the source and integrity of the message can be ensured. This model is just the simplest signature authentication model, and an actual digital authentication scheme is much more complex. The course of signature authentication is as follows.

(1) The sender uses the key generation algorithm to generate a pair of keys SK_A and PK_A for encryption and decryption, where SK_A is the private key of the sender and PK_A is the public key of the sender.
(2) The sender makes the public key PK_A which is used for decryption public and then transfers it to the receiver, while keeps the private key SK_A which is used for encryption secret.
(3) The sender uses his/her own private key SK_A and the public key encryption algorithm to encrypt the message M and produce the ciphertext C ($C = E(SK_A, M)$).

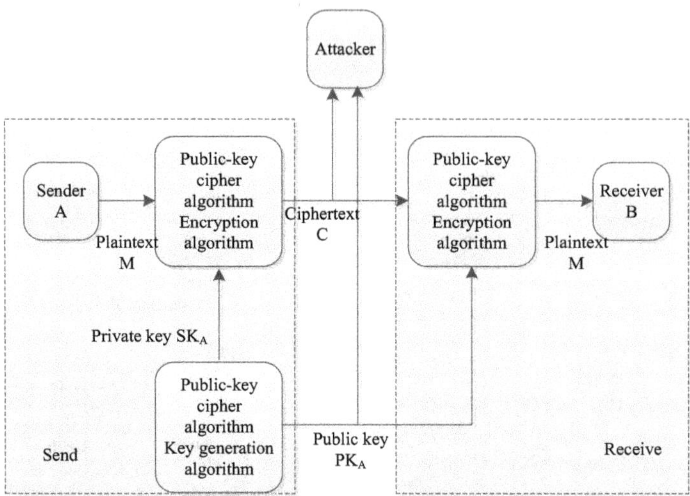

Fig. 2.7 Signature model of the public-key cipher algorithm

(4) After receiving the ciphertext *C*, the receiver uses the public key PK_A of the sender and the public key decryption algorithm to decrypt the ciphertext *C* to obtain the plain text *M* ($M = D(PK_A, C)$).

In the signature model above, as everybody can get access to the public key of the sender and use the public key to decrypt the message, the attacker can listen to the message though he/she cannot tamper the message. This is also a security problem and not desired in many application scenarios. This problem can be solved by using a dual-encryption/-decryption digital envelop model, which is shown in Fig. 2.8. In this digital envelop model, the sensor *A* uses his/her own private key SK_A to encrypt the message first and create digital signature for the message. Then, the sender *A* will use the public key PK_B of the receiver *B* to carry out secondary encryption for the signed message *Z* and produce the ciphertext *C*. During decryption, the receiver *B* will first use his/her own private key SK_B to decrypt ciphertext *C* to obtain ciphertext *Z* and then uses the public key PK_A of the sender to carry out secondary decryption for ciphertext *Z* to obtain the message *M* for authentication. Without the private key of the receiver, the attacker, however, cannot transform ciphertext *C* into ciphertext *Z* and thus cannot obtain plaintext *M*. Therefore, the attacker cannot listen to the content of the plaintext even if he/she has the public key of the sender. This ensures the secrecy and integrity of messages.

Next, we will analyze the symmetric cryptographic algorithm, hash algorithm, and public key cipher algorithm, respectively.

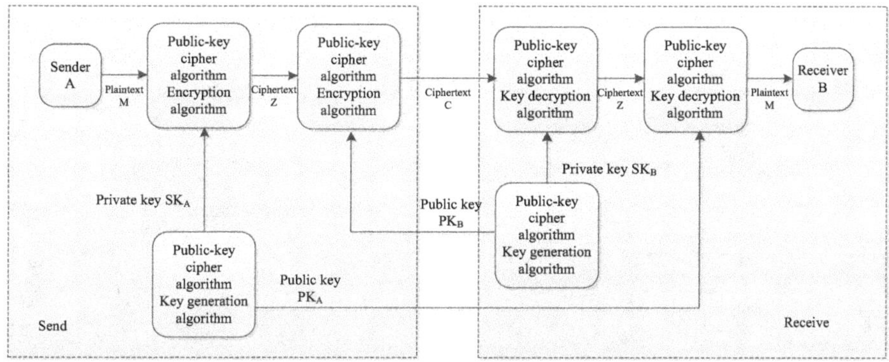

Fig. 2.8 Digital envelope model of the public-key cipher algorithm

2.2 Symmetric Cryptographic Algorithm

2.2.1 Block Cipher Algorithm

1. Introduction to Block Cipher Algorithm

As the most common cipher algorithm, the block cipher algorithm is has the widest application including DES, 3DES, AES, SM4, SERPENT, IDEA, RC6, Mars CAST-256, CRYPTON, SAFER+, Twofish. Some algorithms have many variants. For example, AES can be divided into AES-128, AES-192, and AES-256 according to the key length.

The encoding method of the block cipher algorithm will be introduced by taking AES [5], the most commonly used block encryption algorithm, as an example. AES is a block encryption algorithm established by US National Institute of Standards and Technology in 2001. AES supports keys of three lengths: 128-bit, 192-bit, and 256-bit. For the sake of brevity, AES algorithms mentioned later all refer to the AES algorithm with a key length of 128 bits.

AES adopts typical SPN structure, and its encryption/decryption process is shown in Fig. 2.9. One block of AES is 128 bits long, and 10 rounds of iteration will be carried out for each block. During encryption, such operations as SubBytes, ShiftRows, MixColumns, and AddRoundKey will be carried out in turn in the first nine rounds of iterations, which are completely consistent. The last round of iteration is slightly different from the first nine rounds. It does not contain the operation of MixColumns. The decryption process is similar to the encryption process except the operation sequence. During decryption, such operations as inverse ShiftRows, inverse SubBytes, AddRoundKey, and inverse MixColumns are carried out in turn and the round keys are used in a reverse order, too. In addition, the initial round of both encryption and decryption is AddRoundKey.

The operation of each step will be introduced as follows.

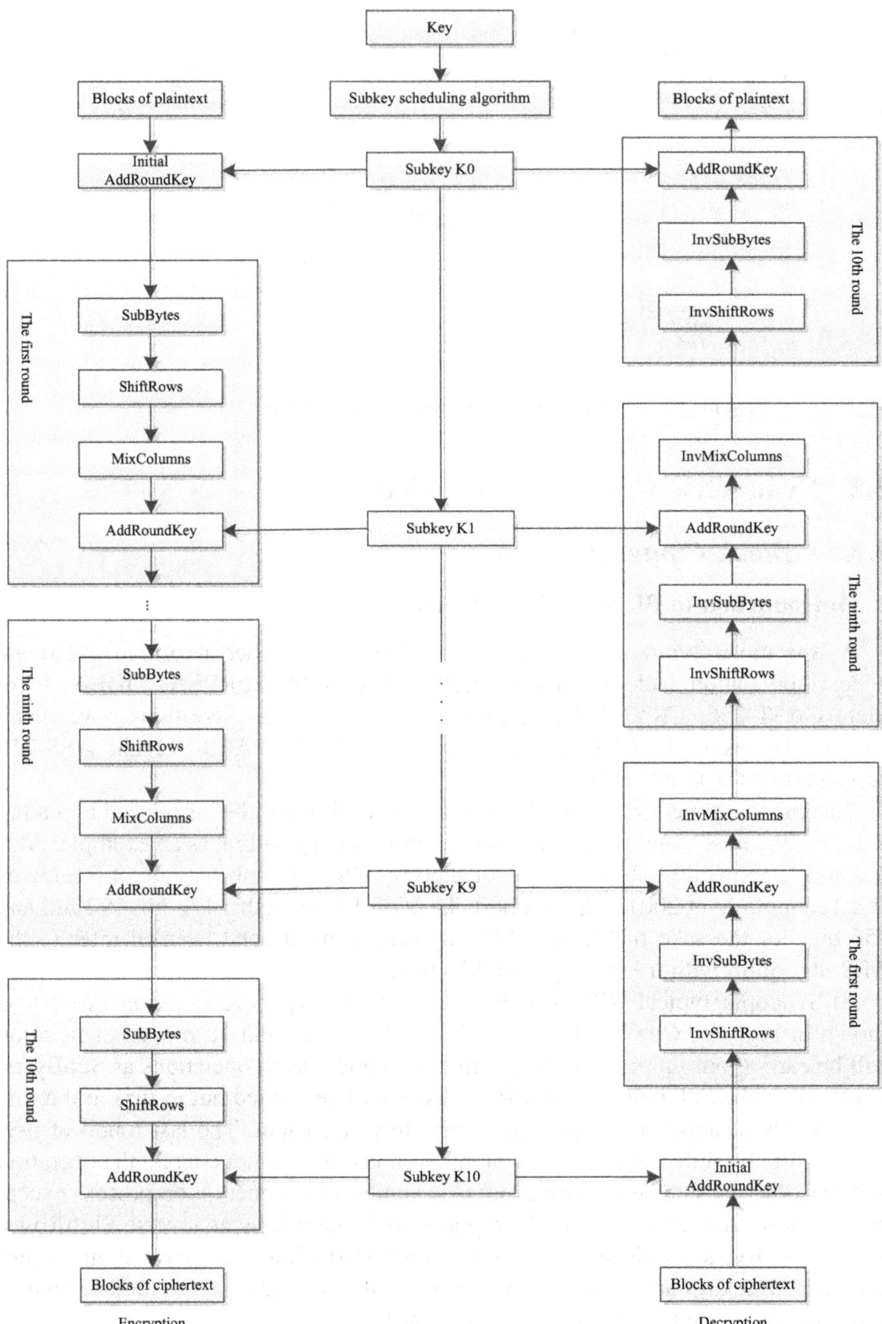

Fig. 2.9 Process of AES encryption/decryption

SubBytes is dividing the 128-bit data into 16 bytes, and each byte is substituted, respectively, via S-Box, as shown in Fig. 2.10. The principle of inverse SubBytes is similar, and the only difference is that the content of the substitution table is different.

The operation of SubBytes can also be constructed mathematically. First, conduct inverse operation for the bytes input in finite field GF(2^8), expand the result by bit to 8 bits, and then conduct the transformation shown in Fig. 2.11. This transformation includes a matrix vector multiplication and a vector addition. Matrix vector multiplication is the multiplication of a matrix, and a multiplier vector to produce a result vector will be obtained. In the result vector, each element is the sum of the products of the corresponding elements in the corresponding line and the corresponding element of the multiplier vector in the matrix. As the elements in the matrix and the multiplier vector are all 1-bit, the multiplication of the elements is equal to the AND operation and the addition of the elements is equal to exclusive OR operation. Actually, each bit in the result vector is obtained through the exclusive OR operation of not more than 8-bit data. As shown in Fig. 2.11, the transformation result is obtained with another vector addition, and each bit is obtained through the exclusive OR operation of not more than 9-bit data, with inverse SubBytes is similar to SubBytes and the only difference is that in inverse SubBytes, matrix transformation is carried out before the inverse operation in finite field GF(2^8), and the contents of the constant matrix and constant vector in the

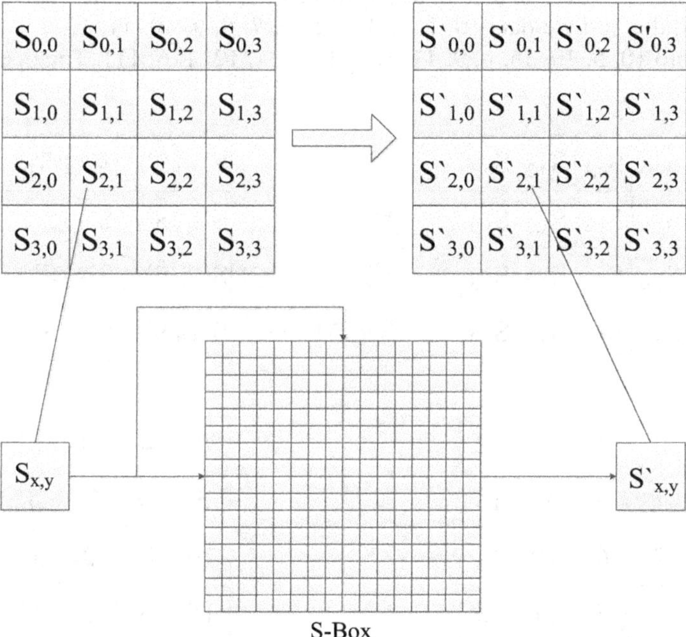

Fig. 2.10 Schematic diagram of SubBytes

$$\begin{bmatrix} y_0 \\ y_1 \\ y_2 \\ y_3 \\ y_4 \\ y_5 \\ y_6 \\ y_7 \end{bmatrix} = \begin{bmatrix} 1 & 0 & 0 & 0 & 1 & 1 & 1 & 1 \\ 1 & 1 & 0 & 0 & 0 & 1 & 1 & 1 \\ 1 & 1 & 1 & 0 & 0 & 0 & 1 & 1 \\ 1 & 1 & 1 & 1 & 0 & 0 & 0 & 1 \\ 1 & 1 & 1 & 1 & 1 & 0 & 0 & 0 \\ 0 & 1 & 1 & 1 & 1 & 1 & 0 & 0 \\ 0 & 0 & 1 & 1 & 1 & 1 & 1 & 0 \\ 0 & 0 & 0 & 1 & 1 & 1 & 1 & 1 \end{bmatrix} \begin{bmatrix} x_0 \\ x_1 \\ x_2 \\ x_3 \\ x_4 \\ x_5 \\ x_6 \\ x_7 \end{bmatrix} + \begin{bmatrix} 1 \\ 1 \\ 0 \\ 0 \\ 0 \\ 1 \\ 1 \\ 0 \end{bmatrix}$$

Fig. 2.11 Matrix operation of SubBytes

matrix transformation are different, which will not be described here. Actually, the constant matrix in SubBytes is the inverse of that in inverse SubBytes, and the product of them is a unit matrix.

The operation of ShiftRows does not have any value change and is a typical permutation operation. First, the 16-byte data are arranged to form a 4-row and 4-column matrix by row and then column, and the data in each row of the matrix are rotated in byte. For encryption, the data on each row are shifted to the left by 0 byte, 1 byte, 2 bytes, and 3 bytes, respectively, as shown in Fig. 2.12. After ShiftRows, the 16-byte data {0, 1, 2, 3, 4, 5, 6, 7, 8, 9, 10, 11, 12, 13, 14, 15} is changed into {0, 5, 10, 15, 4, 9, 14, 3, 8, 13, 2, 7, 12, 1, 6, 11}. The operation for

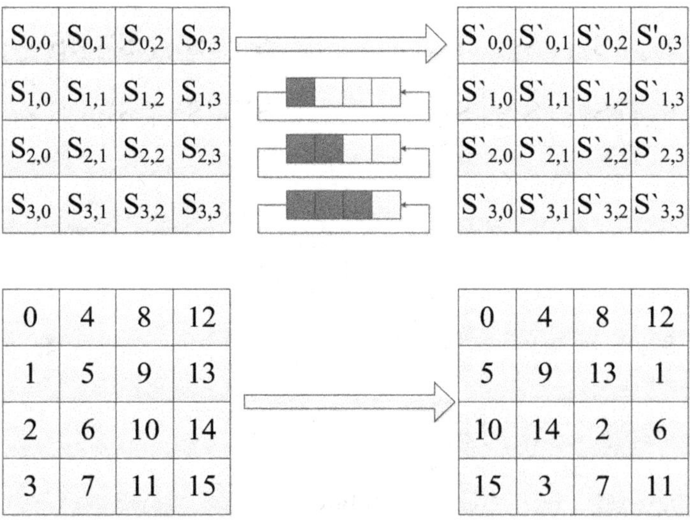

Fig. 2.12 Schematic diagram of ShiftRows operation

decryption is similar, and the difference is that the left shift operation is changed to right shift operation accordingly.

MixColumns is that the four bytes in each column is multiplied with a fixed 4×4 matrix in the finite field, and the product is the result of MixColumns, as shown in Fig. 2.13. On the whole, it can also be seen as the multiplication of a fixed 4×4 matrix and input matrix. The decryption is similar to encryption, and the only difference is that the fixed matrix for multiplication is different. The matrix multiplication in MixColumns is similar to that in SubBytes. Each element in the result vector is obtained by multiplying the corresponding element in a row in the constant matrix and the corresponding element in a column in the state matrix, and then, the product is XORed by bit. However, the elements of the former matrix are bytes and the multiplication is in the finite field GF(2^8), while the elements of the latter matrix are bits and the product is obtained through simple AND operation.

The operation of AddRoundKey is very simple. It is just the exclusive OR operation between the 128-bit plaintext and 128-bit subkey by bit.

In addition to the round transformation described above, AES also includes the key expansion which expands the 128-bit key into subkeys for AddRoundKey in 10 rounds of iterations and preround transformation. There are 44 32-bit words (or 11 128-bit words), as expressed as $W[j]$, $j = 0, 1, \ldots, 43$. The input key is directly used as the first subkey, and each subkey later is generated by transforming the previous subkey using the key expansion. The key expansion is shown in Fig. 2.14. First, the initial key is transformed into four words by column, and these four words expressed as $W[0]$, $W[1]$, $W[2]$, $W[3]$ are used as the first subkey. Then, for the case where j is not an integral multiple of four, $W[j] = W[j-4] \oplus g(W[j-1])$, and for the case where $j\%4 = 0$, $W[j] = W[j-4] \oplus g(W[j-1])$. Repeat this process until the subkeys of all 44 words are generated, that is, j = 43. The g mentioned above is a slightly complex function, and its structure is shown in Fig. 2.14b. W which is composed of four bytes is shifted to the left by one byte, and S-box is applied on each byte for mapping. Finally, the obtained results are XORed with the round constant to produce the final subkey. The 32-bit round constant is (RC[i], 0, 0, 0), where RC is a one-dimensional and RC = {00, 01, 02, 04, 08, 10, 20, 40, 80, 1B,

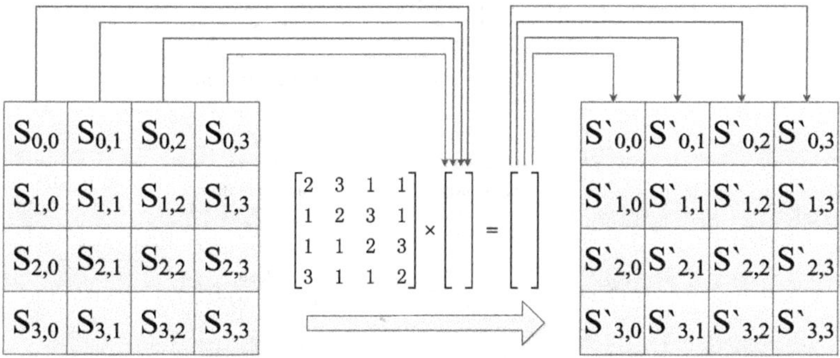

Fig. 2.13 Schematic diagram of the MixColumns operation

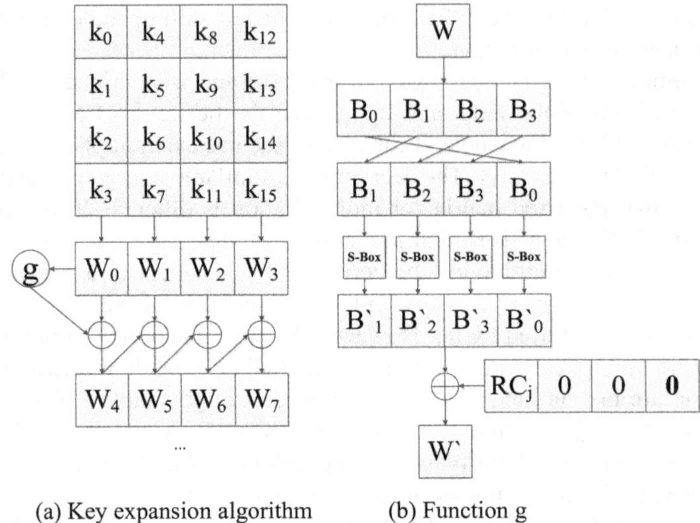

Fig. 2.14 Key expansion

36}. In fact, 10 RC values are enough and RC[0] has not been used in operation. Eleven RC values are used here only for convenient expression. What should be noted is that the subkeys should be used according to the order of the subkeys which are generated by the key expansion algorithm during the encryption and should be used in the inverse order during the decryption.

2. Features of Block Cipher Algorithm

The overall structure features of the block cipher algorithm are analyzed firstly. The typical structure of the block cipher algorithm is shown in Fig. 2.15. It is mainly composed of several rounds of identical iterations, pretransformation,

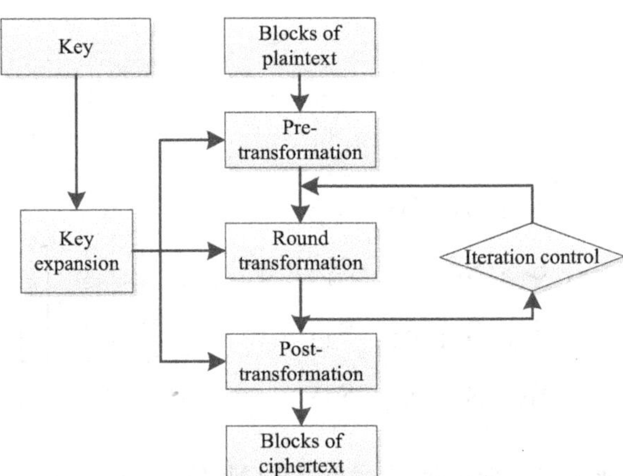

Fig. 2.15 Typical structure of block cipher algorithm

post-transformation, and key expansion. Pretransformation and post-transformation are executed only once but not every algorithm has both transformations. For example, the pretransformation in AES is the preround transformation, that is, initial AddRoundKey, and there is no post-transformation in AES. Round process is executed many times under iteration control and the structure of the last round is slightly different. For the AES algorithm, it is the case that MixColumns transformation is changed in the last round. Due to many executions, round transformation has a relatively complex structure, which is the most important structure in the block cipher algorithm.

Different block cipher algorithms have different round transformation structures, and Feistel network structure and SP network structure are the two main structures in block cipher algorithm. The round transformation of the Feistel network structure is shown in Fig. 2.16, and both the input and output of the round transformation are 2w-bit data. The input data of round transformation can be divided into two parts: left part and right part. Each part is w bits long. The right part of the data is processed using the round function F and then XORed with the left part of the data to produce the right part of results of round processing. Meanwhile, the left part of round processing is simply from the right part of the input data. At the end of the last round of iteration, the left and right parts are exchanged, so that the encryption and decryption have the same structure. The only difference between them is that the order of using subkeys during encryption is exactly the opposite of that during decryption. The decryption of the Feistel cipher is completely identical to the encryption, which is one of the most important features of the Feistel cipher. Both the DES and SM4 (Chinese commercial cipher) are of typical Feistel network

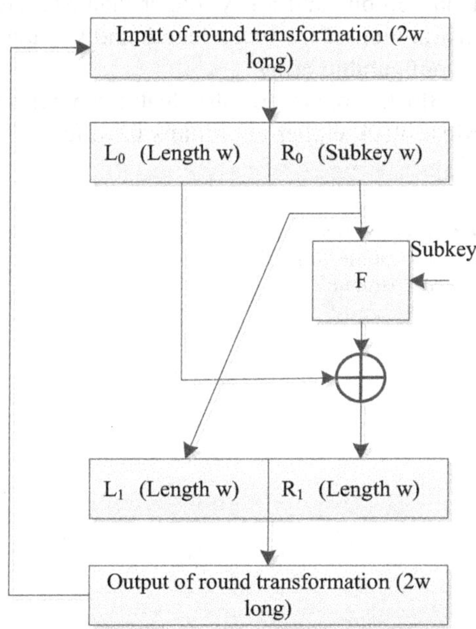

Fig. 2.16 Feistel network structure

structure. Block ciphers based on SP network structure include AES, CRYPTON, SAFER, RIJNDEAL, SERPENT. The SP network structure is based on two basic operations in cryptology: substitution and permutation. Substitution is often called S-Box. It is used for message confusing, which is usually performed under the control of the subkey. Permutation is often called permutation box (P-Box). It is used for message diffusion, which is usually reversible linear transformation. A typical round transformation of the SP network is shown in Fig. 2.17. The input of round transformation is first applied to the reversible substitution controlled by the subkey, and then the permutation or the reversible linear transformation. As the round transformation structure of a cipher on the SP network is more unified than that of the Feistel network structure, and the input subblocks of round transformation are processed identically, it is easier for parallel implementation of algorithms.

The features of the block cipher algorithm in terms of block length, key length, and number of iterations are summarized and analyzed. Though these data vary for different algorithms, they are distributed in a limited range with minor difference and can be categorized easily. For a public block cipher algorithm, a block is usually 64 bits or 128 bits long. The length of a key is usually 64–256 bits and is an integral multiple of 64. This is mainly to facilitate processing by computer software. The number of iterations of round transformation is mainly in range of 8–32. The block length, key length, and number of iterations of some public block cipher algorithms are listed in Table 2.1 [6, 7]. Actually, such special fields as military, national defense, aerospace, and energy and power where national security is involved raise even higher requirements for the security of the cipher algorithm. In these fields, the block length and key length are even greater, but a block longer than 256 bits and a key longer than 512 bits are never seen. Relatively centralized distribution of the block length and key length makes it easy to select the scale of a reconfigurable array.

Block ciphers are also featured with easy control and intensive computation. Most block cipher algorithms have few computing branches or conditional jumps.

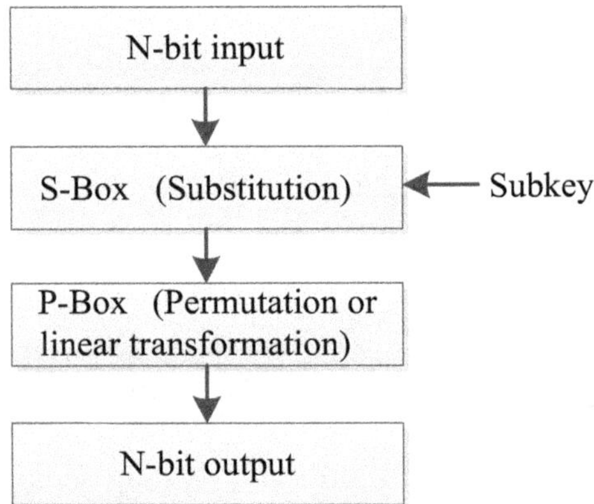

Fig. 2.17 Round transformation of the SP network structure

Table 2.1 Block length, key length, and number of iterations of block cipher algorithm

Algorithm	Block length	Key length	Number of iterations
AES	128	≥ 128	10–14
SM4	128	128	32
DES	64	64	16
IDEA	64	128	8
Blowfish	64	Uncertain	16
NSSU	64	256	32
TEA	64	128	16–32
FEAL	64	64	≥ 8
GOST	64	256	32
SAFER	64	64	8–10
SAFER+	128	≥ 128	8–16
LOKI91	64	64	≥ 15
LOKI97	128	≥ 128	16
CAST	64	64	8
CAST256	128	≥ 128	12
CRYPTON	128	≥ 128	12–16
DEAL	128	≥ 128	6–8
E2	128	≥ 128	12
RC5	64	128	12–16
RC6	128	≥ 128	20
SERPENT	128	≥ 128	31
TWOFISH	128	≥ 128	16
MARS	128	≥ 128	16

The sequence, type, and execution time of operations are all independent of the data to be processed. If the round computation of the block cipher algorithm is expanded, pretransformation, the first round, the second round, etc., and post-transformation are performed in turn without any branches. Even though the round computation is not expanded, there are only conditional statement of the number of iterations and the corresponding branches. Despite the easy control, the block cipher algorithm is compute-intensive. It usually takes ten to twenty rounds of operation for each block of data, and each round involving several substitutions and permutations is quite complicated. The features of few branches, few feedbacks, and intensive computing make it easy to for the block cipher algorithm to be controlled and partitioned by function and timing. Especially, these features make the block ciphers suitable for pipeline implementation. The pipeline implementation is also related to the block cipher mode of operation, which will be discussed later.

Though the block cipher algorithm has a block length as long as 128 bits, the computation granularity is not always equal to the block length. In fact, normally there are multiple computation granularities. A block is an unsigned long integer. For example, a block is usually 64–128 bits long, and a key is usually 64–256 bits long. After key expansion, multiple subkeys, whose lengths are equal to or related to the block length, are generated and used in round operations. For the convenience of round processing, each block of data is usually divided into multiple parts with narrow width, mainly including the whole block, subblock, subword, and bit. Operations with the whole block mainly include permutation, shift, and logic operation, and the block length is usually 64 bits, 128 bits, 192 bits, or 256 bits. Operations with subblocks mainly appear in modular addition, modular multiplication, permutation, shift, and logic operation, and the subblock length is usually 32 bits or 64 bits. Operations with subwords mainly involve S-Box, finite field multiplication, modular addition, modular multiplication, and shift, and the subword length is usually 4 bits, 6 bits, 8 bits, or 16 bits. Operations with bits mainly include register tap and logic operation. Due to a very low execution efficiency, operation software with a granularity of bit is seldom used in modern block ciphers.

3. Degree of Parallelism for the Block Cipher Algorithm

Having a high parallelism, the block cipher algorithm is specially suitable for concurrent execution of reconfigurable arrays which have a lot of hardware resources. The parallelism analysis will be given in three dimensions: intra-block parallelism, inter-block parallelism, and key scheduling and inter-block operation parallelism.

(1) Intra-block parallelism

Each block of data is usually divided into multiple smaller blocks, and many operations in an algorithm are executed in parallel based on these data blocks. This parallelism occurs in the vertical direction of the encryption/decryption data stream and is called horizontal parallelism. The feature of the block cipher algorithm in terms of computation granularity enables dividing block of data into several parts with narrow width for the convenience of data processing. Dividing blocks into fine-grained data has two major advantages in parallelism. One is parallel computing. Specifically, blocks can be divided into several subblocks which can be processed in parallel; some of the subblocks can be subdivided into multiple subwords, which can then be processed in parallel. On the other hand, the fine subdivision makes parallel storage necessary. During cipher processing, S-Boxes, usually implemented through memories (SRAM or register), are utilized extensively. Fine-grained data make it possible to load/store data to memories through multiple memories in parallel.

(2) Inter-block parallelism

Inter-block parallelism, reflected in the same direction of the encryption/decryption data stream, is called vertical parallelism. In ECB and CTR modes, there is no feedback among blocks, and the input of blocks does not rely on the output of previous blocks. Therefore, the computation of each block is independent of each other; that is, inter-block parallelism exists. This parallelism can be realized by using multiple circuits in parallel. This, however, causes exponential increasing area consumption and requires complicated scheduling control circuits. More commonly used method is using pipeline to implement inter-block parallelism for easy control. By adding few registers, further pipeline within the round computation can be implemented. The intra-block and inter-block parallelisms of the AES algorithm are shown in Fig. 2.18.

(3) Key scheduling and inter-block operation parallelism

During encryption, subkeys are generated by the key expansion algorithm and are then used in block round transformation in turn. As a result, subkey scheduling and block operation can be executed in parallel on the whole. That is, subkeys can be generated in parallel with round processing, which can start without waiting for all subkeys generated.

In addition to a high parallelism, block cipher algorithm also has a very close data dependency, including data dependency within a round transformation, between two round transformations, between two blocks, and between subkey scheduling and block operation.

(1) Data dependency in a round transformation

There are linear operations in serial in the round transformation, and there is very intensive read-after-write (RAW) data dependency between two operations. For example, during the round transformation in the AES, the operation of ShiftRows relies on the result of SubBytes, the operation of MixColumns relies on the result of ShiftRows, and the operation of AddRoundKey relies on the result of MixColumns.

(2) Data dependency between two round transformations

In cipher algorithms, there is iteration relation among different rounds of transformations. Therefore, obvious RAW data dependency exists among different rounds of transformations, and the next round of transformation needs to wait for the result from the previous round. As a result, the rounds of operations must be

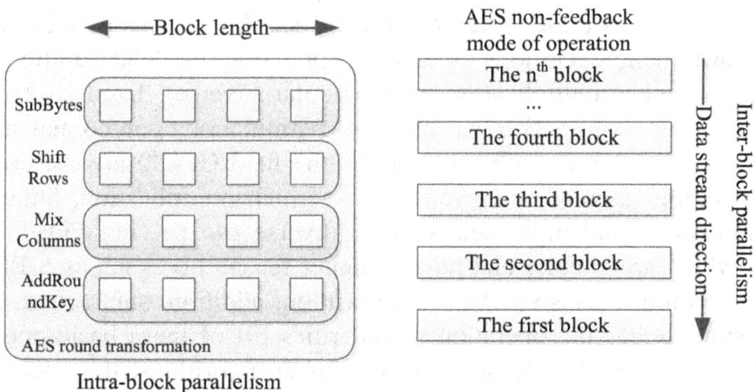

Fig. 2.18 Intra-block parallelism and inter-block parallelism of the AES algorithm

executed in serial, and round operation of multiple iterations is the common feature of all block cipher algorithms.

(3) Data dependency between subkey scheduling and block operation.

These data dependency can be divided into two levels. Microscopically, subkey scheduling and block operation can be executed in parallel on the whole during encryption, but the corresponding subkey should also be computed first during each round of iteration. Therefore, data dependency is generated. Macroscopically, during decryption, a round of operation can be executed only after all subkeys have been computed because the subkeys are used in a reversed order. On the whole, subkey scheduling and block operation are executed in serial and data dependency exists.

(4) Data dependency between two blocks

In such modes of operation as CBC, CFB, and OFB, there is feedback between two adjacent blocks, and the input of the block often relies on the output of the previous block. Therefore, the next block can not be computed until the previous block is processed. This makes direct parallel execution or pipeline execution of multiple blocks rather difficult. If a cipher is executed in a feedback mode, interleaving technology [8] can be used in some application scenarios to separate the feedbacks, so that they are in two blocks far enough from each other. The operation of adjacent blocks can be executed in parallel, but this is no longer a working mode in the normal sense.

4. Common Logic of Block Cipher Algorithm

Many block cipher algorithms are based on similar design theories. This enables these algorithms to have similar structure and operations and brings a lot of common logic for different block cipher algorithms. As mentioned earlier, main structures include the Feistel network-based and SP network-based structure. By analyzing the encryption/decryption round operation and the key expansion of commonly used algorithms including AES, SM4, SERPENT, 3DES, IDEA, RC6, Mars, CAST256, CRYPTON, SAFER+, Twofish, these algorithms show various common logic such as basic logic operations (including exclusive OR, AND, OR, and NOT), arithmetic operations (addition, subtraction, modular addition, modular subtraction, modular multiplication, and modular inverse), fixed shift operation, variable shift operation, S-Box substitution, permutation, polynomial multiplication, finite field GF(2^n) multiplication, and finite field GF(2^n) inverse operation.

Such operations as modular inverse, polynomial multiplication, finite field GF (2n) multiplication, and finite field GF(2n) inverse can be subdivided into basic operations which are simpler and have a higher reusability, such as S-Box substitution, permutation, exclusive OR, multiplication, addition, subtraction, and shift. Table 2.2 summarizes the operation granularities [9] of these basic operations in different block cipher algorithms. In block cipher algorithms, the operation granularities of arithmetic operations and logic operations are mainly byte, half byte, word, or integer multiples of bytes, half bytes, words, and are usually within the range between 8 and 128 bits. Multiplication is usually modular multiplication with a bit width no more than 32 bits, and the bit width of addition is no more than 32 bits. There are few cases with 64-bit addition such as LOKI97 and E2. A great amount of fixed logic shift and rotation operations are used in block cipher algorithms. Usually, 8 or 16 bits of data (a byte or a word) are shifted, but there are also unfixed shifts. S-Box is a very important and commonly used structure in block cipher algorithms. For example, some algorithms use the same S-Box in each round of operation, and S-Box is fixed for each cipher. These algorithms include DES, SAFER, and NSSU. Some algorithms use the same S-Box in each round of operation, and S-Box is fixed for the same key. These algorithms include Blowfish. Some algorithms use different S-Boxes in each round of operation, but S-Box is fixed for each algorithm. These algorithms include SERPENT. The input granularities of S-Box cover a wide range from 4 to 13 bits, with the granularity of 8-bit in the majority. The output granularities of S-Box cover a narrow range. There are only three output granularities, that is, 4-, 8-, 32-bit, and the first two granularities are in the absolute majority. The size of S-Box also covers a wide range from 512b to 80 kb. A block cipher algorithm often needs to access the same or different S-Boxes in parallel. As a result, it raises a high requirement for the number of read ports of the memory which stores S-Box. Keeping a balance between distributed storage and the number of ports is the key to supporting S-Box. Analysis of these common logics, basic operations, and their granularities is an important basis for designing reconfigurable processing units.

2.2.2 *Stream Ciphers*

1. Introduction to Stream Ciphers

The stream ciphers are also a common kind of symmetric ciphers. The stream cipher algorithms which are widely used include ZUC, RC4, SEAL, and A5. Just like in Sect. 2.2.1, a representative stream cipher algorithm, ZUC algorithm, will be first selected to introduce in this section [10]. The ZUC algorithm was included in the Long-Term Evolution (LTE) standards for broadband wireless mobile communication system in 2011 and is the first Chinese algorithm which has become an international cipher standard.

The ZUC algorithm is composed of three parts including the linear feedback shift register (LFSR), the bit-reorganization (BR), and nonlinear function F, and its architecture is shown in Fig. 2.19. An LFSR is composed of sixteen 31-bit register units (S0, S1, …, S15), with each register unit defined on the prime field GF($2^{31} - 1$). Bit-reorganization (BR) is a transition layer, and it is mainly used to extract a total of 128-bit content from the 8 register units of the linear feedback shift register to constitute four 32-bit words (*X*0, *X*1, *X*2, *X*3) for the lower-layer nonlinear function F and key stream output logic. Nonlinear function *F* has 2 32-bit storage units R1 and R2, whose input is *X*0, *X*1, *X*2, *X*3, and whose output is 32-bit word *W*. The S-Box in

Table 2.2 Basic operation granularity of block cipher algorithm

Algorithm	S-Box	Permutation	OR	Shift	Addition (subtraction)	Multiplication	
AES	8 in and 8 out	32 kb		32/ 128	Rotation shift 8 and 32	8	8
SM4	8 in and 8 out	40 kb		32	Rotation shift 2, 10, 18 and 24		
DES	6 in and 4 out	2 kb	32/ 48	32/ 48	Rotation shift		
IDEA				16		16	16
Blowfish	8 in and 32 out	32 kb		32		32	
NSSU	4 in and 4 out	512 b		32	Rotation shift 32		
TEA				32	Shift 32	32	
FEAL				8/32	Rotation shift 8	8	
GOST	4 in and 4 out	512 b		32	Rotation shift 32	32	
SAFER	8 in and 8 out	4 kb		8	Rotation shift 8	8	
SAFER+	8 in and 8 out	4 kb		8	Rotation shift 8	8	
LOKI91	12 in and 8 out	32 kb	32	32/ 48	Rotation shift		
LOKI97	13 in and 8 out						
11 in and 8 out	80 Kb	64	64		64		
CAST	8 in and 32 out	48 kb		32			
CAST256	8 in and 32 out	32 kb	32		Rotation shift 32	32	
CRYPTON	8 in and 8 out	8 kb		8			
DEAL	6 in and 4 out	2 kb	32/ 48	32/ 48			
E2	8 in and 8 out	2 kb		8		64	32
RC5				32	Rotation shift 32 and unfixed value	32	
RC6				32	Rotation shift 32 and unfixed value	32	32
SERPENT	4 in and 4 out	512 b	128	32/ 128	Rotation shift 32		
Shift 32							
TWOFISH	4 in and 4 out	512 b		32/ 128	Shift 32	8/32	8
MARS	9 in and 32 out	16 kb		32	Rotation shift 32 and unfixed value	32	32

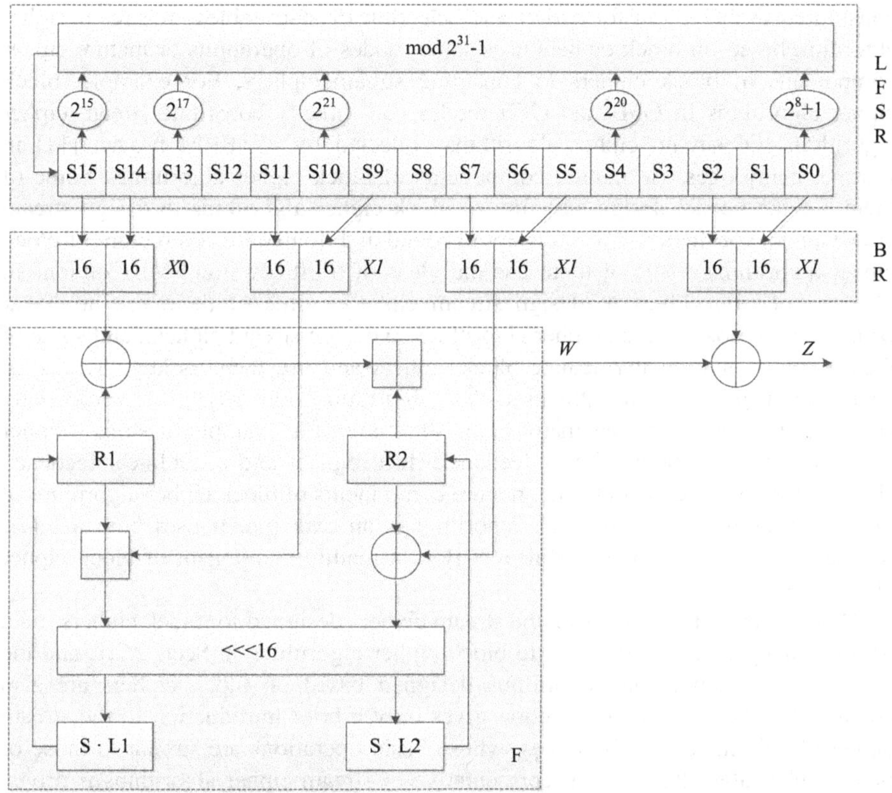

Fig. 2.19 Structure of the ZUC algorithm

nonlinear function *F* is composed of 4 parallel 8 × 8 S-Box (S0, S1, S2, S3), where S0 and S2 are equal, and S1 and S3 are equal. Both linear transformation functions L1 and L2 have 32-bit word inputs and 32-bit word outputs, and their inputs are both from the outputs of S-Box. The combination of S-Box and linear transformation function L1 can be expressed as $S \cdot L1$, and the combination of S-Box and linear transformation function L2 can be expressed as $S \cdot L2$. The outputs of linear transformation functions L1 and L2 are, respectively, reported to storage units R1 and R2.

Currently, stream ciphers adopt such processing structures as stream cipher algorithms based on the linear feedback shift register, the stream cipher algorithms based on the nonlinear feedback shift register, stream cipher algorithms driven by table, and stream cipher algorithms based on block ciphers. In addition to the core, that is, the linear feedback shift register, a stream cipher algorithm based on linear feedback shift register also contains other control and output logics. A stream cipher algorithm based on a nonlinear feedback shift register has a similar structure, but it uses a nonlinear feedback shift register as the core. A stream cipher algorithm driven by table has its core idea originating from the RC4 algorithm, which is constituted through transformation and selection of state tables. A stream cipher algorithm based on block ciphers uses some modes of operations or mature cipher components of block ciphers to constitute stream ciphers. For example, block cipher algorithms in OFB and CFB modes can directly constitute stream cipher algorithms. Of stream cipher algorithms collected by eSTERM project [11] in Europe, many ones use mature components of block cipher algorithms. Some of them use the Feistel design structure of block cipher algorithms to design stream ciphers, some of them use the concept of round and double helix structure of block cipher algorithms, some of them use the ideas of ShiftRows and MixColumns in block cipher algorithms to design stream ciphers, some of them use nonlinear component S-Box in block cipher algorithms to design stream ciphers, and some of them directly output intermediate block ciphers and use them as key streams. In actual applications, some stream cipher algorithms not only use a processing structure, but also combine many other structures; for example, a stream cipher algorithm may use both a linear feedback shift register and a nonlinear feedback shift register and even adopt some mature components of block cipher algorithms at the nonlinear filtering part. ZUC algorithm is an example. It uses both a linear feedback shift register and nonlinear S-Box, a mature component of block cipher algorithms.

This section will not focus on the stream ciphers designed for block ciphers. As a detailed analysis has been given to block cipher algorithms in Sect. 2.2.1, and the features of stream cipher algorithms designed based on block ciphers are also contained in Sect. 2.2.1, this section gives only a brief introduction to the stream cipher algorithms driven by tables, whose basic operations are similar to those of block cipher algorithms. As a representative of stream cipher algorithms is driven by tables and a part of the wireless local area network (WLAN) standard protocol, RC4 algorithm is one of the most widely used stream cipher algorithms. RC4 algorithm itself is very simple. Unlike most stream cipher algorithms which are based on feedback shift registers, it is designed based on dynamic arrays. RC4 algorithm is a stream cipher algorithm oriented to bytes, which has a variable key length of 1–256 bytes. It contains key scheduling algorithm (KSA) and pseudo-random generation algorithm (PRGA). Its first part is KSA (Algorithm 2.1), which performs permutation for initial vectors S{0, 1, …, 254, 255} under the control of the key. Its second part is PRGA (Algorithm 2.2), which is the output part of key stream. This part uses the permutation generated by KSA to generate pseudorandom bytes. In each round of iteration, it generates an output value and performs another permutation for vector S for generating the next key. Vector S in RC4 algorithm is the so-called table. In case of software implementation, tables in this kind of algorithm are implemented through dynamical arrays. In case of hardware implementation, tables are implemented through memories. This is similar to S-Box in block cipher algorithms. S-Boxes in stream cipher algorithms and in block cipher algorithms, however, are not accessed in completely the same mode. The value of S-Box in block cipher algorithms does not change usually, and only the read request needs to be considered. For table-driven stream cipher algorithms, as table permutation needs to be performed in each iteration, the content of table is changed, and both read request and write request need to be considered. In RC4 algorithm, in addition to the operation of table permutation, another main operation is 8-bit modular addition. In other table-driven stream cipher algorithms, other operations except table permutation are all very simple, covered in operations of block cipher algorithms, and have nothing special. In this section, it will not be the analysis emphasis.

Algorithm 2.1 Key scheduling algorithm

```
for i from 0 to 255
    S[i] := i
endfor
j := 0
for i from 0 to 255
    j := (j + S[i] + key[i mod keylength]) mod 256
    swap values of S[i] and S[j]
endfor
```

Algorithm 2.2 Random subkey generation algorithm

```
i := 0
j := 0
while GeneratingOutput:
    i := (i + 1) mod 256
    j := (j + S[i]) mod 256
swap values of S[i] and S[j]
    K := S[(S[i] + S[j]) mod 256]
    output K
endwhile
```

2. Features of Stream Ciphers

This section will mainly analyze two kinds of stream cipher algorithms: stream cipher algorithms based on linear feedback shift registers and stream cipher algorithms based on nonlinear feedback shift registers. As these two algorithms have a lot in common, they are collectively referred to as stream cipher algorithm based on feedback shift registers. Just as described in Sect. 2.1, the core of the stream cipher algorithm is a key stream generator. Key stream generators based on feedback shift register can be divided into two parts: the driving part and the nonlinear combination part. The driving part is usually composed of long-cycle feedback shift registers and uses the feedback shift registers to expand the key to a long-cycle state sequence. The nonlinear combination part, however, uses the stream generated by the former part to generate the final key stream sequence and hide the relationship between the state sequence and the key. The mechanism of the key stream generator conforms to the two basic cryptographic principles: Shannon diffusion and Shannon confusion.

What are usually used to design a key stream generator include the filter generator and the clock-controlled generator. A filter generator is also called a feedforward generator and is usually composed of two parts: the feedback shift register (s) and the filtering function. The output of the feedback shift register is output to the filtering function. According to the number of feedback shift registers, filter generators can be divided into nonlinear filter generators and nonlinear combination generators. A nonlinear filter generator filters the state of a feedback shift register in nonlinear mode, and its structure is shown in Fig. 2.20a, where function f_0 is called nonlinear filtering function. A nonlinear combination generator combines the output of multiple feedback shift registers in the nonlinear mode, and its structure is shown in Fig. 2.20b, where function f_0 is called a nonlinear combination function. If the feedback shift register is a linear feedback shift register according to the structure, the functions selected for these two kinds of generators must be nonlinear functions.

Geffee algorithm is a typical filter generator, and its structure is shown in Fig. 2.21. It uses three linear feedback shift registers, and its nonlinear composite part is a composite function $g(x)$ whose value is x_1, $x_2 + x_3$, $x_2 + x_3$. Geffee algorithm is a typical stream cipher algorithm based on linear feedback shift register. Many other algorithms are roughly the same in terms of principle, except that the number of linear feedback shift registers, position of tap, control of the nonlinear combination part, and function of the nonlinear combination part are different.

A clock-controlled generator is a clock which controls a feedback shift register by using the stream output by one or multiple feedback shift registers. There are two kinds of clock-controlled generators: feedforward clock-controlled generator and feedback-controlled generator. Feedforward clock-controlled generators include alternating step generators, cascade generators, and shrinking generators and usually use a clock-controlled shift register to control the clock of the shift register. An alternating step generator is composed of a clock-controlled register

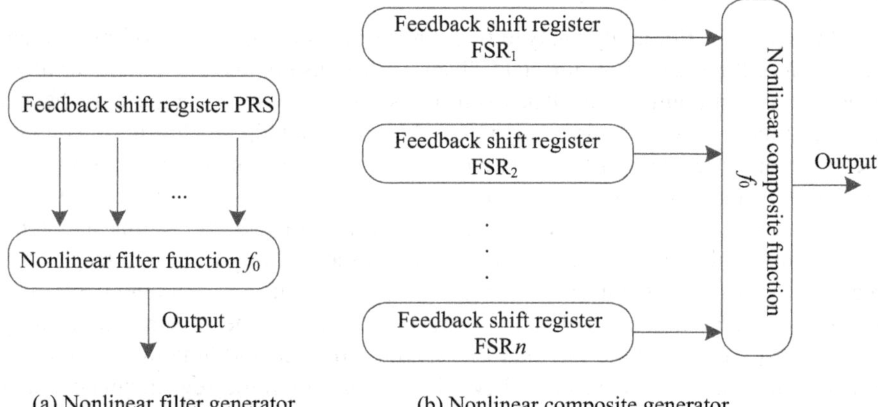

Fig. 2.20 Filter generator

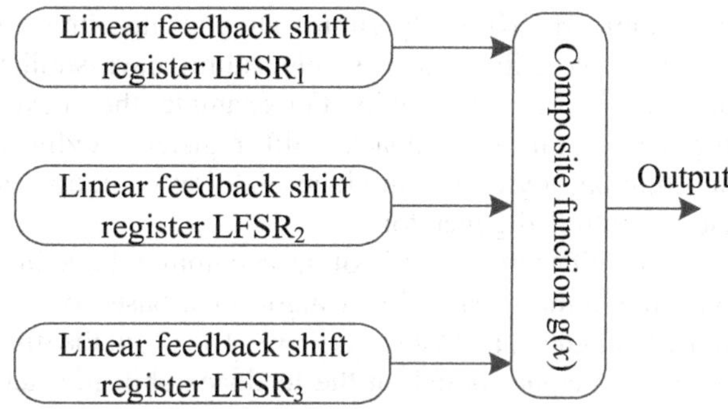

Fig. 2.21 Structure of Geffee algorithm

and two stop generators, and each stop generator uses a linear feedback shift register to control the clock pulse of the other linear feedback shift register. A cascade generator is composed of a series of feedback shift registers, and the clock of each feedback shift register is controlled by the previous feedback shift register. A shrinking generator is composed of two feedback shift registers and uses a feedback shift register to select the output of the other feedback shift register to generate key streams. A feedback clock-controlled generator is so-called in comparison with a feedforward clock-controlled generator. It indicates a clock which uses a shift register to control itself. A5 algorithm is a typical clock-controlled generator, and its structure is shown in Fig. 2.22. It is controlled by using three linear feedback shift registers and the corresponding clock-controlled stop components.

3. Common Logic of Stream Ciphers

According to the analysis above, it can be seen that many stream cipher algorithms are just slightly different in terms of structure. Though these algorithms are subdivided into many categories, they have very similar basic composition and they have much in common in terms of basic implementation logic. In addition, there is a lot of common logic among different stream cipher algorithms. By analyzing such stream cipher algorithms as ZUC algorithm, Geffe algorithm, A5 algorithm, SNOW algorithm, Grain algorithm, SOBER algorithm, Gifford algorithm, Achterbahn

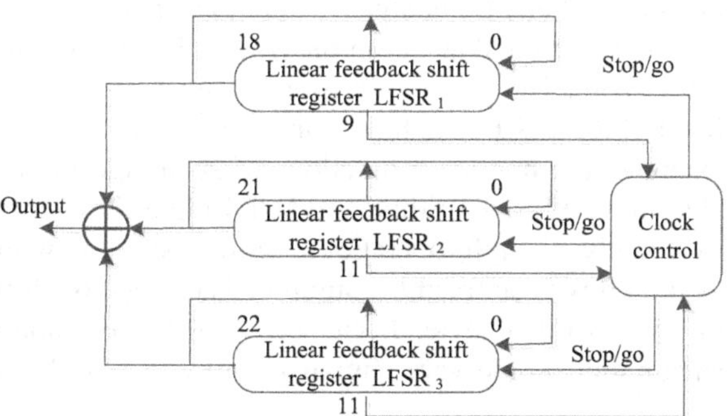

Fig. 2.22 Structure chart of A5 algorithm

algorithm, MICKEY algorithm, Dragon algorithm, and NLS algorithm, we can make a conclusion that these algorithms have similar cipher processing structure and function operation units. For example, they have similar linear feedback shift register, nonlinear feedback shift register, nonlinear Boolean function, S-Box substitution, logic unit, modular addition operation unit, shift operation unit, and clock-controlled generator.

To facilitate the analysis of these common logic in terms of computing grain, we will divide the stream cipher algorithms based on the feedback shift register into finite field GF(2) and GF(2^w) ($w = 8, 16, 32$) in the discussion. On finite field GF(2), the stream cipher based on the feedback shift register has the following features.

① These stream ciphers are generated on a binary field and generate a 1-bit key each time.
② Most algorithms are within 256 series, and the number of feedback shift registers usually does not exceed 10.
③ For different stream ciphers, there are various nonlinear functions. The most commonly used function, however, is still nonlinear Boolean function.
④ It is required to select a state of the feedback shift register and use it as the input of the nonlinear function.

On finite filed GF(2^w) (w = 8, 16, 32), a stream cipher based on feedback shift register is a little more complex than an algorithm on GF(2). The computation granularities of the logic operation and arithmetic operation involved in these stream cipher algorithms are usually bytes or an integral multiple of bytes such as 8, 16, and 32 bits, but seldom exceed 32 bits. The operations are usually ones based on unsigned integers, and there is basically no fixed point decimal operation, signed number operation, or floating-point decimal operation. The nonlinear function part of these stream cipher algorithms is usually composed of various nonlinear transformations. For example, the nonlinear functions of such algorithms as Phelix algorithm, SOBER algorithm, and NLS algorithm adopt modular 2^w addition operation. The nonlinear functions of such algorithms as Phelix algorithm, SOBER algorithm, NLS algorithm, SNOW algorithm, and Dragon algorithm adopt nonlinear logic operations. Such algorithms as RC4 algorithm, SOBER algorithm, and NLS algorithm use S-Box as their nonlinear functions. These S-Boxes are used to implement confusion required by algorithms and have similar functions with S-Box in block ciphers. By analyzing multiple stream ciphers which use S-Box operations, it is discovered that common S-Box is of the following structures: 8-in-8-out, 8-in-32-out, 4-in-4-out, 6-in-4-out, 13-in-8-out, 12-in-8-out, 10-in-8-out, 6-in-2-out. Most of these S-Boxes have consistent structure with S-Boxes in block ciphers. For example, the nonlinear functions of such algorithms as Phelix algorithm, SOBER algorithm, NLS algorithm, and Mir-1 algorithm adopt the shift operation. There are two kinds of shift operations: shift operation with fixed bit number and shift operation with unfixed bit number. The range of shift is usually 1–31, and the unit of shift is bit or byte. It can be seen that the nonlinear transformations used in stream ciphers are very similar to those used in block ciphers.

By analyzing the common logic and computing granularities of stream ciphers, you can see that such operations as logic operation, shift, S-Box, modular addition/subtraction are very similar to those in block ciphers and have been covered in the common logic of block cipher algorithms, which are obtained by the analysis in Sect. 2.2.1. Thus, no more discussion will be made about it here. Linear feedback shift registers and nonlinear feedback shift registers are the most widely used common logic in stream ciphers, and they are generally not involved in block ciphers. Next, we will analyze the structure of the feedback shift register (FSR) in detail.

A feedback shift register (FSR) is composed of a n-bit shift register and a feedback function, as shown in Fig. 2.23. Each shift operation the register performs, all bits right shift 1-bit position, the output value of the rightmost register will be used as the register output, and the output value of the feedback function will be filled into the left most register. The feedback function f is the Boolean function of n variables (b_0, b_1, …, b_{n-2}, b_{n-1}). Each shift operation the register performs, and an output is generated. You can enable register to perform m shifts to generate a m-bit output and form an output sequence b'_0, b'_1, …,b'_{m-1}. What correspond to the feedback shift registers are a lot of Boolean function operations and shift operations in stream ciphers. For example, within a shift cycle, extract the data at the corresponding position of the shift register according to the feedback polynomial and perform exclusive OR operation for the data bit by bit, generate 1-bit data and report it to the highest bit, and perform shift operation for the bits of the register step-by-step. A lot of similar operations exist in various stream ciphers and exist throughout the algorithms.

LFSR refers to the feedback shift register whose feedback function is linear. The common linear function is an exclusive OR operation, which conducts exclusive OR for some bits of the register and then uses the result of the exclusive OR as the input of the register and conducts overall shift for the bits in the register. Nonlinear feedback shift register (NLFSR) is so named in comparison with the linear feedback register. These two kinds of algorithms have similar basic structure, and the difference between them is that the feedback logic of LFSR is linear, while the feedback logic of NLFSR is nonlinear. A common nonlinear logic is AND operation. The feedback logic of LFSR includes only exclusive OR gate, while the feedback logic of NLFSR includes both exclusive OR gate and AND gate. Let us look at the difference between LFSR and NLFSR from algebra perspective. First, let

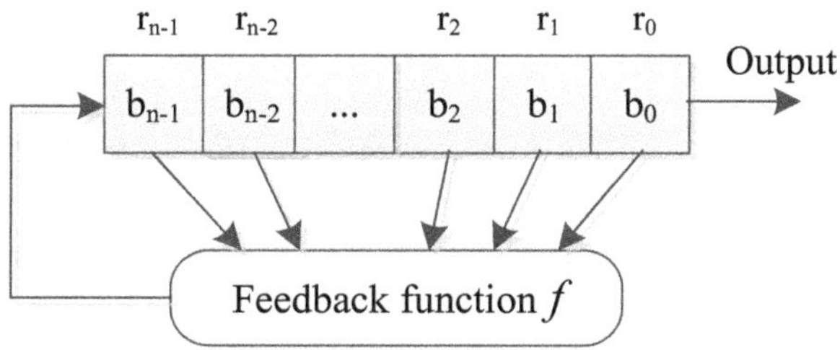

Fig. 2.23 Feedback shift register (FSR)

us express them with a polynomial, specifically addition (+) indicates exclusive OR and multiplication (*) indicates AND. Linear feedback logic contains only addition, and the degree of the highest term of its feedback expression will not increase. However, as nonlinear feedback logic contains multiplication, the degree of the highest term of its feedback expression will increase.

4. Degree of Parallelism for Stream Ciphers

Data dependency can be divided into two levels. First, seen from the encryption/decryption task level of the whole algorithm, the stream cipher executes tasks in serial. Unlike the block ciphers in the ECB or CTR mode which divide a task into multiple subtasks and execute these subtasks in parallel, a stream cipher algorithm can only execute tasks in serial. Second, seen from the internal structure, the read-after-write dependency exists between the components of block ciphers. That is, the operation of the next component starts only after the completion of the operation of the previous component. For stream ciphers, however, write-after-read dependency exists between their components. This is more obvious when two feedback shift registers are cascaded. Only after the lower-level feedback shift register has read the value of its upper-level register and has updated its own state, its upper-level register can update its state. This data dependency has a more obvious reflection in case of software implementation and is inevitable.

Though a stream cipher does not support parallel data encryption/decryption like a block cipher, parallelism exists in its internal components. Unlike block cipher algorithms where the pretransformation, the round transformations, and post-transformations in a block are executed in serial, the components of stream cipher algorithms, that is, the feedback shift registers and nonlinear functions, are executed in parallel. The components need to execute each clock cycle and update their own state simultaneously. This is very similar to the parallelism developed by using the pipeline technology in block ciphers. In case of hardware implementation, the stream cipher parallelism can be developed naturally.

Actually, the research on reconfigurable cryptographic processor lays more emphasis on block ciphers. There are three reasons for it. First, some stream ciphers themselves are from some working modes or some components of block ciphers and have structures similar to those of block ciphers. Second, the structure and operation of the stream cipher are both much simpler than those of block ciphers. Finally, there are not as many public stream ciphers as public block ciphers. Therefore, the demands and challenges brought by stream ciphers are much smaller than those brought by block ciphers.

2.3 Hash Algorithms

2.3.1 Introduction to Hash Algorithms

There are at least over ten commonly used hash algorithms, including SM3, MD5, SHA-1, SHA-2, and SHA-3 [3, 12]. This section introduces the typical hash algorithms: SHA-3 and SM3. The SHA-3 algorithm is a new-generation hash algorithm collected by the National Institute of Standards and Technology (NIST) around the world since 2007. On October 2, 2012, the Keccak algorithm became the standard SHA-3 hash algorithm [13].

The SHA-3 algorithm involves message padding, message expansion, and iterative compression. During the message padding, the input message of any length is padded so that the length of the message is an integral multiple of the message block length. The padding pattern is 10…01 (that is, add a "1" to the end of the message value; then, add several zeros "0"; finally, add a "1"). After the message padding, the message is grouped according to the message block length; the number of bits for each group of messages is the same as the message block length. During the message expansion, "0" expansion is performed for message blocks so that message blocks are expanded to the width of the compression function. An iterative compression part is the iteration process of a compression function. The output of the compression function is called the chain value. At each iteration, the output chain value and message blocks of the last compression function after XOR are loaded to the next compression function for compression. After all message blocks are processed through several rounds of iterations, the output chain value of the compression function at each iteration is directly loaded to the input of the next compression function; the hash value output is bits of the message block length in the chain value until iteration for all hash values ends. The compression function is the key part of the hash algorithm. It involves 24 rounds of iterations. Each round of iteration involves the following steps (permutation and substitution operations): τ, χ, π, e, and θ. The object of the permutation and substitution operations in the preceding steps is data of 1,600 bits. The data are organized into a three-dimensional matrix. The three dimensions x, y, and z correspond to row (five elements), column (five elements), and lane (64 elements). The permutation and substitution functions involved in the preceding steps are introduced below.

θ is a substitution function, as shown in Formula (2.1). This function performs the linear operation (mainly the XOR operation). The input of the θ substitution function comes from message blocks after the message padding or the output of the τ substitution function in the last round of iteration. The θ substitution function processes one column of the three-dimensional matrix with the following steps: 1. Perform the SUM operation for all the five elements in its left column; 2. perform the SUM operation for the five elements in its lower right column; 3. perform the XOR operation for the output of Step 1 and that of Step 2; 4. perform the XOR operation for the result of Step 3 and each element of the column, respectively. The result of Step 4 is the output of the θ function.

$$\theta\text{: } a[x][y][z] \leftarrow$$

$$a[x][y][z] \oplus \sum_{y=0}^{4} a[x-1][y][z] \oplus \sum_{y=0}^{4} a[x+1][y][z-1] \tag{2.1}$$

ρ is a permutation function, as shown in Formula (2.2). This function is also a linear operation. The input of the ρ permutation function is the output of the θ substitution function. The ρ permutation function processes one lane of the three-dimensional matrix. The nature of this function is rotation along the z-axis. The rotation length is the result of the matrix multiplication in Formula (2.2). The bit number of rotation varies according to lane. It mainly depends on the coordinate value of the x-axis, the coordinate value of the y-axis, and algorithm parameters.

$$\rho\text{: } a[\mathrm{x}][\mathrm{y}][\mathrm{z}] \leftarrow \mathrm{a}[\mathrm{x}][\mathrm{y}]\left[\mathrm{z} - \frac{(t+1)(t+2)}{2}\right],$$

$$0 \le \mathrm{t} \le 24, \begin{pmatrix} 0 & 1 \\ 2 & 3 \end{pmatrix}^{\mathrm{t}} \begin{pmatrix} 1 \\ 0 \end{pmatrix} = \begin{pmatrix} x \\ y \end{pmatrix} \in GF(5)^{2\times 2} \tag{2.2}$$

π is a permutation function, as shown in Formula (2.3). This function also performs the linear operation. It performs permutation for cross sections determined by the x-axis of the three-dimensional matrix to change positions of elements on the cross sections. It has the spreading effect. The permutation positions are the results of the matrix multiplication in Formula (2.3). The mapping scheme of the π function is similar to that of the ρ function. The permutation is mainly determined by the x and y coordinates.

$$\pi\text{: } a[\mathrm{x}][\mathrm{y}][\mathrm{z}] \leftarrow a[\mathrm{x}'][\mathrm{y}'][\mathrm{z}], \begin{pmatrix} \mathrm{x} \\ \mathrm{y} \end{pmatrix} = \begin{pmatrix} 0 & 1 \\ 2 & 3 \end{pmatrix} \begin{pmatrix} \mathrm{x}' \\ \mathrm{y}' \end{pmatrix} \tag{2.3}$$

χ is a substitution function, as shown in Formula (2.4). As the only nonlinear transformation function in round functions, this function helps improve algorithm security. This function processes one row of the three-dimensional array. Logical operations mainly consist of NOT, AND, and XOR, wherein the AND operation provides the only nonlinear element in the algorithm. Taking one-row operation of the χ function as an example, the update value of $a[0]$ can be obtained with the following operations: 1. NOT operation for $a[1]$; 2. AND operation for the output of Step 1 and $a[2]$; 3. XOR operation for the output of Step 2 and $a[0]$. The update value of $a[2]$ can be obtained with the following operations: 1. NOT operation for $a[3]$; 2. AND operation for the output of Step 1 and $a[4]$; 3. XOR operation for the output of Step 2 and $a[2]$. The rest may be deduced by analogy. You can obtain the update values of all χ functions.

$$\chi: a[\mathrm{x}] \leftarrow a[\mathrm{x}] \oplus (a[\mathrm{X} + 1]) \oplus 1 \Lambda a[\mathrm{x} + 2] \tag{2.4}$$

τ is a substitution function, as shown in Formula (2.5). This function performs the linear operation. This function adds a constant to the end of each round of operation to destroy the original symmetry of the three-dimensional array, thus increasing security.

$$\tau: a[0][0] \leftarrow a[0][0] \oplus RC[\mathrm{n_r}] \tag{2.5}$$

SM3 is the hash algorithm [14] for commercial cryptography published by the State Cryptography Administration in 2012. This algorithm has advantages such as high security strength, simple construction, and high implementation efficiency of software and hardware. Its overall performance is superior to the SHA-256 hash algorithm that is widely used in the world. Currently, the SM3 algorithm has been widely used as national standard of hash algorithms and cryptography industry standard in China. The SM3 algorithm also involves message padding, message expansion, and iterative compression. During message padding, l bits of 1s are added to the end of the message m with the length of l bits, and then, k bits of 0s are added to the end of the added 1s, wherein l and k shall meet the following condition: $l + k + 1 = 448 \bmod 512$. Finally, the length l is expressed in 64 bits of binary numbers, which are added to the end of the message, thus obtaining the message m'. During message expansion, the expanded message m' is grouped by 512 bits. That is, $m' = B^{(0)} B^{(1)} \ldots B^{(n-1)}$, wherein $n = (l + k + 65)/512$; the length of $B^{(i)}$ is 512 bits. Then, the message block $B^{(i)}$ is expanded to 132 words $W_0, W_1, \ldots, W_{67}, W'_0, W'_1, \ldots, W'_{63}$, which are used as the input of the compression function. During iterative compression, the input is processed by the compression function. That is, $V^{(i+1)} = \mathrm{CF}(V^{(i)}, B^{(i)})$, wherein $V^{(0)}$ is the initial value (IV) of iteration, and $V^{(n)}$ is the result of iterative compression. The compression iteration steps are repeated for n times. Details about the compression function will not be provided here.

2.3.2 *Features of Hash Algorithms*

The preceding section introduces the typical hash algorithms: SHA-3 and SM3. This section analyzes the features of hash algorithms. The operation of all hash algorithms is performed according to the following general principle: The input (such as message and file) can be seen as the sequence of an n-bit block; its output is an n-bit hash value. A hash algorithm processes a block at a time and repeats the process until all blocks are processed. The simplest hash algorithm performs the XOR operation for the corresponding bits of blocks. The generated result is the hash value. The formula is as follows: $C_i = b_{i1} \oplus b_{i2} \oplus b_{i3} \ldots b_{i(m-1)} \oplus b_{im}$. C_i indicates the ith bit of the hash value; b_{ij} indicates the ith bit of the jth block. The operation principle of a slightly improved hash algorithm is as follows: 1. Perform the XOR operation between the hash value and each bit of a block; 2. take a rotation; 3. perform the XOR operation with each bit of the next block until all blocks are processed. In practical application, the construction method for hash algorithms is complex. There are basically two construction methods: construction method based on block cipher and construction method that is not based on block cipher. The simplest method is to use the CBC or CFB working mode of block ciphers and take the key as the initial value to encrypt message blocks one by one. The final ciphertext is the hash value. Earlier hash algorithms are often designed on the basis of mature block ciphers. The hash algorithms using the construction method that is not based on block cipher are called custom hash algorithms. Block cipher algorithms have been discussed in Sect. 1.2.1. The features of hash algorithms constructed on the basis of block cipher are similar to those of block cipher algorithms. This section mainly discusses custom hash algorithms.

All the existing hash algorithms are designed on the basis of the iterative structure of the compression function (or permutation function) [15]. The input of each compression function is a message block of fixed length. Moreover, the operations are the same or similar for each message block. Merkle and Damgård put forward the typical iterative structure (MD structure [15, 16]) for secure hash algorithms. See Fig. 2.24. Many hash algorithms such as SHA-2 and SM3 adopt the MD structure. The MD structure requires iterative use of compression function for several rounds. The input of the current compression function f is the n-bit result of the previous compression function and a b-bit message block. The current compression function generates another n-bit output. If the compression function is collision-resistant, the hash algorithm is also collision-resistant. Thus, it can be seen that the compression function f is the most important and core link in hash algorithms in terms of both the amount of computation and security. The design and attack of hash algorithms mainly start with compression function. Hash algorithms with this type of structure have similar iterative structures. They mainly differ in terms of the iterative operation mode and the internal structure of the compression function.

To solve the problems (such as second collision attack, multicollision attack, length extension attack, and second preimage attack of long messages) of this typical structure, new algorithm structures such as double-pipeline structure [17] (see Fig. 2.25) and HAIFA structure [15] (see Fig. 2.26) are developed. In the

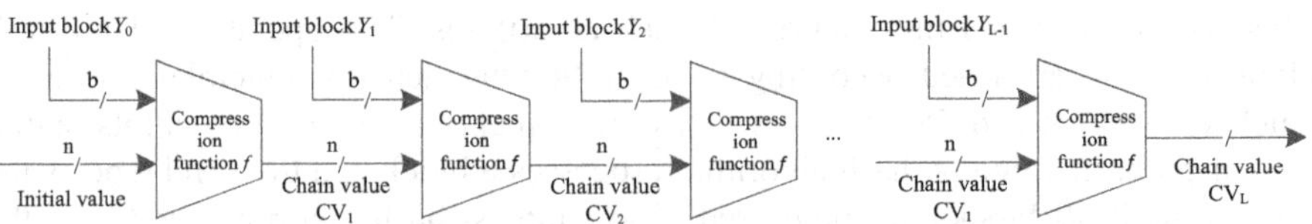

Fig. 2.24 Typical hash algorithm structure

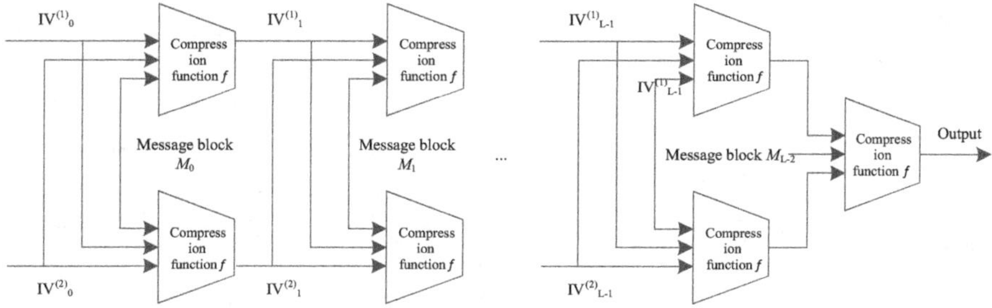

Fig. 2.25 Dual-pipeline structure

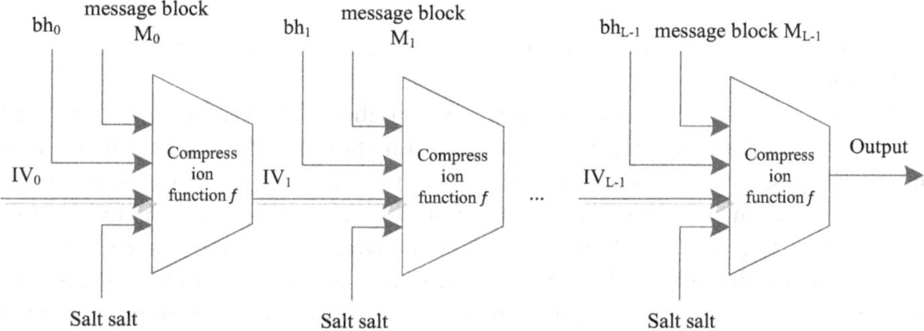

Fig. 2.26 HAIFA structure

double-pipeline structure, the hash algorithm with two parallel narrow pipelines of the same size is used to compress each message block twice in parallel and then compress the final compression result, thus obtaining the hash value. In the HAIFA structure, the random salt and bits that have been processed by the hash algorithm (bh) are added to the input of the compression function.

The most widely used sponge structure (see Fig. 2.27) is applied in SHA-3 and some lightweight hash algorithms such as PHOTON, QUARK, and SPONGENT [18]. The sponge structure is simple and flexible. It allows variable input length and output length. The sponge structure has two parameters: bit rate (r) and capacity (c). The sum of the two parameters determines the permutation width of the compression function f. The sponge structure involves two phases: absorbing and squeezing. In the absorbing phase, the input message is divided into message blocks M_0, M_1, …, M_{L-1} (length: r) after being padded. The XOR operation is performed between each message block and the output value (length: r) of the previous round of compression function. The output with the length of c remains unchanged. The XOR operation result and the output (length: c) are used as the input for this round of compression function. In the squeezing phase, z_0, z_1, …, z_m are extracted from the output of the compression function according to the length of the output

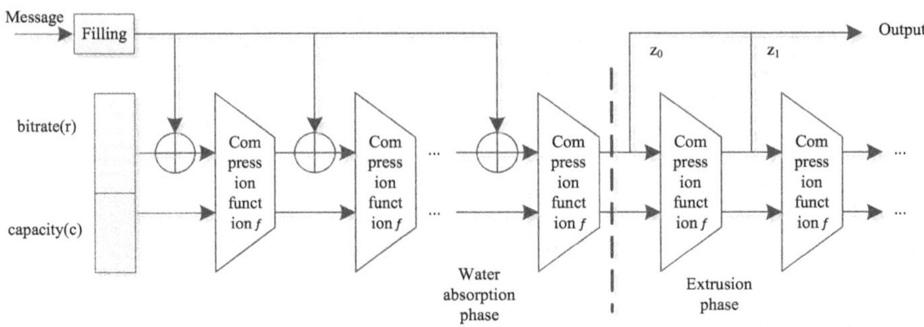

Fig. 2.27 Sponge structure

hash value required by the application; they are concatenated to generate the final hash value.

According to these structures, you can get the operation procedure of hash algorithms based on the iterative compression function: Fill the message in a way so that the length of the message is an integral multiple of the message block length; group the filled message; use the compression function to perform iterative compression for each message block in sequence; convert the output of the end message block to obtain the output of the hash algorithm. Except input of chain values and message blocks, compression functions of some algorithms may require input of parameters such as counter and random salt [19].

Taking SM3, MD5, SHA-1, SHA-2, and SHA-3 algorithms for example, the following section analyzes and summarizes the features of main parameters for common hash algorithms (see Table 2.3). According to the hash value length, SHA-2 can be divided into SHA-224, SHA-256, SHA-384, and SHA-512; SHA-3 can be divided into SHA-3(224), SHA-3(256), SHA-3(384), and SHA-3(512). Although these data vary according to algorithm, there are rules to follow. The grading feature is obvious. Generally, the range of the hash value length is 160–512 bits. SHA-1 is not secure. Therefore, the hash value length mainly falls into four grades: 224, 256, 384, and 512. The maximum message lengths supported by different algorithms are different. The message lengths supported by SM3, SHA-1, SHA-224, and SHA-256 are less than 2^{64}; the message lengths supported by SHA-384 and SHA-512 are less than 2^{128}; the message lengths supported by MD5 and all SHA-3 algorithms are unlimited. Like block cipher algorithms, hash algorithms also process messages block by block. The block processed by hash algorithms is a little longer than that for block ciphers. Generally, the block length for algorithms is not greater than 256 bits. The block length for SM3, MD5, SHA-1, SHA-224, and SHA-256 is 512 bits; the block length for SHA-384 and SHA-512 is 1024 bits; due to the sponge structure, the block length (576–1152 bits) for SHA-3 algorithms varies according to hash value length. The word length is regular. In order to adapt to the word length of mainstream computers, the word length is generally 32 bits or 64 bits. The number of rounds varies from 24 to 80. The number of rounds for SM3, MD5, SHA-224, and SHA-256 is 64; the number of

Table 2.3 Main parameters of common hash algorithms

Algorithm	Length of hash value	Maximum message length	Packet length	Word length	Number of rounds
SM3	256	$2^{64} - 1$	512	32	64
MD5	128	Unlimited	512	32	64
SHA-1	160	$2^{64} - 1$	512	32	80
SHA-224	224	$2^{64} - 1$	512	32	64
SHA-256	256	$2^{64} - 1$	512	32	64
SHA-384	384	$2^{128} - 1$	1024	64	80
SHA-512	512	$2^{128} - 1$	1024	64	80
SHA-3 (224)	224	Unlimited	1152	64	24
SHA-3 (256)	256	Unlimited	1088	64	24
SHA-3 (384)	384	Unlimited	832	64	24
SHA-3 (512)	512	Unlimited	576	64	24

rounds for SHA-1, SHA-384, and SHA-512 is 80; the number of rounds for SHA-3 algorithms is 24.

2.3.3 Common Logic of Hash Algorithms

Correspondingly, you can obtain a general implementation architecture (see Fig. 2.28) according to common features of hash algorithms. The hardware architecture involves the control unit and datapath. The core of the control unit is the control state machine, which generates the internal control signal of the datapath. For the control signal generated by the control unit, information such as the number of iterations and the operation module selection should be specified. In the datapath, the input data is processed by message padding, message expansion, compression function (a chain value is obtained), and output conversion successively. The output data after output conversion is stored in the compression function register. Then, the control unit determines whether the number of iterations has been completed. If the number of iterations has been completed, a hash value is exported; if the number of iterations has not been completed, the data are returned to the compression function for the next round of compression. The constant memory stores the initial hash values and some constants of hash algorithms. For example, T_j = 79cc4519, $0 \leq j \leq 15$; T_j = 7a879d8a, $16 \leq j \leq 63$ in the SM3 algorithm. The SHA-3 and SM3 algorithms that are analyzed in the previous section can be implemented using the hardware architecture as shown in Fig. 2.28. In fact, you can adapt the hardware

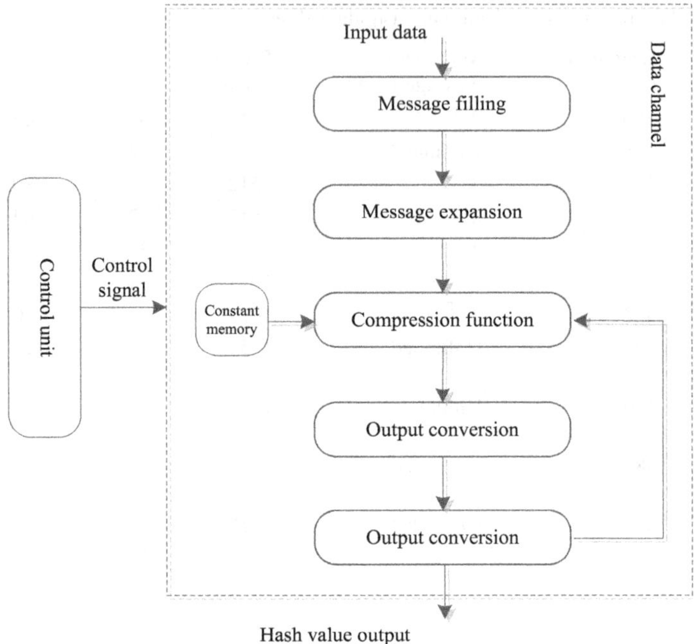

Fig. 2.28 Common hardware architecture of hash algorithm

architecture to different hash algorithms by adjusting parameters such as bit width, compression function in iteration, number of iterations, and contents in the constant memory.

In terms of specific structure, some algorithms have a lot in common. For example, both the SM3 algorithm and the SHA-2 algorithm use the MD structure. The structure of the SM3 algorithm is similar to that of the SHA-2 algorithm in detail. The differences between the SM3 algorithm and the SHA-2 algorithm mainly lie in the data width, operand length, number of rounds, and length of the hash value that is exported in the end. Many other parts are the same or similar. You can even implement the SM3 and SHA-2 algorithms on a hardware circuit, thus achieving flexibility, efficiency, and hardware resource-saving. First, message expansion of the SM3 algorithm is similar to that of the SHA-2 algorithm. Second, both algorithms mainly involve eight steps, wherein computations in four steps are the same (i.e., they are implemented by direct assignment); operations in two steps are similar (you only need to determine whether value is assigned directly or value is assigned after shift); only operations in two steps are quite different for both algorithms. Therefore, the SM3 and SHA-2 algorithms can share common logic such as message expansion and many assignment operations.

Hash algorithms are similar to block ciphers in many aspects. It is true not only for the hash algorithms based on the block cipher structure, but also for the custom hash algorithms. Both hash algorithms and block ciphers process messages block by block. To facilitate understanding of the similarity between hash algorithms and block ciphers, you can make an analogy between the compression function *f* and the block cipher round function and an analogy between iterative structures such as MD structure and sponge structure and feedback operation mode of block ciphers (e.g., CBC and OFB). The compression function *f* performs the round-iteration operation similar to that of block ciphers. The nature of the round-iteration operation is a number of permutation and substitution operations.

Hash algorithms are designed on the basis of similar theories and with similar structures and operation types. Therefore, a lot of common logic exists among different hash algorithms. For example, generally, there are two message padding schemes: 1. One bit of 1 is added to the end of the message; then, several bits of 0s are added; 2. One bit of 1 is added to the end of the message; several bits of 0s are added; then, another bit of 1 is added. Different algorithms differ in the number of filled 0s. Another example is that round transformation is the core operation of the compression function, which is the core operation of hash algorithms. The nature of the round transformation operation is the combination of several substitution and permutation operations, which is exactly the same for block ciphers. The round transformation mainly involves logic operations such as AND, OR, NOT, and XOR and a large number of shift operations such as logic shift and rotation. It seldom involves complex operations such as multiplication and finite field inversion. Therefore, common logic of hash algorithms is basically included in those of block ciphers.

2.3.4 Parallelism of Hash Algorithms

The high structure similarity between hash algorithms and block ciphers determines that they have many common features in parallelism. Hash algorithms process messages block by block. Therefore, parallelism exists among blocks. The block lengths processed by SM3, SHA-256, SHA-384, SHA-512, and SHA-3 are 512 bits, 512 bits, 1024 bits, 1024 bits, and 1600 bits, respectively, which are longer than those of general block ciphers. Horizontally, block ciphers do not perform operations on a block as a whole. Instead, block ciphers perform parallel operations based on a smaller granularity. Generally, the granularity is 32 bits or 64 bits. Compression functions are dependent on each other. Therefore, the parallelism between blocks cannot be developed like block cipher modes such as ECB.

Like block ciphers, hash algorithms involve close data dependency, including the data dependency in the round transformation, the data dependency between round transformations, the data dependency between compression functions, and the data dependency between the message padding and expansion and the compression function. Vertically, the operations in the round transformation are performed in a serial manner. For example, the θ, ρ, π, χ, and τ functions in the round transformation of SHA-3 are executed in sequence. The latter function must wait for the result of its previous function. Intensive RAW data dependency exists between operations. Likewise, the compression function also involves multi-iteration round transformation. The numbers of round transformation iterations for SM3, SHA-256, SHA-384, SHA-512, and SHA-3 are 64, 64, 80, 80, and 24, respectively. Obvious RAW data dependency also exists between these round transformations. Therefore, the round operation of algorithms must be performed in a serial manner. The data dependency between compression functions is the same as that between blocks when block algorithms work in modes such as cipher block chaining (CBC) mode, ciphertext feedback (CFB) mode, and output feedback (OFB) mode. The feedback operation is involved between two adjacent compression functions. Data input of the latter compression function depends on the output of its previous compression function. Therefore, the latter compression function can be computed only after computation of its previous compression function is complete. This directly limits the parallel and pipeline possibilities of compression functions. The data dependency between message padding and expansion and compression functions mainly indicates that message padding and message expansion need to be performed before compression functions are executed, as shown in Fig. 2.28.

According to the preceding analyses, hash algorithms are highly similar to block algorithms in some aspects such as block features, iterative structure, round transformation contents, parallelism, data dependency, and common logic. Therefore, the analysis conclusions on relevant aspects of block algorithms can also be applied to hash algorithms.

2.4 Public-Key Ciphers

2.4.1 Introduction to Public-Key Ciphers

The basic design thought for substitution and permutation operations of public-key ciphers is different from that of symmetric cryptographic algorithms. The substitution and permutation operations of public-key ciphers are designed by using hard problems in mathematics. According to the hard mathematical problems, the public-key ciphers are mainly divided into public-key ciphers (typical representatives: RSA and Rabin–Williams algorithms) based on the integer factorization problem, public-key ciphers (typical representatives: DSA, Diiffe–Hellman, ElGamal, Schnorr and Nyberg–Rueppel algorithms) based on the discrete logarithm problem, public-key ciphers (typical representatives: ECDSA, ECC Diiffe–Hellman, ECC ElGamal, ECC Schnorr and ECC Nyberg–Rueppel algorithms) based on the elliptic curve discrete logarithm problem (ECDLP), public-key ciphers (typical representative: NTRU [20] algorithm) based on the shortest vector problem (SVP) and the closest vector problem (CVP). The following section takes the RSA and ECC algorithms for example to introduce public-key ciphers.

The RSA algorithm was put forward by Rivest et al. from MIT in 1978 [21]. As a simple and easy-to-understand algorithm, RSA can be used for encryption and signature. The RSA algorithm is the most popular and most influential public-key cipher at present. The RSA algorithm is described in Algorithm 2.3.

Algorithm 2.3 RSA algorithm

Randomly select two big prime numbers p and q.

Compute $n=pq$, $\varphi(n)=(p\text{-}1)(q\text{-}1)$.

Randomly select a positive integer d, which satisfies the following conditions: $0<d<\varphi(n)$, $\gcd(d, \varphi(n))=1$.

Compute $e=d^{-1} \bmod \varphi(n)$.

Disclose e and n as the public keys of user A.

Keep secret d, p and q. d is the private key of user A.

Encryption

· For message M, assume $0<M<n$.

· Compute $C=M^e \bmod n$

Decryption

· Compute $P=C^d \bmod n$

The ECC algorithm was put forward by Miller and Koblitz in 1985. ECC is another most important public-key cipher except RSA. With higher performance and security, ECC is supposed to gradually replace RSA and become the most important public-key cipher in this century. ECC has been standardized by organizations such as IEEE, NIST, ANSI, ISO, and China National Commercial Cipher Management Office. The commercial cipher algorithm SM2 in China is a public-key cipher algorithm based on ECC. Compared with the RSA algorithm, the ECC algorithm is more difficult to understand. It is constructed by using an elliptic curve finite group defined in a finite field. Generally, such finite fields include prime field GF(p) and finite field GF(2^m).

First, introduce the elliptic curve defined in the prime field GF(p). Suppose p is a big prime number; $a, b \in \text{GF}(p)$ satisfying $\Delta = -16(4a^3 + 27b^2) \neq 0$; E1 is the solution set of the equation $y^2 = x^3 + ax + b$ on the field GF(p): $E1 = \{(x, y)|x, y \in \text{GF}(p); y^2 = x^3 + ax + b\}$; $E1$ and the special element O constitute a set E: $E = E1 \cup \{O\}$. The operation between E elements enables E to constitute a group, which is called an elliptic curve defined in the finite field GF(p). Generally, the operation in E is expressed by "+"; the identity element in E is O.

Suppose $P1 = (x1, y1)$ and $P2 = (x2, y2)$ are any two elements in $E1$. The group operation rules in E are as follows:

Identity element: $P1 + O = P1$, $O + P1 = P1$.

Inverse element: $-P1 = (x1, -y1)$, $-P2 = (x2, -y2)$, $P1 + (-P1) = O$.

Point addition and point doubling: Suppose P1 $\neq$ −P2. Let $P3 = P1 + P2$, $P3 = (x3, y3)$, then you can calculate P3 according to the values of $P1$ and $P2$ as follows:

Point doubling: When $P1 = P2$

Suppose $\lambda = \frac{3x1^2+a}{2y1}$. Then, $x3 = \lambda^2 - 2x1$ and $y3 = \lambda(x1 - x3) - y1$ in P3 (P3 = P1 + P2).

Point addition: When $P1 \neq P2$,

Suppose $\lambda = \frac{y2-y1}{x2-x1}$. Then, $x3 = \lambda^2 - x1 - x2$ and $y3 = \lambda(x1 - x3) - y1$ in P3 (P3 = P1 + P2).

Then, introduce the elliptic curve defined in characteristic-two finite field GF (2^m). Suppose m is a given positive integer; for finite field GF(2^m), $a, b \in$ GF(2^m) satisfying $\Delta = b \neq 0$. $E2$ is the solution set of the equation $y^2 + xy = x^3 + ax^2 + b$ in the finite field GF(2^m): $E2 = \{(x, y)|x, y \in \mathrm{GF}(2^m); y^2 + xy = x^3 + ax^2 + b\}$. E2 and the special element O constitute a set E: E = E2 ∪ {O}. The operation between E elements enables E to constitute a group, which is called an elliptic curve defined in the finite field GF(2^m). Generally, the operation in E is expressed by "+"; the identity element in E is O.

Suppose $P1 = (x1, y1)$ and $P2 = (x2, y2)$ are any two elements in $E1$. The group operation rules in E are as follows:

Identity element: $P1 + O = P1$, $O + P1 = P1$.

Inverse element: $-P1=(x1, x + y1)$, $-P2 = (x2, x + y2)$. $P1 + (-P1) = O$, $P2 + (-P2) = O$.

Point addition and point doubling. Suppose $P1 \neq -P2$. Suppose $P3 = P1 + P2$, $P3 = (x3, y3)$, then you can calculate P3 according to the values of $P1$ and $P2$ as follows:

Point doubling: When $P1 = P2$

Suppose $\lambda = x1 + \frac{y1}{x1}$. Then, $x3 = \lambda^2 + \lambda + a = x1^2 + \frac{b}{x1^2}$, and $y3 = x1^2 + \lambda x3 + x3$ in P3 (P3 = P1 + P2).

Point addition: When $P1 \neq P2$,

Suppose $\lambda = \frac{y2+y1}{x2+x1}$. Then, $x3 = \lambda^2 + \lambda + x1 + x2 + a$ and $y3 = \lambda(x1 + x3) + x3 + y1$ in P3 (P3 = P1 + P2).

If <*P*> is a cyclic subgroup of elliptic curve E, the prime number n is its order. For any positive integer k satisfying the condition of $0 < k < n$, define $kP = P + P + \cdots + P$ (number of Ps = k). The operation for calculating kP with P and k is called multiple point operation of elliptic curve, point multiplication operation, or scalar multiplication of elliptic curve. Otherwise, suppose $Q = kP$. When P and Q are given and k is unknown, the problem for solving the value of k with P and Q is called discrete logarithm problem of elliptic curve.

The prerequisites for constructing the elliptic curve cryptographic algorithm are a specific elliptic curve E and a prime-order cyclic subgroup of E, <*P*>. Suppose the order of P is the prime number n; the private key of the elliptic curve cipher is a random integer k in the interval $(0, n)$; the public key is a point $Q = kP$ on E and the system parameters. The feature of the elliptic curve cipher is that it involves the multiple point operation. Otherwise, the algorithm cannot be called the elliptic curve cipher. There are many relevant public-key ciphers, key exchange protocols, and digital signature algorithms based on elliptic curve. This section does not provide examples for the relevant algorithms and protocols.

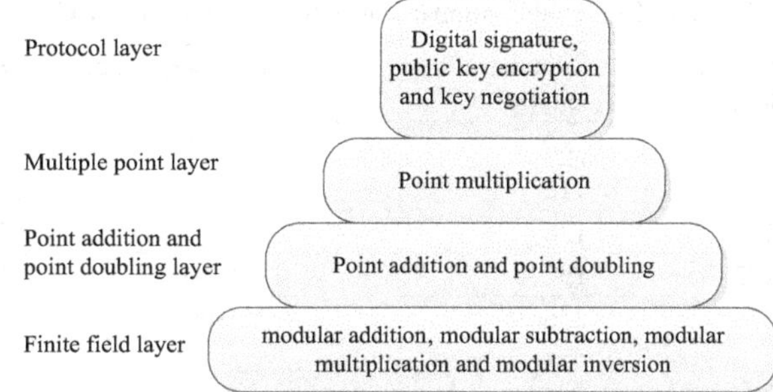

Fig. 2.29 Schematic diagram of layered implementation of the ECC algorithm

The elliptic curve cipher is not as simple and intuitive as the RSA algorithm. Generally, the elliptic curve cipher is implemented layer by layer, as shown in Fig. 2.29. The first layer is the protocol layer, where convention system protocols such as key exchange, public key encryption, and digital signature are defined; the second layer is the multiple point layer, where some algorithms are used to decompose point multiplication into a series of point addition and point doubling operations; the third layer is the point addition and point doubling layer, where the point addition and point doubling formulas are used to convert point addition and point doubling operations into a series of operations in finite fields; the last layer is the finite field layer, where the lowest level of operations is completed in finite field and the multiple point operation is finally completed. Therefore, the ECC algorithm is finally implemented into modular addition, modular subtraction, modular multiplication, and modular inversion of integers.

2.4.2 *Features of Public-Key Ciphers*

To ensure the security, the values of parameters for public-key ciphers such as key are large. First, we analyze ranges of parameters for these public-key ciphers. These parameters determine the size of operands participating in operations of algorithms.

The key lengths of the public key ciphers based on big-integer problem and those based on discrete logarithm problem indicate the number of bits occupied by binary system expression of modulus n. The longer the key length, the higher the security level. Meanwhile, the amount of computation is larger, and the speed is slower. The modulus n of actual RSA algorithms is required to be an integer of big bit width, which is generally 1024 bits or 2048 bits or even can reach up to 15,360 bits. The numbers of bits are close for the decomposed prime numbers p and q of modulus n. The number of bits for p or q is about half the number of bits for modulus n.

Compared with RSA algorithms, the public-key algorithms based on ECDLP can use smaller keys to provide equivalent or higher security, as shown in Table 2.4. The key lengths of ECC algorithms in the prime number field GF(p) indicate the number of bits occupied by binary system expression of p (generally, p

Table 2.4 Key length comparison between RSA and ECC

Secrecy level	Symmetric key length/Bit	RSA key length/ Bit	ECC key length/ Bit	Secrecy year
80	80	1024	160	2010
112	112	2048	224	2030
128	128	3072	256	2040
192	192	7680	384	2080
256	256	15,360	512	2120

is a prime number of bit 160 to bit 571); the key length of ECC algorithms in the characteristic-two finite field GF(2^m) indicates the value of m (generally, m is a prime number of bit 160 to bit 571 [22]).

The key parameter for public-key cipher algorithms based on NTRU is (*N*, *q*) (reference values sorted by security strength in the ascending order: (107, 64), (167, 128), and (503, 256)) [23].

2.4.3 Common Logic of Public-Key Ciphers

First, we will also analyze the typical RSA and ECC algorithms. The computations involved in the implementation of RSA algorithms include generation of random numbers, prime number test, modular inversion operation on big integers, and modular exponentiation operation on big integers. The first three operations are mainly performed in the key generation phase. Their execution frequency is low. According to the layered structure in Sect. 2.4.2, the computations required for implementation of ECC algorithms in key exchange, public key encryption, and digital signature include generation of random numbers and multiple point operation, wherein the multiple point operation is the key part. The multiple point operation is finally converted into big-integer modular multiplication, big-integer modular addition, big-integer modular inversion, polynomial modular multiplication, polynomial modular addition, and polynomial modular inversion.

As a matter of fact, the core operations of main types of public-key ciphers are very close to those of RSA and ECC algorithms. The core operations of public-key ciphers based on integer factorization problems are modular exponentiation, modular multiplication, and modular addition. The public-key ciphers based on discrete logarithm problem perform operations defined in the residue class field. Their core operations are the same as those of public-key ciphers based on integer factorization problem. There are two conditions for public-key ciphers based on ECDLP. The operations defined on the elliptic curve in the prime field GF(*p*) are also converted into modular multiplication and modular addition operations on big integers; the operations on the elliptic curve in the characteristic-two finite field GF (2^m) can be converted into polynomial modular multiplication and polynomial modular addition operations on $Z_2(x)$; the degree of the polynomial is m; the coefficient is 0 or 1. In the characteristic-two field, all data can be seen as polynomials with the coefficient of 0 or 1; the operation of the data is equivalent to that of polynomials and shall follow polynomial operation rules. The coefficient has only one bit. Therefore, addition indicates modular addition by bit, that is, XOR operation by bit without considering the carry bit. Correspondingly, some partial product addition operations in multiplication also indicate XOR operations by bit without the need for carrying. Compared with addition and multiplication of integers, this type of addition and multiplication is much faster. The public-key ciphers based on NTRU perform operations defined on ring $R = Z(x)/(x^N - 1)$. The core operations are polynomial modular multiplication and modular addition on ring R; the degree of the polynomial is N; the coefficient belongs to Z_q.

Therefore, core operations of public-key ciphers come down to modular exponentiation, modular multiplication, and modular addition of big integers or big polynomials. The modular exponentiation operation can be converted into a series of modular multiplication operations and square operations by using square-multiplication algorithm. Therefore, the modular multiplication operation of big integers and polynomials is the most important common logic of public-key ciphers. The big-integer operation and polynomial operation have much in common. The main differences lie in minor aspects such as additive carry chain. Therefore, some architectures are available to help implement integer operations and polynomial operations, thus implementing the dual-field elliptic curve algorithm [24]. For convenience, in the following, big integers and high degree of polynomials are not distinguished.

The modular multiplication operation is one of the most basic and the most frequently used operations for public-key ciphers. Modular multiplication of such great numbers is time-consuming. The computing speed of the modular multiplication operation directly impacts the speed of high-layer protocols in public-key cipher algorithms. The key to implement fast public-key ciphers is how to implement fast modular multiplication. The most direct way for performing the modular multiplication operation is to perform the multiplication and then perform the modulo operation. The most direct way for performing the modulo operation is division, which is hard to implemented on hardware. The Montgomery modular multiplication algorithm can avoid division. In the Montgomery modular multiplication algorithm, the modulo operation is converted into multiplication and shift operations [25], thus greatly reducing the complexity of the modulo operation.

Due to the conversion of the modular exponentiation operation with the square-multiplication algorithm and conversion of the modular multiplication operation with the Montgomery algorithm, the core common operations of public-key ciphers are multiplication, addition, and shift.

2.4.4 Parallelism of Public-Key Ciphers

We analyze the parallelism of public-key ciphers from both the higher level and the lower level. The lower-level parallelism mainly indicates the parallelism of internal implementation for common logic of public-key ciphers. The higher-level parallelism mainly refers to operations above the multiplication operation.

First, we analyze the higher-level parallelism. We analyze the higher-level parallelism of the RSA and ECC algorithms (the higher-level parallelism of other public-key ciphers is similar to that of the RSA and ECC algorithms). The main operation for algorithms such as RSA is modular exponentiation operation. If the modular exponentiation operation is performed directly, the algorithm performs the modular multiplication operation in sequence in a cumulative manner; the modular multiplication operations are data-dependent and must be performed in a serial manner. If the modular exponentiation operation is performed by using the square-multiplication algorithm, the square operation and multiplication operation are also data-dependent and are performed in a serial manner. Other algorithms are similar to the RSA and ECC algorithms. They cannot perform parallel computation, even though they involve many great-number modular multiplication units.

We analyze the ECC algorithm layer by layer according to its layered feature. On the protocol layer, various signature verification protocols and the key negotiation protocol involve two multiple point operations, which can be performed in parallel. On the multiple point operation layer, the parallelism is similar to that of the modular exponentiation operation for RSA; point doubling can be an analogy to square, and point addition can be an analogy to multiplication. The point doubling and point addition operations invoked by scanning from left to right and scanning from right to left in the multiple point operation are data-dependent. The latter operation must rely on the result of its previous operation. On the point addition and point doubling layer, the point addition and point doubling operations are a sequence constituted by operations such as modular multiplications and modular additions. Instruction-level parallelism exists in common programs. As an analogy, operation-level parallelism exists in these operation sequences. Some operations that are not data-dependent can be performed in parallel. There also exist a large number of data-dependent operations; the latter operation replies on the result of its previous operation. On this layer, the method for developing instruction-level parallelism in the computer system architecture can be used to develop operation-level parallelism. The finite field layer mainly involves the modular multiplication, modular addition, and modular inverse operations. The involved parallelism is coincident with the lower-level parallelism.

Then, we analyze the lower-level parallelism. In common logic of public-key ciphers, addition and shift are much simpler than multiplication. Therefore, we mainly analyze the parallelism of the multiplication operation. The operand size of core operations involved in public-key ciphers is determined by the parameter described in Sect. 1.4.2. The bit widths of big integers required by public-key ciphers based on integer factorization problem and discrete logarithm problem are generally 1024–2048 bits. The bit widths of big integers required by ECC algorithms based on the prime number field GF(p) are generally 160–571 bits. The bit widths for degrees of polynomials on $Z_2(x)$ required by ECC algorithms based on the characteristic-two field GF(2^m), which are equivalent to the bit widths of integers in the prime field, are generally 160–571 bits. The degree of polynomial on Z_q required by public-key ciphers based on NTRU is several hundred; the coefficient is generally 64, 128, and 256 bits. This shows that the operand size is a big integer between hundreds of bits and thousands of bits, or the polynomial coefficient is equivalent to the operand size. The bit widths of these data are large. If data with such big bit widths are used to perform computation directly, the number of bits for multiplication is large. This is obviously not good for computation. Especially, the multiplier of such big bit width is seldom directly used in hardware applications. The common practice is to use multipliers of small bit widths to construct a multiplier of big bit width and convert an operation of big bit width into multiple operations of small bit widths. Using multiple multipliers of small bit widths to constitute an array is a common practice. In the array, the multipliers of small bit widths can be executed in parallel; data dependency does not exist between the multipliers. The partial product addition involves carry chain propagation; it is a little more complicated. Moreover, partial product additions vary according to transmission direction of carry chain. The general case is as follows: On the one hand, some partial products are in the same phase of carry chain propagation; operations of the partial products can be performed in parallel; the parallelism is increased with the increase in the array size. Some other partial product additions are data-dependent on the propagation path of the entire carry chain. Addition of a partial product can be performed only after the carry output of the partial product at the previous level is exported. Similarly, additions of big bit widths can also be constructed by multiple adders of small bit widths. Due to addition carry between partial additions, partial additions are data-dependent. Generally, the shift amount for shifts of big bit width data is known. Therefore, the shifts can be performed in parallel, and the granularity of each part can be flexibly specified.

In conclusion, according to the analysis on the reconfigurable feature of various cipher algorithms, the overall structures of various cipher algorithms are relatively consistent; various parameters are distributed in a smaller scope; the algorithm control part is relatively simple; the computation is intensive with rich parallelism; the basic computing units are regular. These features show that it is suitable to implement cipher algorithms with the reconfigurable computing technology. This chapter analyzes and extracts the common logic and computation granularity of cipher algorithms, thus providing the basis for extracting operators, determining the functions and granularities of reconfigurable processing units, and determining the sizes of reconfigurable arrays.

References

1. Shannon CE (1949) Communication theory of secrecy systems. Bell Labs Tech J 28(4): 656–715
2. Williams (2015) Cryptography and network security: principles and practice, 6th ed. Publishing House of Electronics Industry, Beijing, p 554
3. Stallings W (2013) Cryptography and network security: principles and practice international edition, 6th ed. Prentice Hall, Upper Saddle River, pp 121–136
4. Diffie W, Hellman M (1976) New directions in cryptography. IEEE Trans Inf Theory 22(6): 644–654
5. NIST-FIPS (2001) Announcing the advanced encryption standard. Natl Inst Stand Technol 29(8):2200–2203
6. Kitsos P, Sklavos N, Koufopavlou O (2002) Hardware implementation of the SAFER +encryption algorithm for the bluetooth system. In: IEEE international symposium on circuits and systems, pp 878–881
7. Kaicheng L (1998) Computer cryptography: data confidentiality and security in computer networks. Tsinghua University Press, Beijing
8. Schneier (2014) Applied cryptography: protocol, algorithm and C source program. Mechanical Industry Press, Beijing, pp 63–83
9. Jingfei J (2004) Research and design of reconfigurable cryptographic processing structure. Doctoral dissertation of National University of Defense Technology, Changsha
10. Feng X (2011) ZUC algorithm: 3GP LTE international encryption standard. Inf Secur Commun Priv 12:031
11. Robshaw M (2008) The eSTREAM project. Lect Notes Comput Sci 2:1–6
12. Koç ÇK (2009) About cryptographic engineering. Springer, New York, pp 1–4
13. Bertoni G, Daemen J, Peeters M et al (2013) Keccak. In: Annual international conference on the theory and applications of cryptographic techniques, pp 313–314
14. Xiaoyun W, Hongbo Yu (2016) SM3 cryptographic hash algorithms. Inf Secur Study 2(11): 983–994
15. Merkle RC (1989) One way Hash functions and DES. In: International cryptology conference on advances in cryptology, pp 428–446
16. Damgård IB (1989) A design principle for Hash functions. In: International cryptology conference on advances in cryptology, pp 416–427
17. Lucks S (2005) A failure-friendly design principle for Hash functions[J]. Lect Notes Comput Sci 3788:474–494
18. Xiaoyun W, Hongbo Yu (2015) Review of cryptographic hash Algorithms. Inf Secur Study 1(1):19–30
19. Dunkelman O, Biham E (2006) A framework for iterative hash functions: HAIFA. In: The 2nd NIST cryptographic hash workshop
20. Hoffstein J, Pipher J, Silverman JH (1998) NTRU: a ring-based public key cryptosystem. Springer, Heidelberg, pp 267–288
21. Rivest RL, Shamir A, Adleman L (1978) A method for obtaining digital signatures and public-key cryptosystems. Commun ACM 21(2):120–126
22. Menezes AJ (2012) Euiptic curve public key cryptosystems. Springer, Berlin
23. O'Rourke C, Sunar B (2003) Achieving NTRU with Montgomery multiplication. IEEE Trans Comput 52(4):440–448
24. Satoh A, Takano K (2003) A scalable dual-field elliptic curve cryptographic processor. IEEE Trans Comput 52(4):449–460
25. Montgomery PL (1985) Modular multiplication without trial division. Math Comput 44(170): 519–521

Chapter 3
Hardware Architecture of Reconfigurable Cryptographic Processors

The hardware architecture of reconfigurable cryptographic processors is the customization of the generic reconfigurable computing architecture in the cryptographic field. On the basis of the generic architecture described in Sect. 1.4.1, designers need optimize each concrete structure and parameter involved in the architecture framework in the cryptographic field. Different from the hardware architecture design of traditional cryptographic processors, i.e., the hardwired design of the data flow diagram for a single cipher algorithm in ASIC and the design of extended instruction set for specific operators and functions of cipher algorithm in ISAP, the hardware design of reconfigurable cryptographic processors shall integrate features of multiple cipher algorithms to implement the flexible and efficient reconfigurable datapath and reconfigurable controller. The reconfigurable computing unit, interconnection networks, heterogeneous module, data storage, configuration control method, configuration information organization, and storage are designed on the basis of common features of cipher algorithms. This chapter summarizes the basic design methods for the hardware architecture of reconfigurable cryptographic processors from the aspects of reconfigurable datapath and reconfigurable controller, thus helping designers analyze how to perform reasonable architecture designs based on a specified demand.

3.1 Reconfigurable Datapath

The datapath of reconfigurable cryptographic processors follows the basic forms shown in Figs. 1.33 and 1.34. The following sections introduce the reconfigurable datapath from aspects such as the reconfigurable computing unit design, the interconnection network design, the data storage design, and the heterogeneous module design.

L. Liu et al., *Reconfigurable Cryptographic Processor*,
https://doi.org/10.1007/978-981-10-8899-5_3

3.1.1 Reconfigurable Computing Unit

The PE is the core element of reconfigurable computing arrays. The PE structure design directly determines the computing efficiency of reconfigurable datapath. The following sections discuss the core factors (computation granularity, storage units, and computing functions) of PE design in details.

1. Computation Granularity

The PE computation granularity determines the unit width of parallel data processing in reconfigurable arrays. The computation granularity shall match the cipher algorithm set supported by the processor. If the PE computation granularity is too large compared with the basic processing width of cipher algorithms, only some bit width of PE participate in the computation. This results in waste of computing resources and ultimately affects the overall performance. For example, if the basic operation granularity of the cipher algorithm is mainly 16 bits and a reconfigurable computing unit with the computation granularity of 64 bits is used, the residual 48 bits in the unit may be wasted because only one computing function can be synergistically performed for different data bits in one PE at a given time (e.g., AND logical operation or multiplication is configured). If the PE computation granularity is too small compared with the basic processing width of cipher algorithms, although multiple PEs can be joined to complete the computation, this will result in waste of interconnection resources, control resources and configuration resources and ultimately reduces the area efficiency and energy efficiency of overall implementation. As far as interconnection resources are concerned, smaller PE granularity means more complex interconnection topology (i.e., larger interconnection area and longer interconnection delay) with the same array processing width. As far as control resources are concerned, to maintain the same computing power, the decrease in PE computation granularity will result in the increase in the number of PEs. As a result, the array control becomes more complex. As far as configuration resources are concerned, under the precondition of the same configuration granularity (e.g., PE-based configuration), if the PE computation granularity decreases, the arrays of equivalent computing power require a larger size of configuration information to maintain the computation and reconfiguration of computation datapaths, which will consume a larger configuration storage area and result in longer configuration delay.

According to the analysis on features of cipher algorithms, it can be seen that symmetric ciphers, hash algorithms, and public-key ciphers have their own features. As far as the basic operation granularity of algorithms is concerned, the basic operation granularity of symmetric ciphers is close to that of hash algorithms, which is basically about 8 bits to 32 bits; the operand width of public-key cipher algorithms is larger, but it can be further split into small data blocks to enable hardware implementation; the general splitting width is mainly 8 bits to 32 bits. Generally speaking, the PE computation granularity of reconfigurable cryptographic processors is recommended to be a value ranging from 8 bits to 32 bits. Due to PE cooperation, the selected granularity shall be exponential times of two (8, 16, or 32). It is necessary to note that the PE processing granularity shall be adjusted accordingly in actual architecture design to better satisfy application requirements if special requirements are expressed for algorithm sets.

The PE computation granularity also exerts an effect on the size of the reconfigurable computing array. The relationship between the PE computation granularity and the size of the reconfigurable computing array is waxing and waning. Generally speaking, the data processing width of the reconfigurable computing array shall match the data width of the supported algorithm. For example, currently, the data width of the widely used symmetric ciphers is mainly 128 bits; then, the data processing width of the reconfigurable cryptographic processor for the symmetric cipher application can be set to 128 bits (a larger data processing width can be used in case of further requirements for parallel processing). If unidirectional one-dimensional interconnection (interconnection between PE rows) is adopted for such arrays, the PE computation granularity determines the number of PEs in each row. If the PE computation granularity is 8 bits, the number of PEs in each row is 16; if the PE computation granularity is 32 bits, the number of PEs in each row is 4. The array size will ultimately exert an effect on many aspects such as interconnection topology, data interfaces, and configuration control. The specific impact also depends on other structures and parameters of the reconfigurable cryptographic processor.

2. Register Units

The reconfigurable cryptographic processor has the advantage of using the rich computation resources of array structures to perform parallelism and pipelining, thus greatly improving the throughput. To form pipelines between PEs, storage units need to be set to store intermediate processing data. As shown in Fig. 1.34, the general design method is to set a register that matches the PE computation granularity for the output (or input) of each PE to cache the computed result of PE at each stage. Generally, the number of registers matches the number of PE unit output (input) ports. According to the situation, bypass registers can be set to directly cache the input data, thus expanding the data transmission width in arrays. Except basic output registers, a register file can be set in PE units to store data of multiple computations. According to feature of cipher algorithms, the register file does not need to be very deep (appropriate depth: less than 4) so that reasonable computing performance can be implemented in the case of smaller hardware resource consumption.

3. Computing Function

As shown in Fig. 1.34, the ALU is the core component for operation execution in PE. It is necessary to note that the LUT can be used to replace the ALU theoretically, but a lot of additional consumption will be incurred. The FPGA actually uses the LUT to implement computation. The LUT stores the truth table for which operation is to be implemented and uses the input data as the memory address to obtain the computed result of the corresponding storage unit. The storage process does not limit concrete values of the truth table. Theoretically, the LUT can complete random computations but costs are huge. On the one hand, truth table storage occupies a large memory area; moreover, mapping of complex operations to multiple LUTs cannot guarantee that each LUT is utilized efficiently (i.e., many LUTs only implement simple functions). Thus, the overall implementation efficiency may be reduced. On the other hand, as far as the dynamic reconfigurable computing form is concerned, to reconfigure LUT functions, all values of the truth table in the memory must be changed (i.e., writing the memory to overwrite original values in the truth table). Mass configuration information and memory write operations will result in long reconfiguration time and high power consumption. Most reconfigurable cryptographic processors have the requirement for dynamic reconfiguration. Therefore, the vast majority of designs use the ALU as the core computing component. As supplementary, a relatively small-sized LUT can be added as an operator of the ALU or a function module in the reconfigurable computing array to speed up substitution operations in cipher algorithms.

The operator in the ALU is mainly constituted by common logic of cipher algorithms. According to the analysis on cipher algorithms, the same type of cipher algorithms has similar common logic. The basic operations mainly fall into the following categories:

① Logical operations: including AND, OR, NOT, and XOR.
② Arithmetic operations: including multiplication and addition in finite field.
③ Shift operations: including left-and-right shift and rotation (belonging to permutation operations with special rule in nature).
④ Substitution operations: S-Box in various algorithms.
⑤ Permutation operations: out-of-order operations of the bit level.

Designers can add appropriate ALU operators according to the cipher algorithm set to be supported. One operator design example is provided in ALU Opcode Design of Sect. 3.2 "Reconfigurable Controller." The following sections introduce the matters needing attention in operator design, mainly including multi-layer design of logical operations, heterogeneous array design of large-area operators, design of efficient shift logic, design of substitution function, and design of permutation function.

(1) Multi-layer of Logical Operations

Logical operations in the ALU of many reconfigurable cryptographic processors adopt the multi-layer design. Figure 3.1 shows the operator structure of ALU in the reconfigurable cryptographic processor [1]. Except additional operations such as XOR and multiplexer, the LUT unit and arithmetic unit (including the addition and multiplication modules) only involve one layer of core computation. The logic unit has three layers in serial; that is, the logic unit can complete the combination of any three logical operations at a time.

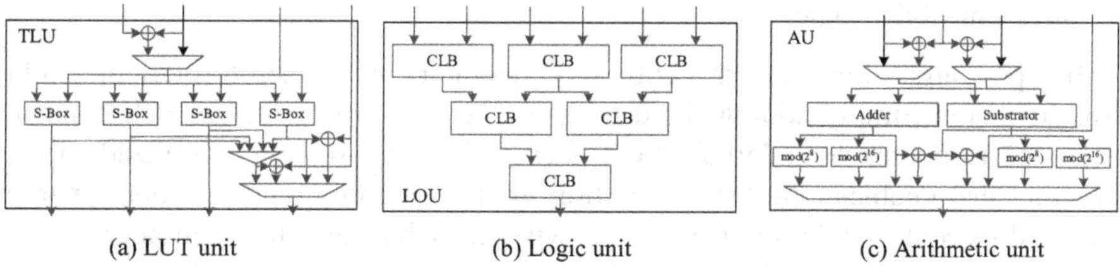

(a) LUT unit (b) Logic unit (c) Arithmetic unit

Fig. 3.1 ALU operator of the reconfigurable cryptographic processor

The multi-layer design specific to logical operations can improve the ALU processing capacity under the premise that the array computing performance is not affected. As far as general synchronous circuits are concerned, one PE corresponds to one stage of registers; that is, operations of each ALU correspond to a clock cycle. The length of the clock cycle is generally determined by the critical path in the ALU. Compared with path delay of other operators in cipher algorithms, the path delay of logical operation is short. If multiple layers of logical operations are reasonably superposed, the length of the critical path and the clock frequency of array computation will not be affected. In actual architecture design, designers can select the number of layers for the logical operation path by means of delay path simulation. If other long-delay function operators exist in the ALU, other simple operators except logical operations may also use the multi-layer design.

(2) Heterogeneous Design of Large-area Operators

The most basic reconfigurable computing arrays adopt the homogeneous array architecture; that is, the operation functions of all PEs are the same. It is true that the array design is very neat, but this will also result in waste of computing resources. In the case that large-area operators exist in the ALU, this waste is particularly obvious. For example, if the ALU involves a multiplier of large bit width (e.g., 16 bits or 32 bits), the circuit area occupied by the multiplier is far larger than that occupied by most other operators in the ALU. The multiplication operation is only part of the operation set for cipher algorithms. Most PE multiplication operators are not used most of the time. This will reduce the overall area efficiency of arrays to a large extent.

In the case that large-area operators with lower frequency of use exist, the heterogeneous array design can be considered. For example, we can add the multiplication operator to the ALUs of some PEs in the array (e.g., some PE rows in the array). When the multiplication operation is performed, the mapping is optimized and transformed to match the PE distribution with the multiplication function. It is true that the algorithm mapping is more complex than the mapping of fully homogeneous arrays and the mapping performance of some algorithms may be degraded. However, considering from the overall area efficiency, it is still a design optimization method that is worth trying.

(3) Efficient Shift Logic

In cipher algorithms, the frequently used operators include shift, rotation, and bit concatenation. Simply adding these operators to the ALU structure will result in waste of resources to a large extent. Figure 3.2 provides an optimized circuit structure that enables efficient completion of shift, rotation, and bit concatenation for fixed or controllable number of bits, wherein N indicates fixed number of shift bits; C indicates controlled number of shift bits; W indicates the width of the input data; in0 and in1 are the input data to be processed. First, the MUX can be used to select the source of shift bits (constant from the fixed shift end or the variable from the controlled shift end). If simple logical left shift or right shift operation needs to be performed, enter the data in in0 or in1 to directly obtain the shift result at the output. If left rotation needs to be performed, enter the data to be shifted in in0 or in1. N or C indicates the number of bits for left rotation; the output after bitwise OR is the rotation result. Similarly, if only the bit concatenation operation needs to be performed, enter two data to be joined in in0 and in1 and then press bitwise OR to obtain the concatenation result.

(4) Substitution Function

The substitution operation is a common operation specific to cipher algorithms. The nonlinear interaction generated in the substitution process can improve the security of cipher algorithms to a large extent. The methods for using the hardware to implement the substitution operation mainly include the arithmetic method and LUT method [2, 3]. The substitution operations of many cipher algorithms are based on some mathematical thoughts. For example, S-Box of the AES algorithm is implemented on the basis of the finite field. For these algorithms, the logic circuit can be directly used to implement the mathematical expression of the substitution operation. The critical path is generally short for the arithmetic-based implementation model, which is good for fast operation. However, only substitution

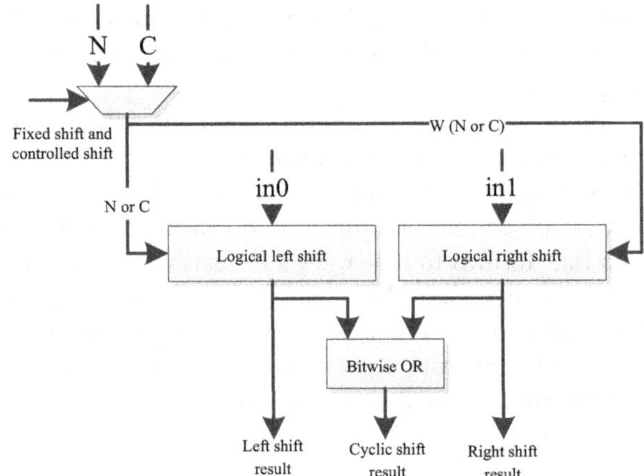

Fig. 3.2 Logic for shift, rotation and bit concatenation

operations of some algorithms have the corresponding mathematical expressions, and expressions of various algorithms differ greatly. For reconfigurable cryptographic processors, the arithmetic implementation model is hard to satisfy the flexibility requirements of various cipher algorithms for the substitution operation. Another method for implementing the substitution operation is LUT. The procedure is as follows: Store the substitution operation list in the LUT; use the input data of the substitution operation as the address input of the LUT; the output data of the LUT is the result of the substitution operation. The LUT method is slower than the arithmetic method for implementing the substitution operation. Moreover, the LUT method involves a larger area than the arithmetic method. The LUT method can satisfy various substitution operations of different algorithms. From the view of resource reuse of multiple algorithms, this method is actually highly efficient.

In reconfigurable cryptographic processors, the LUT can be used to implement the substitution operation in two forms: distributed LUT and heterogeneous LUT module. In the distributed LUT form, a small LUT is set as a special operator in the ALU of each PE; various PE LUTs can be joined and expanded to implement larger LUT. For example, Fig. 3.3 provides an instance of LC (lookup-table cell) in an ALU. The LC consists of four (6:1) LUT subcells. That is, the memory capacity of each subcell is 2^6 bits. Each LC can complete various substitutions such as (4:4) and (6:4) alone. Figure 3.4 provides an instance of (4:4) substitution operation completed by the LC. Connect corresponding bits in the LC input and then import the connected bits as the input data of the substitution operation; the LC output is the result of the substitution operation. Similarly, the substitution operation of a larger size can be completed by joining multiple LCs. Take the LC in Fig. 3.3 as an example. The combination of LCs in eight PEs can be used to complete the (8:8) substitution operation. Figure 3.5 shows how to use LC additional MUX to complete (8:1) substitution. The output values of eight LCs can be further joined to obtain an eight-bit output of (8:8) substitution. Except distributed LUT, the heterogeneous LUT module can also be used to implement the substitution operation of cipher algorithms.

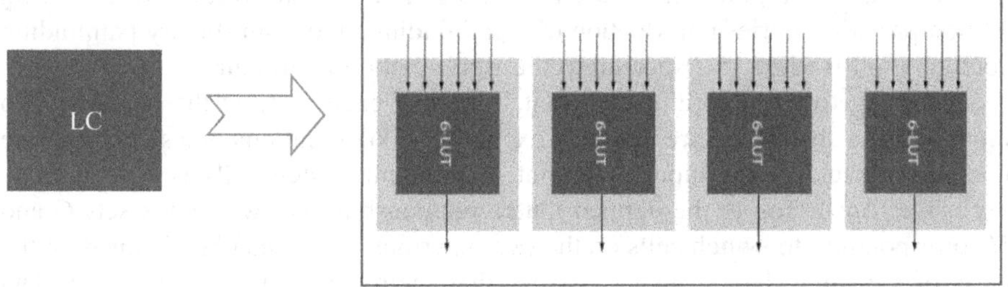

Fig. 3.3 Instance of distributed LUT

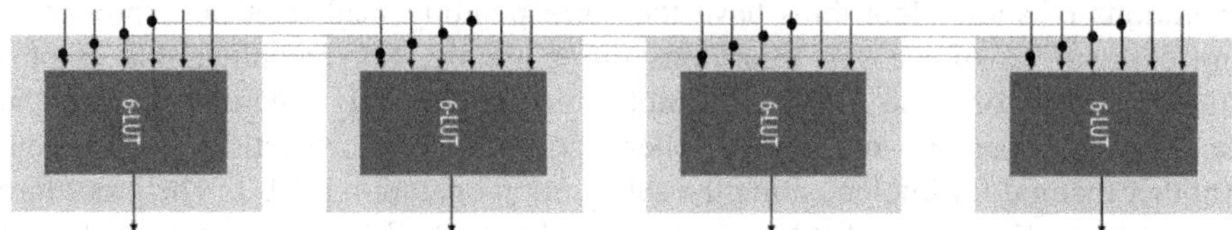

Fig. 3.4 (4:4) substitution operation instance completed by LC

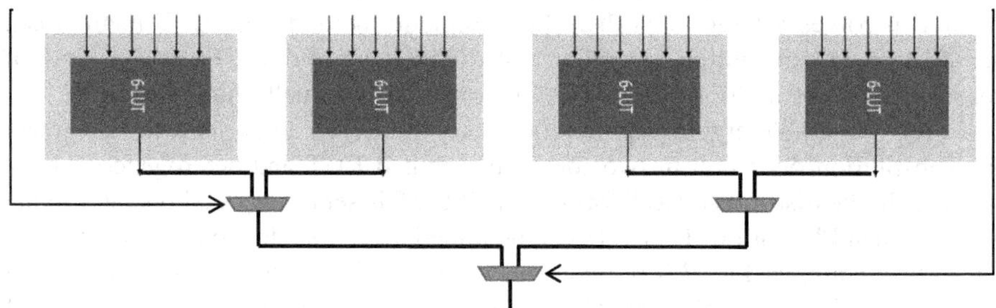

Fig. 3.5 Instance of (8:1) substitution operation instance completed by LC

(5) Permutation Function

The permutation operation is a unique operator in cipher algorithms. The hardware implementation of the permutation operation is generally processed in the form of non-blocking permutation network (PN). Figure 3.6 provides an instance of the network structure (Benes network structure [4]) for implementing the permutation operation. The data for which permutation needs to be performed is used as the network input; a permutation result can be obtained in network output by rationally configuring the function state (cross or direct connection) of each switch cell in the network. In the process of implementing permutation, to ensure that the permutation is not blocked, an appropriate pathfinding algorithm is required to determine the configuration of each switch cell in the network. The following section provides a brief introduction of a pathfinding algorithm (binary pathfinding configuration method [5]) specific to the Benes network structure.

According to the algorithm, for switch cells of each layer in the network, two input bits of a switch cell are mutually exclusive; two output bits of a switch cell are mutually exclusive; the input or output of different switch cells is not mutually exclusive. According to the defined mutex relationship, the two mutex sets G and H corresponding to switch cells on the two outermost layers can be obtained on the basis of the input data sequence and required permutation result. The input data sequence and required permutation result have been given. The input of switch cells on the first layer and the output of switch cells on the last layer are fixed; therefore, only the output of switch cells on the first layer and the input of switch cells on the last layer need to be determined. Attach the connection line of elements in G to the

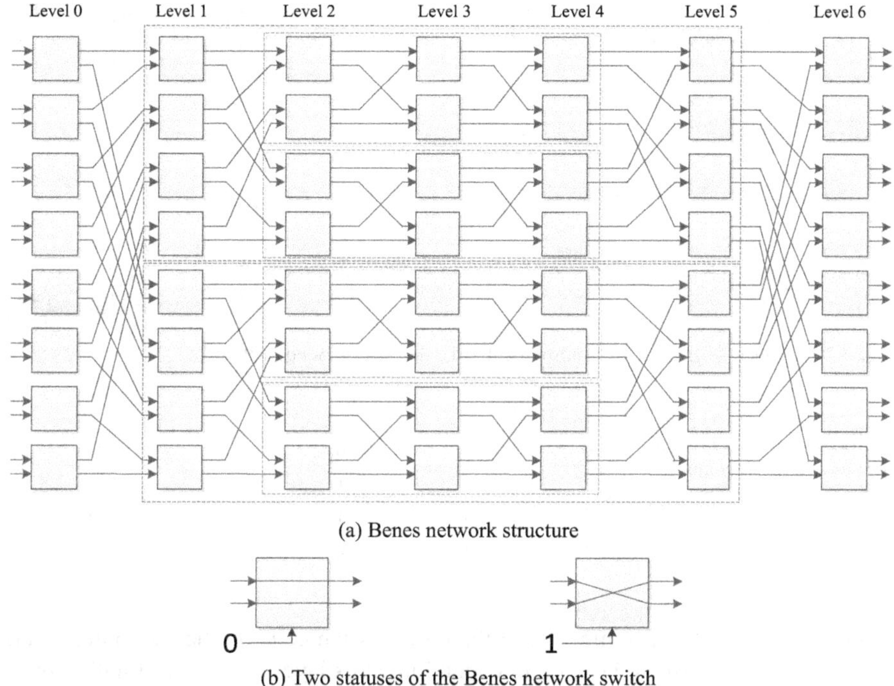

(a) Benes network structure

(b) Two statuses of the Benes network switch

Fig. 3.6 Schematic diagram of Benes PN structure

upper end of both the output of switch cells on the first layer and the input of switch cells on the last layer and attach the connection line of elements in H to the lower end of both the output of switch cells on the first layer and the input of switch cells on the last layer to identify the connection relation between switch cells on the two secondary outermost layers. The rest may be deduced by analogy. Specify the connection relation between switch cells on two layers at a time and expand further to the middle of the network to finally specify the connections of each switch cell in the entire network. Figure 3.7 provides an instance of identifying the connection relation of switch cells in an 8×8 Benes network. The input sequence of permutation is 1, 2, 3, 4, 5, 6, 7, 8; the output permutation sequence is 8, 3, 1, 5, 4, 2, 7, 6; then, the input mutex pairs of switch cells on the leftmost first layer are {1, 2}, {3, 4}, {5, 6}, and {7, 8}; the output mutex pairs of switch cells on the rightmost last layer are {8, 3}, {1, 5}, {4, 2}, and {7, 6}. Construct two mutex sets G and H to ensure that the number of intersection elements between each mutex set and each mutex pair is 1. Sets $G = \{1, 4, 6, 8\}$ and $H = \{2, 3, 5, 7\}$ can be obtained. As previously mentioned, the connection relation between switch cells on the two outermost layers can be identified by connecting the corresponding elements in G set to the upper end of switch cells and connecting the corresponding elements in H set to the lower end of switch cells. The rest may be deduced by analogy.

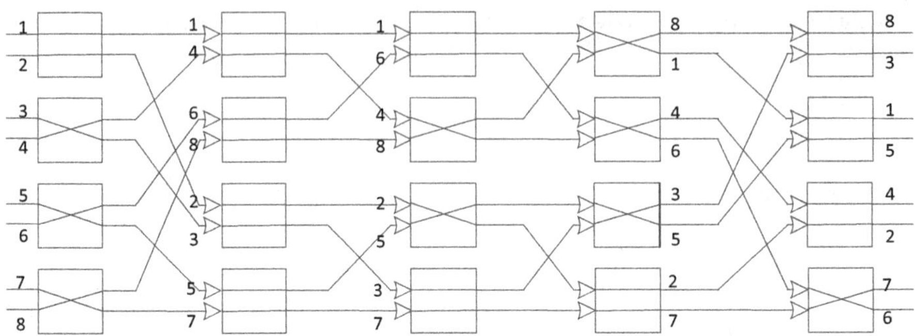

Fig. 3.7 Instance of permutation implemented in the 8 × 8 Benes network

Table 3.1 Benes network configuration information array

0	0	0	1	0
1	0	1	1	0
1	0	1	1	0
1	0	0	0	1

Continue to identify the connection relation of switch cells on the two outer layers except the two outermost layers and expand inwards layer by layer to finally obtain the connection relations of all switch cells. If **0** indicates that the function status of the switch cell is configured as direct connection and **1** indicates that the function status of the switch cell is configured as cross, the list of the corresponding configuration information matrix for implementing the permutation is shown in Table 3.1.

Similar to the substitution function design, the permutation function can also be implemented by using the distributed PN or heterogeneous permutation module in the reconfigurable cryptographic processors. With the distributed PN method, a small-size PN of appropriate width is added to each PE as a permutation function operator in the ALU. If the permutation exceeds the processing width of the operator, multi-level permutation is required to be performed for multiple PEs. The heterogeneous permutation function module is independent of PE. It can complete the permutation operation matching the data processing width of the reconfigurable computing array. For details, see the section of heterogeneous module design.

To enable readers to have a more intuitive understanding of the ALU design, an instance of ALU operator design is provided. Table 3.2 provides the ALU opcode list for the instance. We can see that it involves the logical operations, arithmetic operations, and shift operations mentioned above. It is necessary to note that this opcode list has its own design specificity. For example, the permutation operation and substitution operation implemented in the heterogeneous module do not appear in ALU opcodes. In addition, the opcodes include some application-specific operations. These details are not covered.

Table 3.2 Example of ALU opcode definition

<table>
<tr><th>Signal</th><th>Type</th><th colspan="2">Value</th><th>Function (out0=)</th></tr>
<tr><td>Opcode_in</td><td>Output all zeros</td><td>000000</td><td>0</td><td>Output ‘b0
(out0 = out1 = out2 = 0)</td></tr>
<tr><td></td><td>Boolean operation with two operands</td><td>000001</td><td>1</td><td>A^B</td></tr>
<tr><td></td><td></td><td>000010</td><td>2</td><td>A&B</td></tr>
<tr><td></td><td></td><td>000011</td><td>3</td><td>A|B</td></tr>
<tr><td></td><td>Boolean operation with three operands</td><td>000100</td><td>4</td><td>A&B&C</td></tr>
<tr><td></td><td></td><td>000101</td><td>5</td><td>A&B|C</td></tr>
<tr><td></td><td></td><td>000110</td><td>6</td><td>A&B^C</td></tr>
<tr><td></td><td></td><td>000111</td><td>7</td><td>A|B&C</td></tr>
<tr><td></td><td></td><td>001000</td><td>8</td><td>A|B|C</td></tr>
<tr><td></td><td></td><td>001001</td><td>9</td><td>A|B^C</td></tr>
<tr><td></td><td></td><td>001010</td><td>10</td><td>A^B&C</td></tr>
<tr><td></td><td></td><td>001011</td><td>11</td><td>A^B|C</td></tr>
<tr><td></td><td></td><td>001100</td><td>12</td><td>A^B^C</td></tr>
<tr><td></td><td>Fixed shift, rotation and bit concatenation</td><td>001101</td><td>13</td><td>A ≪ N</td></tr>
<tr><td></td><td></td><td>001110</td><td>14</td><td>B ≫ N</td></tr>
<tr><td></td><td></td><td>001111</td><td>15</td><td>(A ≪ N)|
(B ≫ (W − N))</td></tr>
<tr><td></td><td>Controlled shift, rotation and bit concatenation</td><td>010000</td><td>16</td><td>A ≪ C</td></tr>
<tr><td></td><td></td><td>010001</td><td>17</td><td>B ≫ C</td></tr>
<tr><td></td><td></td><td>010010</td><td>18</td><td>(A ≪ C)|
(B ≫ (W − C))</td></tr>
<tr><td></td><td>Arithmetic Operation</td><td>010011</td><td>19</td><td>A + B (carry in the group)
2-input unsigned addition</td></tr>
<tr><td></td><td></td><td>010100</td><td>20</td><td>A + B (no-carry)
2-input unsigned addition</td></tr>
<tr><td></td><td></td><td>010101</td><td>21</td><td>A + B (carry in the group)
2-input signed addition</td></tr>
<tr><td></td><td></td><td>010110</td><td>22</td><td>A*B (unsigned)</td></tr>
<tr><td></td><td>Controlled selection</td><td>010111</td><td>23</td><td>C?A: B</td></tr>
<tr><td></td><td>Comparison</td><td>011000</td><td>24</td><td>if(A == B) out0 = 0;
else out0 = l;
(Disconnected output)</td></tr>
<tr><td></td><td></td><td>011001</td><td>25</td><td>if(A == B) out0 = 0;
else out0 = l;
(Connected output)</td></tr>
<tr><td></td><td>Parity check</td><td>011010</td><td>26</td><td>out0 = Cout = ^A;
(carry in the group)
Value of the least significant bit of out0</td></tr>
<tr><td></td><td></td><td>011011</td><td>27</td><td>out0 = ^A;(no-carry)</td></tr>
</table>

(continued)

Table 3.2 (continued)

Signal	Type	Value		Function (out0=)
	Direct connection			A (out0 = A)
	GF(2^8) multiplies by 2	011101	29	For irreducible polynomials $m(x) = x^8 + x^4 + x^3 + x + 1$ out0 = $\begin{cases} b_6b_5b_4b_3b_2b_1b_00 & b_7 = 0 \\ (b_6b_5b_4b_3b_2b_1b_00) \oplus 00011011 & b_7 = 1 \end{cases}$ Wherein, $\{b_7b_6b_5b_4b_3b_2b_1b_0\} = in_0$

3.1.2 Interconnection Network

The interconnection network is an important component of the datapath in the reconfigurable cryptographic processor. The interconnection network rationally organizes computing units in the reconfigurable array to complete mapping of data flow diagrams for cipher algorithms. According to spatial connection, the interconnection in the reconfigurable array can be classified into uni-dimensional interconnection and two-dimensional interconnection. The two-dimensional interconnection can further be classified into island interconnection, hierarchical interconnection, and point-to-point interconnection; the uni-dimensional interconnection can further be classified into interlayer full interconnection and interlayer partial interconnection. It is necessary to note that bus interconnection in the array is also a type of interconnection. The bus interconnection is not suitable for implementing fast local dynamic reconfiguration in the case of large-data-size computation. Therefore, it is not introduced too much. The following section describes various interconnection network structures involved in reconfigurable computation and then provides suggestions on the interconnection network architecture suitable for reconfigurable cryptographic processors.

As a very common interconnection form in reconfigurable arrays, island interconnection is widely used in fine-grained generic reconfigurable architecture FPGA. Figure 3.8 takes the island interconnection in the FPGA as an example to show the basic forms of island interconnection [6]. Like islands, configurable logic blocks (CLBs) in the FPGA scattered through the ocean of the interconnection network. Horizontal and vertical meshed lines are regularly distributed in the island interconnection, which also involves two interconnection unit modules: switch box (SB) and connection box (CB). The SB is responsible for connecting the horizontal and vertical lines; the CB is responsible for connecting the CLB to the interconnection network. The CB flexibility can be expressed by *Fc*, which indicates the proportion of the adjacent channels that can be connected to each CLB to the total number of channels in the CB; corresponding to the input and output of the CLB, *Fc*(in) indicates the input flexibility of the CB and *Fc*(out) indicates the output flexibility of the CB. The SB flexibility is expressed by *Fs*, which indicates the

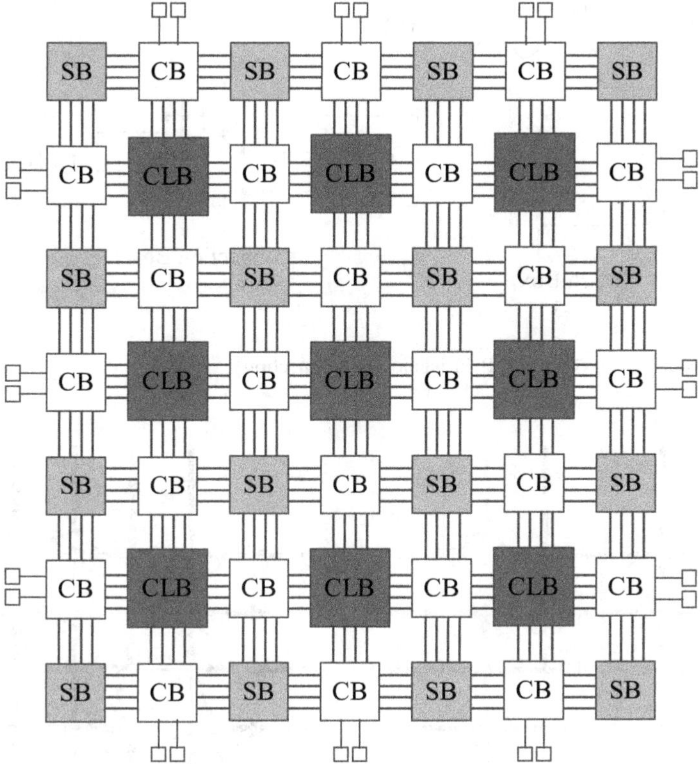

Fig. 3.8 Schematic diagram of island interconnection structure

number of SB output ends that can be connected by each connection input in the SB. The data width (expressed by *W*) of each SB horizontal and vertical line group is called the channel width of the island interconnection. Generally, the data widths shall be the same for line groups to facilitate design of interconnection modules. Figure 3.9 illustrates the value of each parameters involved in island interconnection. Wherein, $W = 4$. Corresponding to the SB in Fig. 3.9a, $Fs = 3$ (i.e., the input of each SB can be connected to three different SB output ends); corresponding to the CB in Fig. 3.9b, $Fc = 0.5$ (i.e., each CLB that is connected to the CB can only be connected to 50% of adjacent channels). In the island interconnection, the lines connected by the SB can be unidirectional or bidirectional. Figure 3.10 provides an instance of unidirectional interconnection and bidirectional interconnection. In the two types of interconnections, $Fs = 3$. The left subfigure corresponds to bidirectional interconnection. Each line can be interconnected with lines of three other directions in two directions. The right subfigure corresponds to unidirectional interconnection. Although the SB flexibility is unchanged, each line can only bear the data flow in the specified single direction. According to the SB structure of

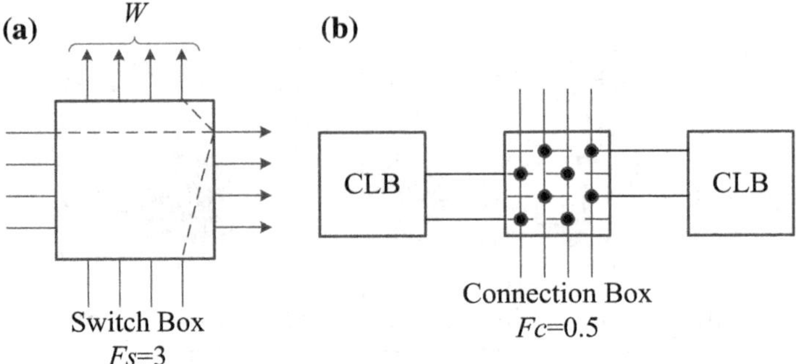

Fig. 3.9 Examples of SB and CB in island interconnection

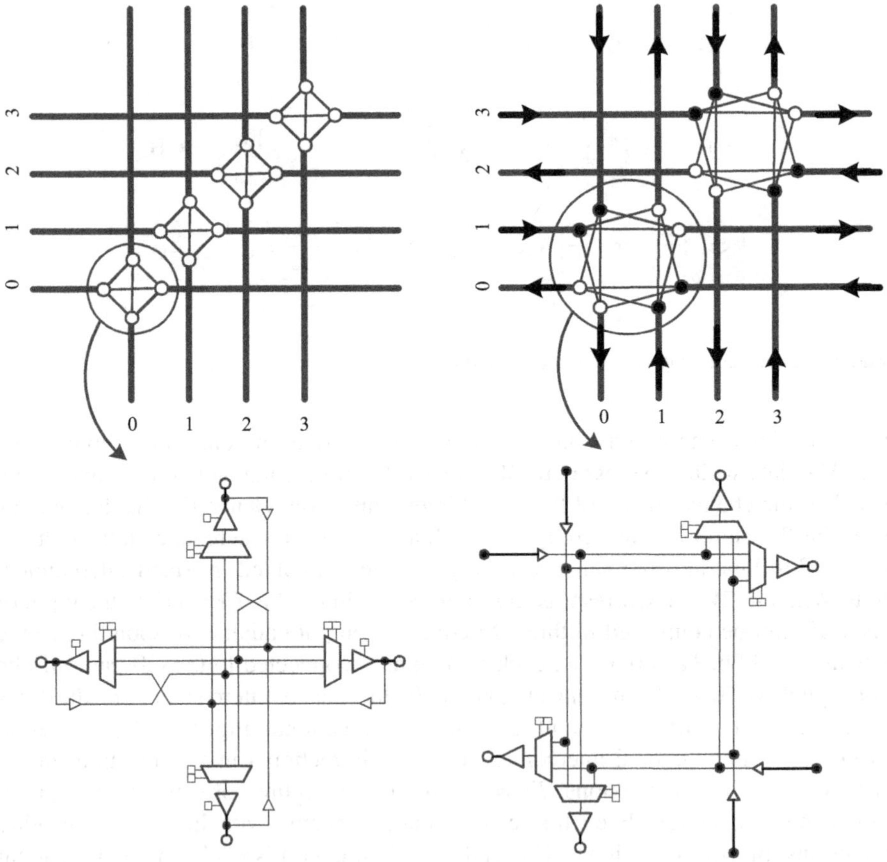

Fig. 3.10 Examples of unidirectional interconnection and bidirectional interconnection

unidirectional interconnection, the switch structure of unidirectional interconnection can be easily built only when the data width is an even number.

Hierarchical interconnection is another two-dimensional interconnection form that is commonly used. The hierarchical interconnection uses the data connection locality of the logical computing unit to divide computing units into multiple groups; the layer-by-layer recursion mode is used among groups to form the overall hierarchical connection of the array. The connection relation of hierarchical interconnection is similar to the familiar tree network topology. The internal units of each subgroup can be interconnected by using the local interconnection; subgroups require to be connected by using the upper-layer interconnections. The basic channel width c and interconnection growth rate p can be used to describe the structure of hierarchical interconnection. The basic channel width c indicates the data width of the bottom leaf node; the interconnection growth rate p indicates the ratio of the data channel width that is increased for the upper-layer connection compared with its lower-layer connection. The interconnection growth can be achieved by two types of switch units: non-compression-type unit and compression-type unit. In the non-compression-type unit, the connection data width of the upper-layer is the sum of the data widths of its connections to its lower layer. In the compression-type unit, the connection data width of the upper layer is the same as that of its lower layer; it is not increased. Figure 3.11 provides an instance of hierarchical interconnection. The switch units between level 1 and level 2 and between level 3 and level 4 are non-compression-type units; the switch unit between level 2 and level 3 is a compression-type unit. Therefore, $c = 3$ and $p = 0.5$.

The commonly used point-to-point two-dimensional interconnection in the reconfigurable array is mainly based on interconnection of nearby units. Figure 3.12 provides two typical interconnection forms. Figure 3.12a shows the neighboring-unit-type interconnection (each unit is connected to its four nearest units). Figure 3.12b shows the neighboring and subneighboring-unit-type interconnection (except its four nearest units, each unit is also connected to its subneighboring units). Such point-to-point interconnection based on nearby units takes full advantage of data locality in array processing. Such kind of point-to-point interconnection only involves local data transmission. Therefore, its hardware consumption will not increase with the array size. If data needs to be transmitted among multiple PEs, the data transmission needs to consume multiple clock cycles. This results in long delay. That is, if the data flow diagram of the target algorithm does not match this type of interconnection form (mass data need to be transmitted among multiple PEs), the final throughput will be greatly reduced.

Except two-dimensional interconnection, some uni-dimensional interconnection forms are also frequently used in reconfigurable computing arrays. In arrays constructed by uni-dimensional interconnection, data follow the one-way flow style (i.e., data only flow from one side of the array to the other side). Interlayer full interconnection is a commonly used uni-dimensional interconnection form. As shown in Fig. 3.13, in interlayer full interconnection, the output of each upper-layer unit can be connected to the input of any unit in the lower layer. This greatly increases the data transmission flexibility in uni-dimensional interconnection, but it

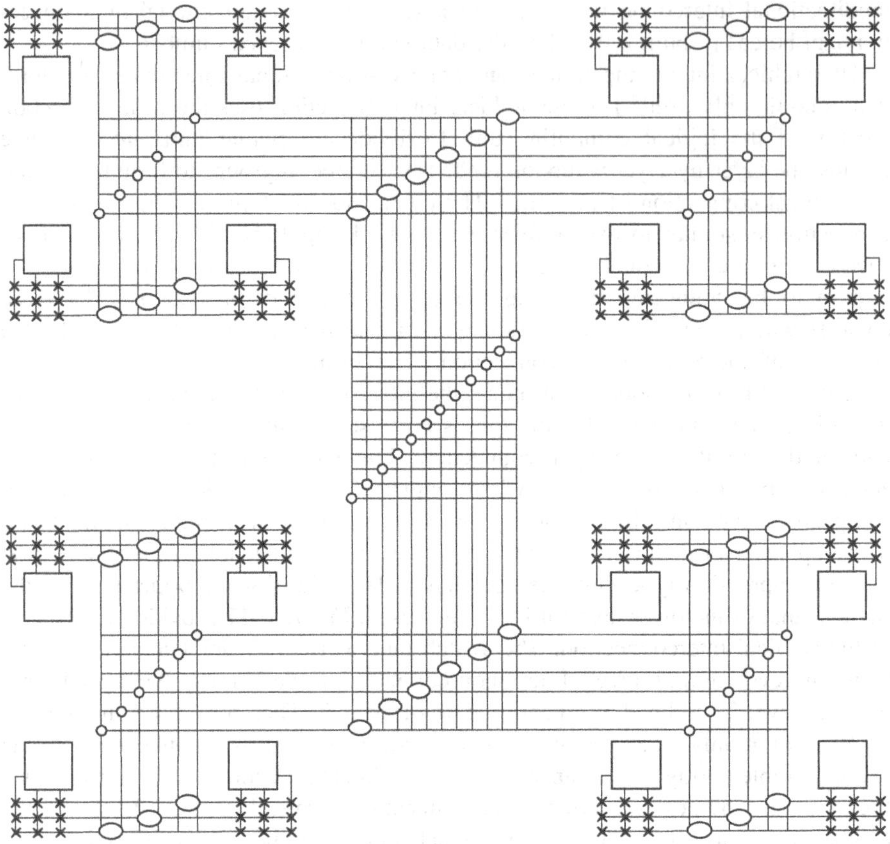

Fig. 3.11 Examples of hierarchical interconnection forms

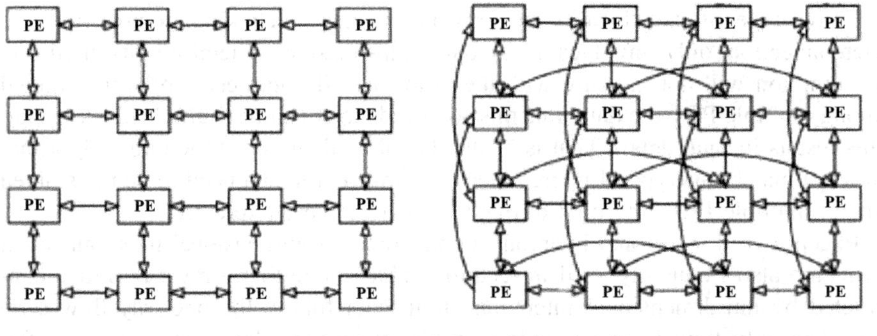

(a) Neighboring type interconnection

(b) Neighboring and sub-neighboring type interconnection

Fig. 3.12 Example of two-dimensional point-to-point interconnection

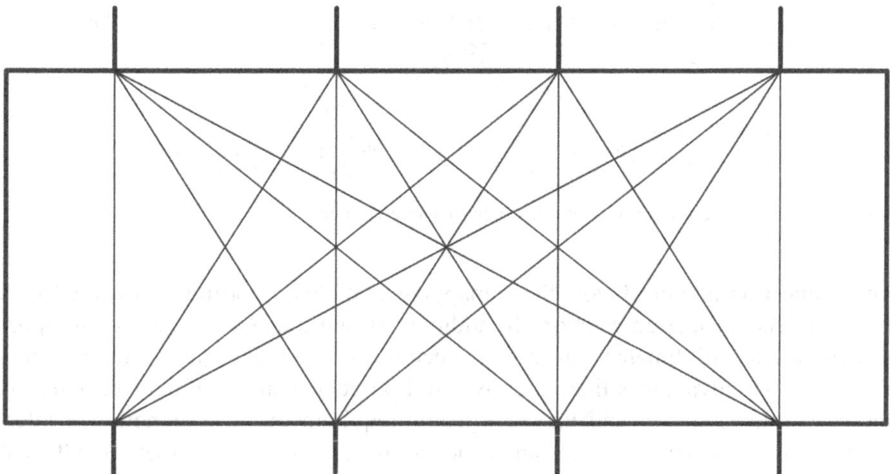

Fig. 3.13 Schematic diagram of interlayer full interconnection

will also result in large hardware consumption. If the number of input and output ports on each layer is large, the interconnection consumption that is obtained by the square of the number of data ports will become unbearable. This greatly limits the extensive use of interlayer full interconnection in large-size arrays.

The interlayer partial interconnection covers the shortages of interlayer full interconnection in interconnection consumption by utilizing the data locality theory. In interlayer partial interconnection, each upper-layer unit is only connected to the neighboring units on the lower layer. As shown in Fig. 3.14, each upper-layer unit is connected to the nearest nine units on the lower layer. Compared with connection with all units on the lower layer, a lot of interconnection area consumption can be saved. What should be noted is that the decrease in the area consumption of partial interconnection is at the cost of decreased interconnection flexibility. In practical application, the topology for interlayer partial interconnection shall be rationally selected according to algorithm features so as to minimize the loss of algorithm mapping efficiency caused by flexibility. We can obtain some variants from interlayer partial interconnection through further extension. For example, units on each layer can be divided into several groups; full interconnection is adopted between each upper-layer unit and its nearest groups on the lower layer; partial interconnection is adopted between the upper-layer unit and the other groups on the lower layer. The group-based interconnection with both tight coupling and loose coupling may further improve the efficiency of interlayer partial interconnection.

In the datapath design of reconfigurable cryptographic processors, an interconnection form matching the cipher algorithm shall be rationally selected to improve the computing efficiency of reconfigurable arrays. According to the analysis on cipher algorithms, mappings of cipher algorithms in reconfigurable arrays mostly adopt unidirectional data flow diagrams. In this case, the uni-dimensional

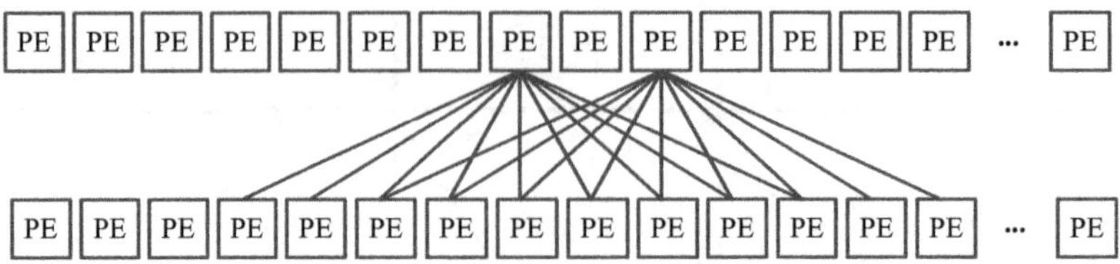

Fig. 3.14 Schematic diagram of interlayer partial interconnection

interconnection is more favorable to mapping of cipher algorithms compared with two-dimensional interconnection. In addition, strong data locality exists in cipher algorithms; in uni-dimensional interconnection, the interlayer partial interconnection has more advantages than interlayer full interconnection. Of course, based on the target algorithm set and various specific requirements of designers, the interconnection mode can be determined according to the actual situation; efficient mapping of most algorithms shall be used as the basis of interconnection design.

3.1.3 Data Storage

In datapath of reconfigurable cryptographic processors, corresponding data storage modules shall be designed to facilitate data transmission and exchange between internal units of the reconfigurable computing array and between the reconfigurable computing array and the array external.

The data transmission between the reconfigurable computing array and the array external mainly involves structured plaintexts, ciphertexts, keys, and computation parameters. Computation of cipher algorithms can be basically completed in the reconfigurable computing array independently. Therefore, the data interactions between the reconfigurable computing array and the array external mainly involve input of plaintext computational data, output of ciphertext computational data and transmission of keys and computation parameters required by the computation process. These data are structured with fixed data block size. Scattered small-width data with frequent interactions basically does not exist. In this case, an FIFO memory can be set on the I/O interface between the reconfigurable computing array and the array external. The I/O FIFO can play the role of buffering the data interaction between the reconfigurable computing array and the array external, so as to avoid data read and write errors caused by mismatch between the data I/O rate and the computing rate, thus simplifying the design of the control state machine for data interaction. What should be noted is that the data FIFO depth shall match the actual computation of the reconfigurable cryptographic processor. If the depth is too small, data overflow may occur or the processor may read the empty; if the depth is too large, the processor resources will be wasted. Designers can use the prototype of the processor architecture to simulate FIFO read and write in the earlier stage of design to determine an appropriate FIFO depth.

Most data interactions between internal units of the reconfigurable computing array can be completed by interconnection in the array; additional data transmission tasks mainly include transmission of intermediate results between round computations and result transmission between multiple data subgraphs. For example, for reconfigurable datapath using interlayer interconnection in the array, data can only be transmitted downward between units of adjacent layers; to enable data transmission from a lower-layer unit to a upper-layer unit, data storage is required to schedule the data transmission. Generally speaking, to implement such intra-array data transmission, the shared memory oriented to array units can be used. The memory can be constructed by using register files. In addition, multiple interfaces are set between the memory and different layers of the array as required to ensure the parallel interaction rate of data.

Figure 3.15 provides an instance of the design for data storage of the reconfigurable computing array. The data interaction between the reconfigurable computing array and the array external is implemented by using the I/O FIFO; the interlayer data interaction in the array is implemented by using the shared memory. Finally, what needs to be pointed out is that the cache mechanism of general-purpose processors can also be used to design data storage of reconfigurable cryptographic processors [7]. The cache mechanism mainly uses the time and spatial locality in data access to store copies of the memory data based on hit and miss. After the cache mechanism is applied to general-purpose processors, the average access speed can be successfully improved, thus reducing the bottlenecks incurred by memory read and write delay. However, the cache mechanism is generally complex. Except the control consumption, frequent content rewrite will also result in large power consumption. Data read and write and storage are regular and are of high predictability for the cipher application. Therefore, complex cache

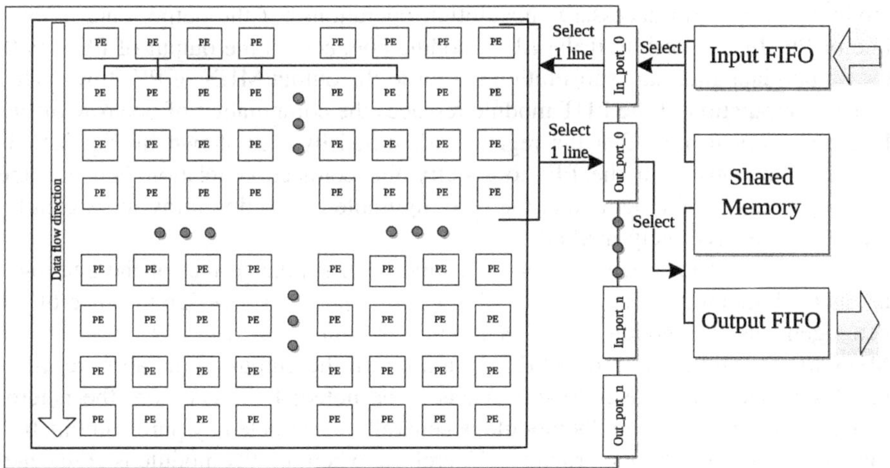

Fig. 3.15 Instance of design for data storage of reconfigurable computing array

scheduling mechanism is not appropriate to the cipher application. The simplified cache mechanism can be customized for cipher algorithms as required so as to reach the expected memory performance with the lowest consumption.

3.1.4 Heterogeneous Module

Except the structured computing unit array, the heterogeneous function module can also be added to the reconfigurable datapath to help the PE complete computation. The following section takes the heterogeneous LUT module and heterogeneous PN module as examples.

The heterogeneous LUT module can take the place of the distributed LUT to implement the substitution operation in cipher algorithms. The heterogeneous LUT utilizes the parallelism of substitution operations in cipher algorithms and does not require addition of small LUTs in each PE; instead, substitutions are performed in a uniform manner by a structured standard RAM that is added to the computing array. In cipher algorithms, substitution operations generally feature high parallelism; that is, multiple substitution operations need to be concurrently processed. For example, the byte substitution operations of the AES algorithm consist of 16 S-Boxes. A RAM of big bit width can be added to the reconfigurable computing array to complete data substitution of multiple PEs in a unified manner. Figure 3.16 provides an instance of the heterogeneous LUT function module. Multiple rows of PEs in a reconfigurable computing array can be considered as a LUT sharing group; PE rows in the group share the same LUT module (four rows of PEs share a LUT module). To enable sharing of the LUT module, data I/O in each row of PEs is connected to the LUT module. In algorithm implementation, the MUX can be used to select the data source and destination of the LUT module. That is, the LUT module can perform necessary substitution operations for the output data in each line of PE. In this design, if the LUT module connects to the output of PE Row0, the output data after data substitution is sent to the output MUX of PE Row1. That is, the computation of the LUT module replaces the computation of one row of PE. It is necessary to note that the register (identified by reg) shown in Fig. 3.16 is actually the register in the PE; to specify the connection relation between the heterogeneous LUT module and the reconfigurable computing array more clearly, the register is separately marked.

The design of the heterogeneous PN module is similar to that of the heterogeneous LUT module in many aspects. Figure 3.17 provides a design instance of the heterogeneous PN module. The PN module is connected to the PE array by the MUX to receive the input of a PE row and return the output permutation result to the designated row. The following needs to be noted: Different from the heterogeneous LUT module, the PN module is generally constructed by pure combination logic (e.g., Benes network structure). Therefore, when the PN module is connected to the PE array, the original register logic of the PE is not required to be bypassed; only ALU computation of the corresponding PE is replaced.

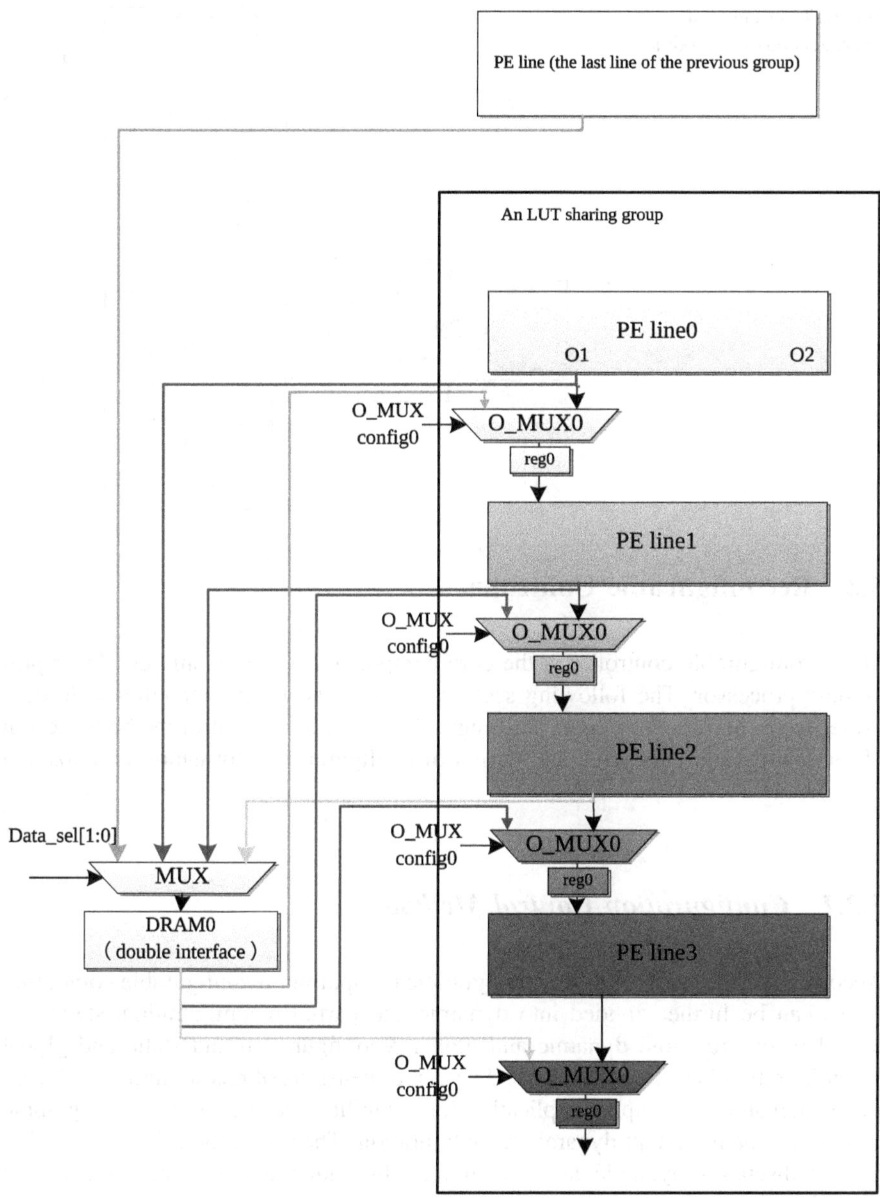

Fig. 3.16 Instance of the heterogeneous LUT function module

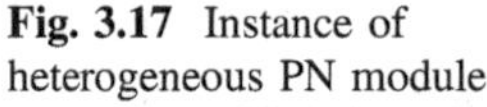

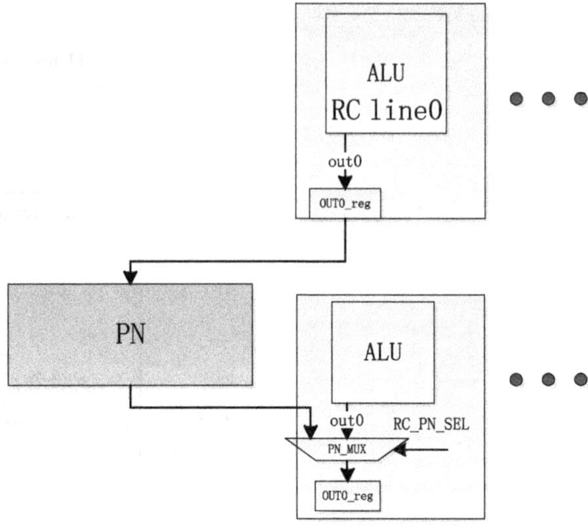

Fig. 3.17 Instance of heterogeneous PN module

3.2 Reconfigurable Controller

The reconfigurable controller is the core component of the reconfigurable cryptographic processor. The following sections mainly discuss the reconfigurable controller from the following aspects: design of configuration control methods, design of the control state machine, and design of configuration information organization and storage.

3.2.1 *Configuration Control Methods*

According to the categories of reconfigurable computing, reconfigurable computing forms can be further divided into dynamic and partial reconfiguration, static and partial reconfiguration, dynamic and global reconfiguration and static and global reconfiguration based on the time and spatial dimensions of reconfiguration. Driven by the demand of the cipher application for flexibility, reconfigurable cryptographic processors mainly adopt dynamic reconfiguration. Therefore, the following section mainly discusses dynamic and global reconfiguration and dynamic and partial reconfiguration.

In reconfigurable cryptographic processors with dynamic and global reconfiguration, the reconfiguration controller mainly uses the centralized control method to complete resource scheduling for computing units. The reconfigurable controller configures data flow diagrams for algorithms based on the reconfigurable computing array; that is, the configuration switchover is performed on the basis of data flow subgraphs in the array. Figure 3.18 shows the global reconfiguration process between subgraphs. For most algorithms, the size of the data flow diagram is greater than the size of the reconfigurable computing array; the data flow diagram needs to be divided into multiple subgraphs matching the array based on certain rules and mapping shall be performed according to the sequence of data dependency. In centralized configuration control, only array-based global reconfiguration is performed. After all computations of a subgraph are completed, the intermediate computation result data is exported and the next subgraph is configured to the array for corresponding computations. As shown in Fig. 3.18, the overall data flow diagram is cut into three interdependent subgraphs. The array will be reconfigured twice to enable switchover between subgraphs. Data interactions occur on edges between subgraphs. Except configuration of the subgraph function for the array, the reconfigurable controller also needs to be configured to enable control over data interaction and storage, thus implementing complete functions of the entire algorithm.

In the centralized configuration control, the overall running time and its components are shown in Fig. 3.19. First, the configuration controller configures the array as a whole to complete functions in the first data flow subgraph. Then, the array performs data input according to the configuration context. After the input data is ready, the array starts to perform computations in the first subgraph. To improve throughput, the reconfigurable array generally uses the pipelining mode to complete computations (performing computations corresponding to the jagged lines in the computation part of the diagram). After computations are completed, the array outputs the intermediate results. The rest may be deduced by analogy. The corresponding configurations and computations of the next subgraph are performed.

The advantage of dynamic and global configuration under the centralized configuration control mainly lies in simplification of the reconfigurable configuration controller. Configuration of all units in the array is performed as a whole. Therefore, as for the sequential control, the controller only needs to coordinate the input, output, and the global configuration; complex synchronization of intra-array units does not need to be considered. The array-based overall reconfiguration also has many problems such as waste of computation resources, additional configuration costs, and performance loss of reconfiguration. As far as computation resources are concerned, data transmission and time sequence control of each unit are completed in a unified manner in one reconfiguration for a subgraph; the computing units that are not occupied by the subgraph in the array are hard to be concurrently used to perform other computations; that is, the computing units are wasted in this subgraph execution. As far as configuration costs are concerned, in the centralized configuration mode, the units that fail to perform effective functions are also configured, which results in increase in the configuration consumption. As far as the algorithm execution performance is concerned, after completing the computations corresponding to a subgraph, each unit in the array cannot start the computations corresponding to the next subgraph until all other units complete the computations and overall configuration is implemented. A large interval exists between each computation and configuration for most units, which results in decrease in the overall computation performance of the algorithm.

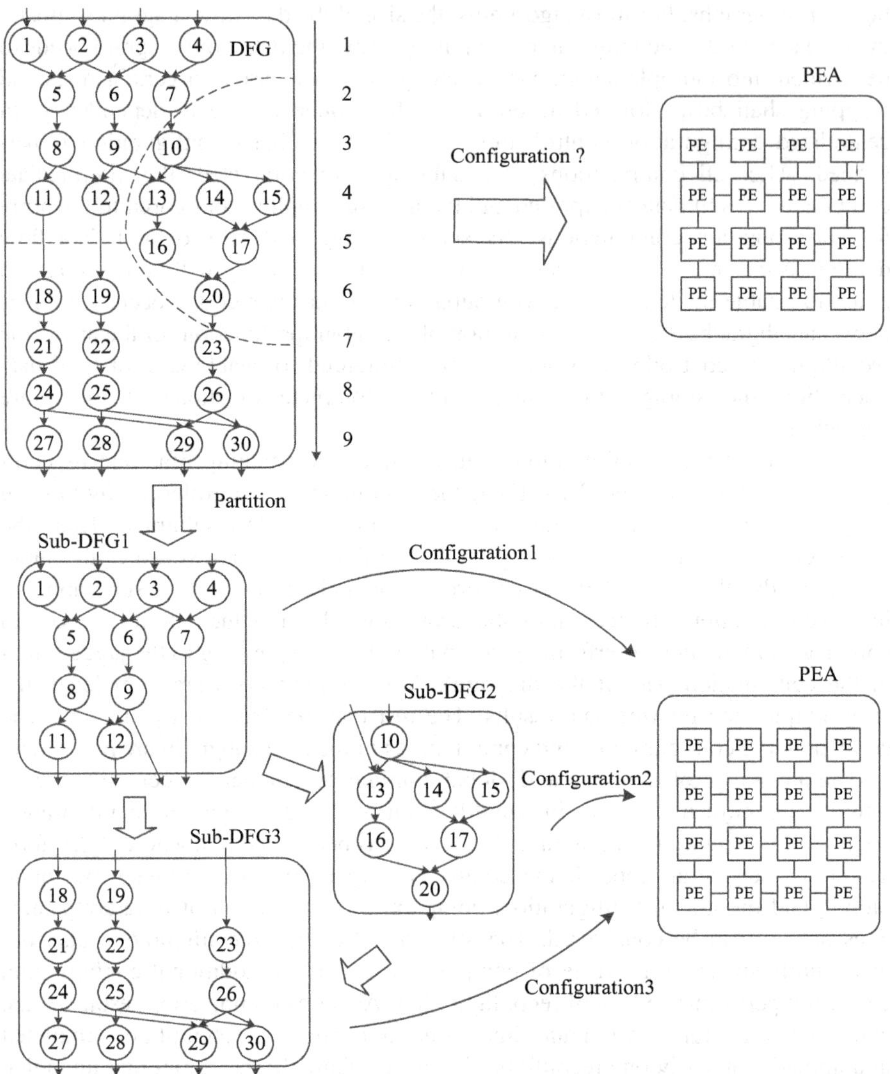

Fig. 3.18 Global configuration switchover between subgraphs

The dynamic and partial reconfiguration is widely used in reconfigurable cryptographic processors. In such processors, the reconfigurable controller mainly adopts the distributed control mode to perform resource scheduling for the computing array. Most of such reconfigurable cryptographic processors use the configuration control method based on PEs; that is, each PE has a separate controller, which enables the PE to perform computation and configuration autonomously while maintaining synchronization with other PEs. Figure 3.20 shows a PE under

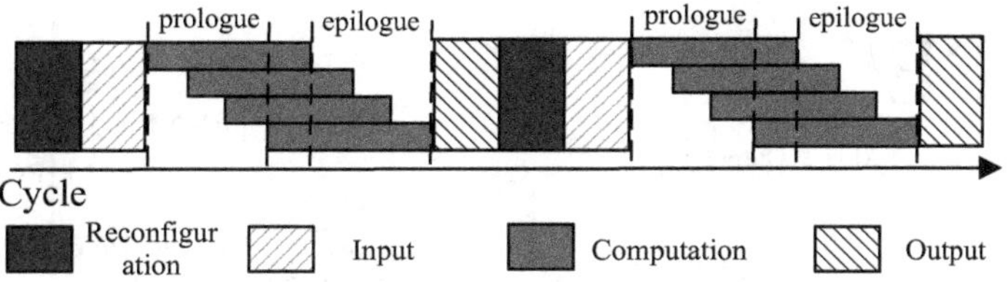

Fig. 3.19 Running time of the array in centralized configuration control

distributed configuration control and its controller. The PE includes the controller (PE_Controller), ALU, output register (Output_regs), unit interconnection (Router), register file (Register_File), and configuration memory (Conf_MEM). As the internal scheduling core of the PE, the controller controls data, computation, and configuration. Various enable signals and completion signals are used between PEs to perform synchronization and communication. On the upper layer of the PE controller, the PEA controller receives the completion signals of each PE controller in the entire array and distributes the enable signals after synchronization to each PE controller. The PE controls its own computations while ensuring synchronization with other PEs; after completing related tasks, the PE returns a completion signal to notify other PEs. If the PE needs to be configured, the PE controller sends a configuration enable signal (Conf_EN) and address information (Conf_ADDR) to the PE configuration memory and receives the configuration information (Conf_DATA) returned by the configuration memory. If the PE needs to perform data access, the PE controller sends enable signals (RF_EN, SM_EN, and FIFO_EN) and address information (RF_ADDR, SM_ ADDR, and FIFO_ ADDR) to the PE register file, array shared memory, and array FIFO; then the PE controller reads or writes corresponding data (RF_DATA, SM_ DATA, and FIFO_ DATA) to the data memory. If the PE needs to perform computation, the PE controller sends the input data (ALU_INPUT) to the ALU, receives the computed result of the ALU (ALU_OUTPUT), and sends the enable signal (OUTPUT_REG_EN) to the PE output register to store the computed result.

Contrary to the situation of the centralized configuration control, the dynamic and partial configuration under the distributed configuration control can effectively resolve problems such as waste of computation and configuration resources and performance loss of reconfiguration interval at the cost of complex configuration controllers. Each PE separately performs time sequence control over itself; in different application scenarios, idle computing units can choose not to perform any configuration to save power consumption or organize to perform other operations to improve the overall resource utilization. In addition, configuration of each PE does not need to wait for completion of computations by other PEs in the array; switchover between computation and configuration can be performed at any time; the time interval that may degrade the performance does not exist any more. The main cost of the distributed configuration control mode is the complex control part. On the one hand, each PE corresponds to its own controller, which generates

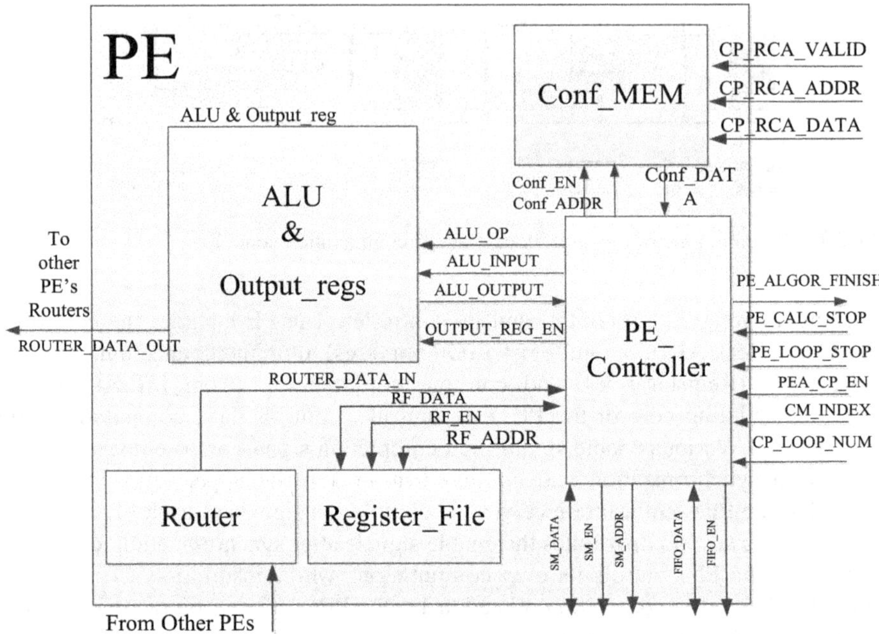

Fig. 3.20 Instance of the PE and its controller design under the distributed configuration control

certain area consumptions; on the other hand, operations such as inter-PE synchronization result in more complex controller design compared with the controller design in the centralized control mode. Generally speaking, the benefits from distributed control far outweigh its consumptions. This is also the reason why many modern new-type reconfigurable cryptographic processors adopt local dynamic reconfiguration under distributed control.

What needs to be pointed out is that some compromise proposals between array-based control and PE-based control exist except the familiar PE-based distributed configuration control mode. In these proposals, reconfiguration and configuration are performed on the basis of some PEs in the array (e.g., one row or several adjacent PEs in the array) according to the specific architecture of the reconfigurable datapath. The controllers are designed on the basis of multiple PEs. The specific design method for such controllers is similar to that for PE-based control.

3.2.2 Control State Machine

As the core element of the reconfigurable controller, the control state machine is also the final implementation carrier for various configuration control methods. The control state machine corresponding to the distributed control mode is relatively complex. Here provides a brief introduction to the design method for state machines taking the state machine corresponding to the controller in Fig. 3.20 as an example. Figure 3.21 provides the schematic diagram of the state machine for the PE controller (ignoring some secondary signals). The state machine mainly involves four states.

① _PE_IDLE: PE idle state, waiting for the PE configuration enable signal (CP_EN_in).
② _PE_TOP: PE configuration initiating state, confirming the PE configuration.
③ _PE_CONF: PE configuration state, parsing the configuration line and fetching data from the PE register file, array shared memory and other PEs (data from other PEs through interconnection).
④ _PE_CALC: PE computing state, performing computation; exporting the computation results to the PE register file, array shared memory or storing the computed results to the PE output register.

Jumps between these states correspond to ten types of situations.
0: Indicates jump from _PE_IDLE to _PE_TOP.
Jump condition: Configure enable (PEA_CP_EN) of the array is valid.

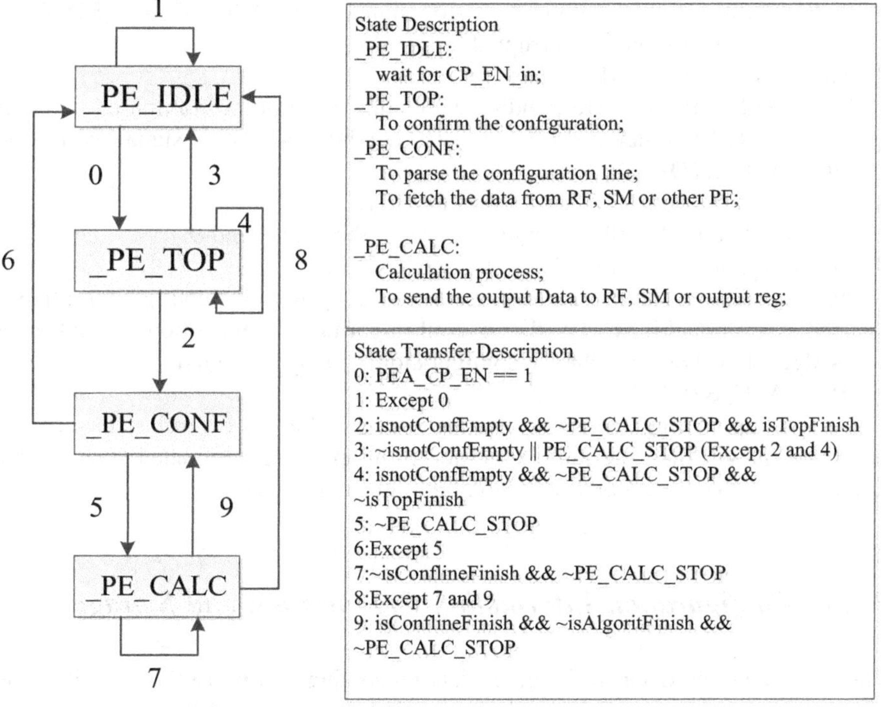

Fig. 3.21 PE_Controller state machine

1: Maintains the _PE_IDLE state.
Jump condition: Configure enable (PEA_CP_EN) of the array is invalid.
2: Indicates jump from _PE_TOP to _PE_CONF.
Jump condition: The configuration memory corresponding to the current pointer is not empty (isnotConfEmpty=1); the configure stop signal is invalid (PE_CALC_STOP=0); the top-level configuration initialization is completed (isTopFinish=1).
3: Indicates jump from _PE_TOP to _PE_IDLE.
Jump condition: The configuration memory corresponding to the current pointer is empty (isnotConfEmpty=0) or the configure stop signal is valid (PE_CALC_STOP=1).
4: Maintains the _PE_TOP state.
Jump condition: The configuration memory corresponding to the current pointer is not empty (isnotConfEmpty=1); the configure stop signal is invalid (PE_CALC_STOP=0); the top-level configuration initialization is not completed (isTopFinish=0).
5: Indicates jump from _PE_CONF to _PE_CALC.
Jump condition: The configuration stop signal is invalid (PE_CALC_STOP=0). That is, if no special interrupt control exists, the PE enters the computing state by default after configuration is completed.
6: Indicates jump from _PE_CONF to _PE_IDLE.
Jump condition: The configuration stop signal is valid (PE_CALC_STOP=1). That is, configuration is interrupted.
7: Maintains the _PE_CALC state.
Jump condition: The corresponding computing cycle of the configuration line is not over (isConflineFinish=0) and the configuration stop signal is invalid (PE_CALC_STOP=0).
8: Indicates jump from _PE_CALC to _PE_IDLE.
Jump condition: All other conditions except those in 7 and 9
9: Indicates jump from _PE_CALC to _PE_CONF.
Jump condition: The corresponding computing cycle of the configuration line is over (isConflineFinish=1); the overall algorithm computation is not over (isAlgoritFinish=0); the configuration stop signal is invalid (PE_CALC_STOP=0).

To sum up, during the array running process, the PE control state machine constantly judges the state jump condition to jump to the designated state in time and performs the corresponding configuration or computation.

3.2.3 Configuration Information Organization and Storage

The reconfigurable datapath involves data processing and transmission. Likewise, the reconfigurable controller mainly parses and distributes configuration information. Execution of cipher algorithms in reconfigurable cryptographic processors is

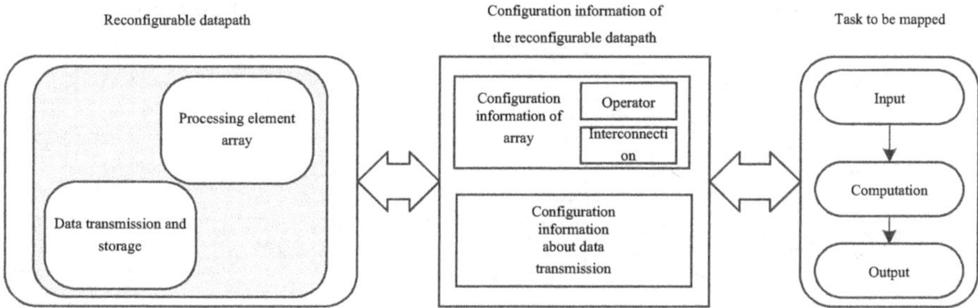

Fig. 3.22 Relations among a reconfigurable cryptographic processor, a cipher computing task and the configuration information

driven by configuration information. Figure 3.22 shows the relations among a reconfigurable cryptographic processor, a cipher computing task, and the configuration information. Through control over array operators, interconnections and data transmission in the processor datapath, the configuration information controls the data input, output and data flow diagram of the cipher task to be mapped in a reconfigurable computing array.

Generally speaking, affected by the size of the reconfigurable array and the complexity of cipher application computation, the configuration information is huge in the reconfigurable cryptographic processor. The core problem to be solved for design of configuration information organization is how to integrate the huge amount of configuration information so as to efficiently complete the storage, parsing, and distribution of configuration information. The configuration information organization directly influences hardware logic of the reconfigurable controller such as configuration information parsing logic and distribution logic. Generally, the configuration information organization scheme varies according to the configuration control method. The following section discusses the configuration information organization methods for centralized configuration control and distributed configuration control. In addition, the configuration information that is organized in a reasonable manner finally needs to be stored in the configuration memory in a certain form. Therefore, design of storage schemes for configuration information is required. The following section introduces the design of configuration organization hierarchy and the design of storage forms of configuration information in the memory.

To organize configuration information, the information with different control functions shall be rationally integrated. Figure 3.23 provides the classification of main configuration information in a reconfigurable cryptographic processor. The configuration information for controlling reconfigurable computing unit functions is one type of the most basic configuration information. It corresponds to the function of each operator in the algorithm data flow diagram. In addition, the configuration information for controlling interconnections between reconfigurable computing units is also very important. It relates to transmission of data between units, that is,

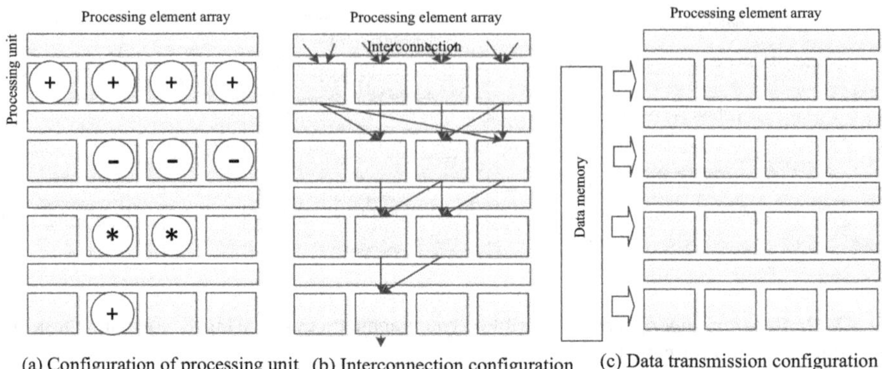

Fig. 3.23 Classification of main configuration information in a reconfigurable cryptographic processor

internal data flow in the data flow diagram. It is necessary to note that configuration of interconnections will be finally implemented into the level of computing units, that is, input source and output direction of each unit. Defining of the data input source or data output direction for each reconfigurable computing unit is equivalent to defining of the specific interconnection form. For example, in a configuration diagram, if the input of a unit in the lower line is the output of a unit in the upper line, the output of the unit in the upper line is in fact connected to the input of the unit in the lower line. That is, interconnection configurations can be incorporated into configurations of a reconfigurable computing unit. The configurations related to data transmission mainly control data interactions between the data memory and the reconfigurable computing array. For example, plaintext and ciphertext input and output from FIFO or transmission of intermediate results by using the shared memory. Generally, the configuration information related to data transmission also includes the control information about array time sequence, that is, inter-unit synchronization control. Synchronization between reconfigurable computing units is performed to ensure the correctness and consistency of data read and write. Closely related to data transmission control, this part of configuration information can be classified as configuration information for data transmission control. Actually, as far as the definition of specific configuration bits is concerned, the time sequence control and data transmission control are also overlapped to a large extent.

Figure 3.24 provides an instance of configuration information organization under centralized configuration control. Configuration initialization and switchover are performed on the basis of array. Generally, the configuration information about overall time sequence control over an array is integrated with the configuration information related to data transmission control. Interconnection configurations can be incorporated into I/O configurations of computing units or be organized in the array form separately. According to the data flow diagram of an algorithm, configurations about data transmission and time sequence control can be used to implement iterations and pipeline operations.

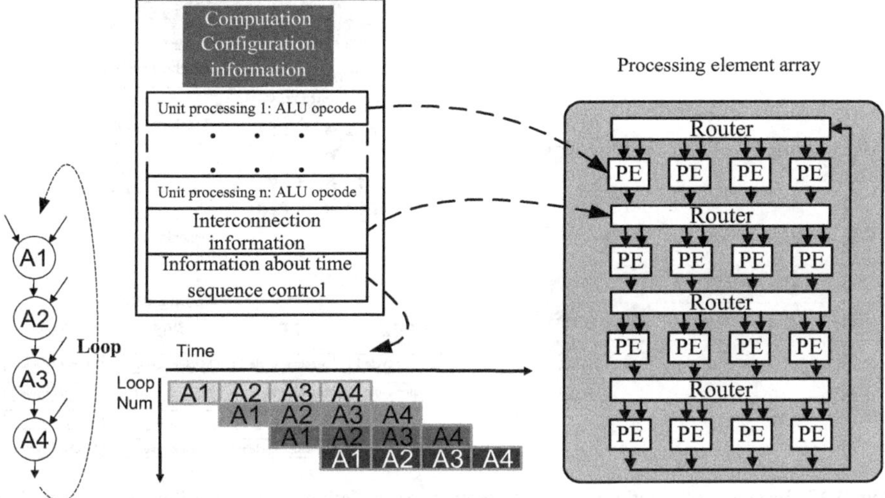

Fig. 3.24 Instance of configuration information organization under centralized configuration control

Figure 3.25 provides an instance of configuration information organization under distributed configuration control. In this case, configurations of processing units are relatively independent. Therefore, the configuration information is generally organized on the basis of processing units. Now the concept of machine cycle can be introduced. The multiple clock cycles involved in the process of performing the same function by a processing unit are defined as a machine cycle. The configuration information about each processing unit can be described in the timeline mode on the basis of the machine cycle. The operator function, data input source, and data output direction of the processing unit can be defined in the configuration information about each machine cycle. Actually, the processing unit-based description method with the machine cycle concept is logically equivalent to the previous computing array-based description method. However, the two description methods have different starting points. In addition, organization of configuration information based on the machine cycle can be considered to have implemented compression of configuration information to some extent. The configuration information for implementing the same function in multiple continuous clock cycles can be compressed into a piece of information, thus reducing the total amount of configuration information to some extent.

In a reconfigurable cryptographic processor, storage of configuration information is multi-layered. On the outer layer of the processor, generally a level of external memory is set to store the configuration information about multiple algorithms. On the inner layer of the processor, at least one level of internal configuration information buffer is set. The implementation mode of the internal configuration buffer varies according to the configuration control mode. In the distributed configuration

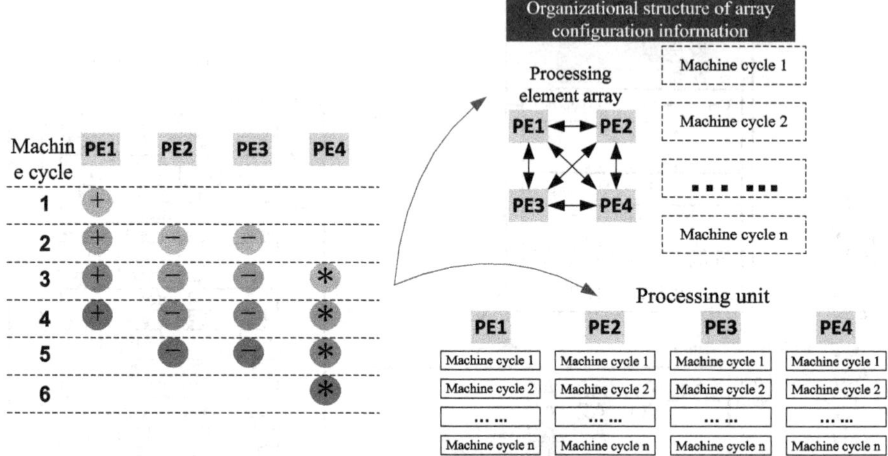

Fig. 3.25 Instance of configuration information organization under distributed configuration control

control mode, the internal configuration buffer can be implemented by multiple distributed memories corresponding to the processing units; in the centralized configuration control mode, the internal configuration buffer is implemented by an integrated memory that supports unified read. Generally, the external configuration memory (ECM) of a reconfigurable cryptographic processor has big data read width; Fig. 3.26 shows a storage form example of the configuration information in the external configuration memory. First, for a group of configuration information, the header of the configuration package should give the basic information of the package, for example the length of the configuration package (when the configuration package has an unfixed length) and whether the configuration package contains the configuration of the heterogeneous modules. Then, store the text of the configuration package. Based on the width of the configuration memory and the width of the basic unit of the configuration information (here, the configuration information under the control of distributed configuration is used as an example), store the configuration information of a specified number of processing unit in each line of the configuration memory. For example, each line of the configuration memory in the figure above can store the configuration information of i processing units. In addition to conventional configuration of the reconfigurable processing element array, an external configuration memory can also store the configuration of heterogeneous functional modules, such as PN and S-Box modules.

In configuration memories at different levels, the configuration information needs to be transferred accordingly. Thus, the problem of delayed transfer of configuration information is generated. A reconfigurable cryptographic processor raises a high requirement for the configuration speed. Therefore, it is recommended to minimize the use of serial configuration transfer mode in the traditional structure, but use the parallel configuration transfer mode which provides a great

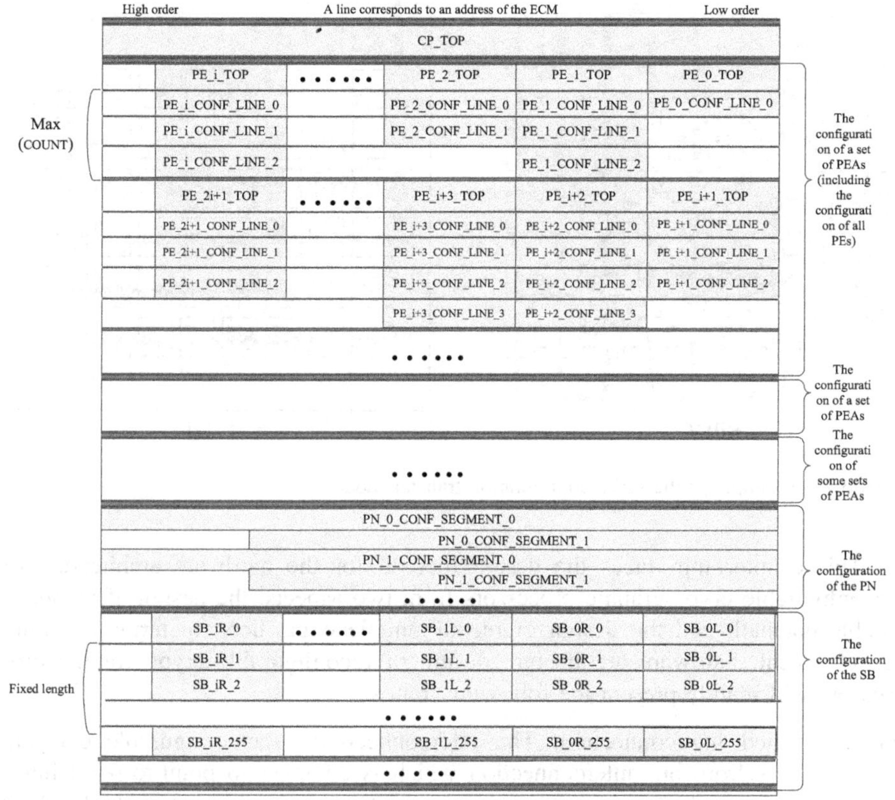

Fig. 3.26 Storage form example of the configuration information in an external configuration memory

configuration width instead. Figure 3.27 gives the design examples of two serial configuration transfer modes, BilRC [8] and XPP-III [9]. In the BilRC processor, the configuration information of all PEs enters the PEA from the customized configuration interface of the PEA and is then transferred in the PEA in the shape of the letter S until it arrives the configuration cache of the destination PE. In this case, the transfer delay of configuration information depends on the position of a PE in the PEA, and the maximum transfer delay is xy cycles, where x and y, respectively, indicate the number of lines and number of columns of the PEA. In the XPP-III processor, the configuration chain is even more complex, which improves the configuration delay to $x + y$ cycles accordingly. Serial configuration, however, still causes a great configuration delay to the far end of the PEA. To sum up, in a reconfigurable cryptographic processor which raises a high requirement for the transfer delay of configuration information, it is recommended to use the PE-level parallel configuration transfer mode to shorten the configuration transfer time of the overall PEA with a higher parallel configuration bandwidth.

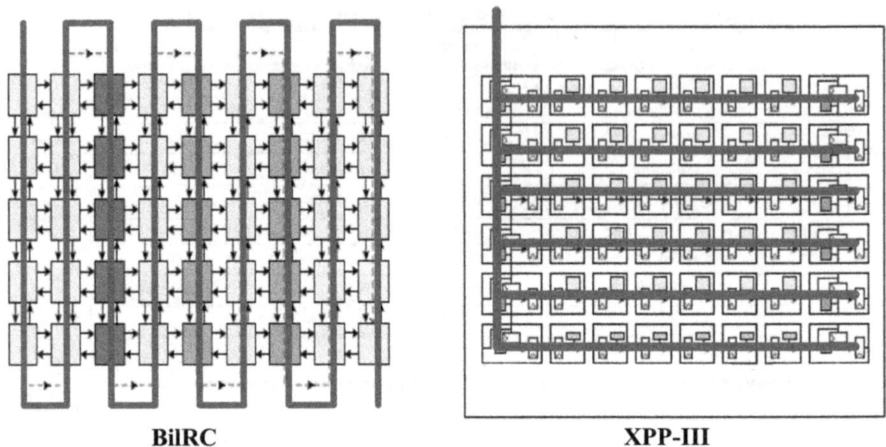

Fig. 3.27 Example of the serial configuration transfer mode

This chapter introduces the design method for the hardware architecture of reconfigurable cryptographic processors from two aspects: the design of reconfigurable datapath and the design of reconfigurable controllers. In terms of future development, hardware architecture design of reconfigurable cryptographic processors will mainly present the following trends.

(1) Simplified interconnection. The interconnection in reconfigurable datapath develops from bus interconnection which is complex to point-to-point interconnection which is simpler, ensuring data processing at a high clock frequency.
(2) Hierarchical data storage. With the rapid improvement of the computing performance of processors, data transmission becomes a bottleneck. Hierarchical data storage, however, can improve the efficiency of data transmission and thus finally ensure the execution performance of the algorithm.
(3) Distributed configuration control. Configuration control develops from centralized configuration at PEA level to distributed configuration at PE level, improving the resource utilization and configuration efficiency.

References

1. Chiou D (2014) Cryptoraptor: high throughput reconfigurable cryptographic processor. In: Proceedings of the 2014 IEEE/ACM international conference on computer-aided design, pp 154–161
2. Mathew S, Sheikh F, Agarwal A et al (2010) 53 Gbps native GF $(2^4)^2$ composite-field AES-encrypt/decrypt accelerator for content-protection in 45 nm high-performance microprocessors. In: 2010 IEEE symposium on VLSI circuits (VLSIC), pp 169–170

3. Hodjat A, Verbauwhede I (2006) Area-throughput trade-offs for fully pipelined 30 to 70 Gbits/s AES processors. IEEE Trans Comput 55(4):366–372
4. Nassimi D, Sahni S (1981) A self-routing Benes network and parallel permutation algorithms. IEEE Trans Comput 30(5):332–340
5. XiangNan Dai Zibin, Jinsong Xu (2007) A reconfigurable bit permutation system design based on the Benes network. Comput Eng 33(22):178–180
6. Farooq U, Marrakchi Z, Mehrez H (2012) FPGA architectures: an overview. Tree-based heterogeneous FPGA architectures 7–48
7. Hwang K, Jotwani N (2011) Advanced computer architecture. McGraw-Hill, New York
8. Atak O, Atalar A (2013) BilRC: an execution triggered coarse grained reconfigurable architecture. IEEE Trans Very Large Scale Integr VLSI Syst 21(7):1285–1298
9. Campi F, König R, Dreschmann M et al (2009) RTL-to-layout implementation of an embedded coarse grained architecture for dynamically reconfigurable computing in systems-on-chip. In: International symposium on system-on-chip, pp 110–113

Chapter 4
Compilation Method of Reconfigurable Cryptographic Processors

As an implementation of reconfigurable computing processors in specific fields, a reconfigurable cryptographic processor inherits the basic compilation framework of reconfigurable computing processors: The algorithm is described in high-level programming languages; the hardware and software partition is made through the static or dynamic analysis; then, the hardware part is transformed into the universal intermediate representation through the front-end compilation tools, which is then optimized through the middle-end compilation tools; finally, the mapping is implemented through back-end compilation tools including the synthesis tool, placement and routing tool, and the configuration information of the reconfigurable computing structure is generated. This chapter will be based on this framework and consider the particularity of the compilation method of reconfigurable cryptographic processors. As a cipher algorithm has many obvious code features such as the fixed-boundary loop, loop-carried data dependency, simple control flow, and quite different data granularity, the compilation method of the compiler of a reconfigurable cryptographic processor needs to be optimized based on these features. This chapter will start with general reconfigurable computing processors and introduce their universal compilation technologies and methods, including the main steps throughout compilation process. Then, this chapter will discuss the compilation methods of reconfigurable cryptographic processors, focusing on the steps which are very important for cipher application, such as code transformation and optimization, division and mapping of intermediate representations. Finally, this chapter will give examples about compilation and implementation of different cipher algorithms.

L. Liu et al., *Reconfigurable Cryptographic Processor*,
https://doi.org/10.1007/978-981-10-8899-5_4

4.1 General Compilation Methods for Reconfigurable Computing Processors

Figure 4.1 shows the general compilation process of a reconfigurable computing system, which starts with the algorithms described in high-level programming languages and ends with the generation of configuration information of the

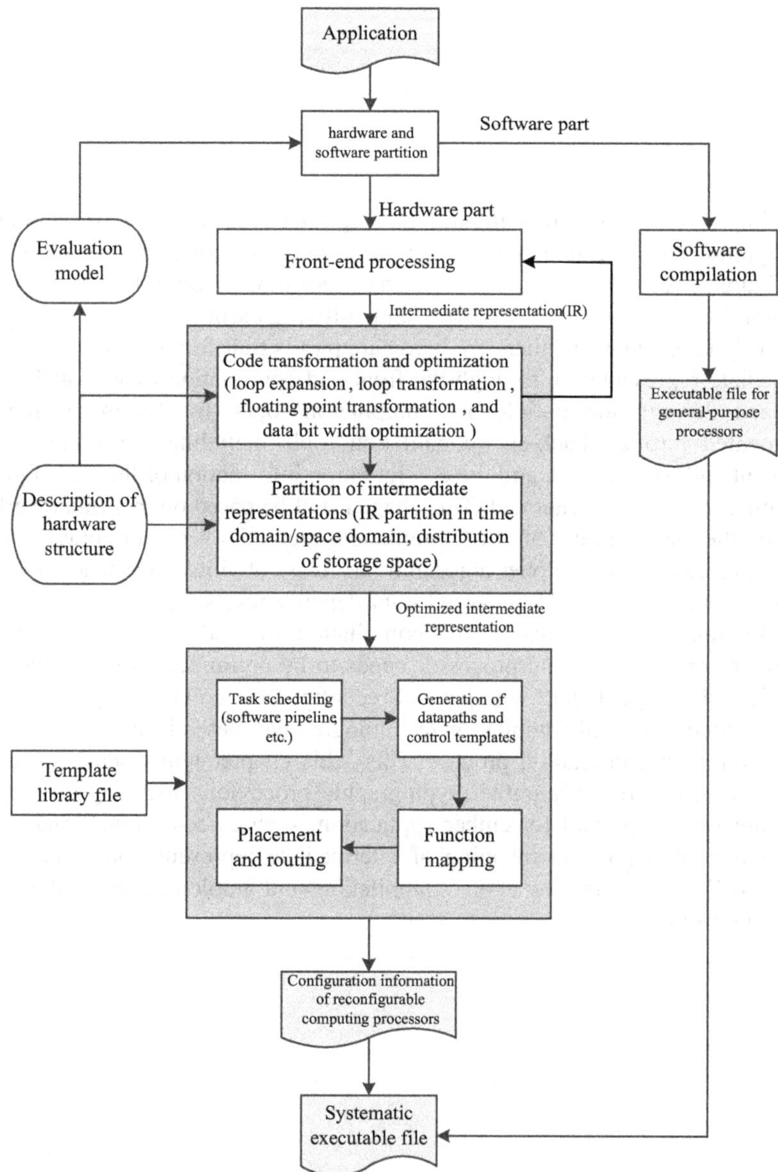

Fig. 4.1 General compilation process of a reconfigurable computing system

reconfigurable processor and of the executable file of the general-purpose processor. Next, the steps and elements in Fig. 4.1 will be introduced [1].

1. Programming Language

As reconfigurable computing processors of different architectures support different programming languages and paradigms, a target application needs to be programmed using a specific language. At present, a main obstacle that a reconfigurable computing processor faces is that there is a lack of an efficient programming language and process which provides a higher level of abstraction than the current hardware description language (HDL). FPGA uses a mature HDL, but it requires that the programmer should have a certain knowledge of the hardware structure. In addition, it needs a long development period. A compilation tool which supports high-level programming languages can greatly shorten the development period of a reconfigurable computing system and accelerates the migration of mature algorithm and applications to a reconfigurable computing system. There has been no, however, mature automatic compilation technologies which are completely based on high-level languages.

2. Programming Paradigm

Programming paradigm is a key feature of a programming language, including the execution feature of the programming language, and the organization form and basic syntax of the codes. There are two main programming paradigms: imperative one and declarative one. An imperative programming paradigm uses imperative statements to describe the functions of a program, and the processor just needs to execute these commands in sequence. Many classic programming languages which are oriented to procedure or object are imperative ones, such as Fortran, Basic, C, C++, Perl, Python, Java, and PHP. A declarative programming paradigm uses descriptive statements to describe the logical structure of the target computation function, but does not specify the execution sequence of the computation explicitly, and the processor needs to complete the execution through analysis. Such function programming languages as Common Lisp, Scheme, Haskell, Erlang, F# and such logical programming languages as Prolog belong to declarative programming languages. It should be noted that the programming paradigm is closely related to the hardware structure. The machine codes used by the underlying hardware of an ISA processor themselves are imperative ones, so an imperative programming language is more suitable for a processor of such an architecture. However, with the quick development of systems of such spatial parallel computing architectures as multicore, many-core, and FPGA, declarative programming languages gradually gain wide attention of the academia and industry. This is because underlying hardware is no longer limited to imperative machines codes on the one hand, and the declarative programming paradigm can greatly simplify parallel programming on the other hand. Just like the research on programming languages, there has been no mature discussion on the programming paradigm for reconfigurable computing structures currently. Obviously, a new programming paradigm is required for a reconfigurable computing structure due to its particularity [2].

3. Front-End Processing

For algorithm and applications which are described in different programming languages, what should be done first is the front-end processing, so as to form a general intermediate representation independent of the concrete programming language. The front-end processing of the HDL based on communicating sequential processes (CSP) is more mature and efficient than that of the high-level language based on the command execution model. The front-end processing based on an imperative programming language spends more time on parallelism extraction, or relies on the hardware developer to provide a lot of mature library function templates and programming interfaces.

4. Intermediate Representation

The main function of an intermediate representation (IR) is to explicitly represent all data dependencies and control dependencies in applications and to give support for exploring parallelism at different levels and of different types. A traditional control data flow graph (CDFG) is a natural intermediate representation. It establishes a control flow structure for the application, maps the control flow to the controller or software implementation, distributes the operators to basic blocks with no internal control, and then maps the basic blocks to the reconfigurable computing arrays. A control flow graph (CFG), however, incorporates some of the control dependencies in the basic blocks to form hyperblocks [3], so as to represent and explore more parallelism. A task flow graph, for example, a hierarchical task graph (HTG) [4], is an intermediate representation which is mainly used to parse application structure, explore task-level or function-level parallelism. Various intermediate representations can be combined to get better effect.

5. Hardware and Software Partition

Hardware and software partition can be conducted by a compiler according to the composition of the reconfigurable computing system. Usually, reconfigurable computing systems include reconfigurable computing processors and general-purpose processors. Therefore, a compiler needs to divide the intermediate representation of the application into two parts accordingly: hardware and software, with the former executed on the reconfigurable computing processor and the latter executed on the general-purpose processor. Meanwhile, a compiler also needs to generate extra configurations or instructions to perform data synchronization between hardware and software. The basis for hardware and software partition is the evaluation model of the reconfigurable computing system, and the main goal is to optimize the system performance and power consumption. It should be noted that a general-purpose processor is not a must in a system. For a system which contains no general-purpose processor, the software compilation part in Fig. 4.1 can be omitted. Compilation of the software part can be directly completed by the compiler of the general-purpose processor, and no detail will be given here. Compilation of the hardware part can be divided into two main parts: middle-end processing and back-end process.

(1) Middle-end processing

The middle-end processing of a compiler includes code optimization and transformation and task division, with an aim to enable the generated control data flow graph to provide a higher parallelism and to better adapt to the mapping implementation on the target hardware structure, so that a better result can be obtained in back-end processing of the compiler. Code optimization includes not only methods related to hardware structure (such as data bit width optimization and loop optimization), but also methods independent of the structure (such as constant folding and memory access optimization). Task division falls into two categories: temporal partition and spatial partition. The former is to break down and assign the tasks to different time slices of the reconfigurable processing array and it uses the principle of time multiplexing for computation. The latter is to break down and assign the tasks to multiple reconfigurable processing arrays and it uses the principle of spatial parallelism for computation.

(2) Back-end processing

The back-end processing of a compiler is mainly to synthesize or map the control data flow graph to the configuration information of a concrete system, including function synthesization of task scheduling, datapaths, and control units, mapping of control data flow graph, and placement and routing. In task scheduling, optimization measures such as software pipelining should be performed for tasks according to the hardware structure. In synthesization, mapping, and placement and routing, the features such as the granularity, interconnection, and storage of the hardware structure should be considered.

6. Description of Hardware Structure

Description of hardware structure provides a basis for the optimization techniques of the compiler at each stage. As the hardware structure of a reconfigurable processor has a high flexibility and variance, an optimized result can be obtained only after a complex design space exploration (DSE) has been performed. The performance and power evaluation model of the hardware structure provides a basis for hardware and software partition. Usually, the most time-consuming modules in algorithms are loop structures, which can provide a high computation and data parallelism and greatly improve the performance when executed on the hardware structure. With the help of the evaluation model, a compiler can distribute the algorithm modules which have great potential improvement for performance or power consumption to the hardware implementation, and distribute the other algorithm modules to the software implementation.

7. Library of Function Template

A reconfigurable processor can provide a template library file to improve the efficiency and performance of the compiler. Similarly to other compilers, a template library file is optimized based on the specific algorithm structure in specific field, and efficient synthesization and mapping of critical codes can be quickly implemented through library calling.

The above content covers each step of the general compilation process for reconfigurable computing processors. Methods for code transformation and optimization and methods for intermediate representation mapping will be introduced in the following section. Next, we will choose some representative reconfigurable computing compilers for introduction. Table 4.1 lists some typical reconfigurable computing compilers, and Tables 4.2 and 4.3 summarize the main features of these compilers. What should be

Table 4.1 Basic information about a typical reconfigurable computing compiler

Compiler name	Published in	Organization
PRISM-I, II	1992	Brown University
Transmogrifier-C	1995	University of Toronto
CoDe-X	1995	University of Kaiserslautern
Handel-C	1996	University of Oxford
SPC	1996	Karlsruhe Institute of Technology, Imperial College London
NAPA-C	1997	Sarnoff Corporation
RaPiD-C	1997	University of Washington
Galadriel & Nenya	1998	INESC-ID, University of Algarve
SPARCS	1998	University of Cincinnati
CAMERON	1998	Colorado State University
Garpcc	1998	University of California, Berkeley
DIL	1998	Carnegie Mellon University
DEFACTO	1999	University of Southern California
DeepC	1999	Massachusetts Institute of Technology
MATCH	1999	Northwestern University in the USA
Maruyama	2000	University of Tsukuba
Stream-C	2000	Los Alamos National Laboratory, Sarnoff Corporation, Adaptive Silicon Corporation
CHIMAERA-C	2000	Northwestern University in the USA
HP-Machine	2001	HEWLETT-PACKARD
XPP-VC	2002	PACT
ROCCC	2003	University of California, Riverside
DRESC	2007	Interuniversity Microelectronics Centre
REMUS-C	2009	Tsinghua University, Shanghai Jiaotong University
TRIPS compiler	2009	University of Texas, Austin
DySER-LLVM	2012	University of Wisconsin

Table 4.2 Main features of a typical reconfigurable computing compiler-1

Compiler	Programming language	Programming model
Transmogrifier-C	C language subset	Software, command
PRISM-I, II	C language subset	Software, command
Handel-C	C language subset extension	CSP model, delay model
Galadriel & Nenya	JAVA subset	Software, command
SPARCS	VHDL task	VHDL, task
DEFACTO	C language subset	Software, command
SPC	C, Fortran language subset	Software, command
DeepC	C, Fortran language subset	Software, command
Maruyama	C language subset	Software, command
MATCH	MATLAB	Software, command
CAMERON	SA-C	Software, function
NAPA-C	C language subset extension	Software, parallel command
Stream-C	C language subset extension	Software, stream processing-based, multi-process
Garpcc	C language	Software, command
CHIMAERA-C	C language	Software, command
HP-Machine	C++ extension	Multi-thread models updated periodically
ROCCC	C language subset	Software, command
DIL	DIL	Delay model
RaPiD-C	RaPiD-C	RaPiD structure specific
CoDe-X	C language subset	Software, command
XPP-VC	C language subset extension	Software, command
REMUS-C	C language subset extension	Software, command
TRIPS compiler	C/FORTRAN	Software, command
DySER	C/C++	Software, command
DRESC	C language	Software, command

noted is that what are listed in the tables are all compiler schemes which mainly adopt high-level languages. According to the differences in hardware structure, reconfigurable computing compilers can be divided into three categories:

(1) The compiler of fine-grained reconfigurable computing structures, for example, the compiler of FPGA.
(2) The compiler of a reconfigurable computing system tightly coupled with a general-purpose microprocessor.
(3) The compiler of a coarse-grained reconfigurable computing processor.

Their main features include whether the programming model is a parallel one (e.g., CSP), or a serial one, whether the programming language is a high-level one or a low-level one, and whether the compiler is oriented to a specific structure or a specific field.

FPGA is the main representation of fine-grained reconfigurable computing structures. Early compilers (e.g., Transmogrifier-C [5]) use low-level languages for programming, including subsets and special syntax structures of C language. Schemes PRISM-I and PRISM-II [6] both firstly focus on the problem of

Table 4.3 Main features of a typical reconfigurable computing compiler-2

Compiler	Front end	Data type	IR	Parallel mining
Transmogrifier-C	Structure customization	Bit type	Abstract syntax tree (AST)	Operator
PRISM-I, II	Lcc	Basic type	DFG	Operator
Handel-C	Structure customization	Bit type	AST	Operator
Galadriel & Nenya	Structure customization	Basic type	DFG	Operator, intra-basic block, inter-loop
SPARCS	Structure customization	Bit type	USM	Operator, task
DEFACTO	SUIF	Basic type	AST	Operator
SPC	SUIF	Basic type	DFG	Operator
DeepC	SUIF	Basic type	SSA	Operator
Maruyama	Structure customization	Basic type	DDG	Operator
MATCH	Structure customization	Basic type	AST, DFG SN	Operator
CAMERON	Structure customization	Bit type	Hierarchical DDCF+DFG +AHA graph	Operator
NAPA-C	SUIF	Preprocessing (bit type)	AST	Operator
Stream-C	SUIF	Preprocessing (bit type)	AST	Operator
Garpcc	SUIF	Basic type	DFG +hyperblock	Operator, basic block
CHIMAERA-C	GCC	Basic type	DFG	Operator
HP-Machine	Structure customization	Bit type	DFG +hyperblock	Operator, process
ROCCC	SUIF2	Basic type	CIRRF	Operator
DIL	Structure customization	Bit type	Hierarchy acyclic graph	Operator
RaPiD-C	Structure customization	Bit type	Control tree	Operator, function
CoDe-X	Structure customization	Basic type	DAG	Operator, loop
XPP-VC	SUIF	Basic type	CDFG	Operator, basic block, inter-loop parallelization
REMUS-C	SUIF	Basic type	DFG	Operator, basic block, loop
TRIPS compiler	Scale	Basic type	AST	Operator
DySER	LLVM	Basic type	LLVM	Operator
DRESC	IMPACT	Basic type	DDG	Operator, loop

compilation directly from subsets of C language to a reconfigurable computing system, but they mainly optimize underlying logic gates. With the development of technologies, many researchers begin to focus on technologies which provide programming frameworks (such as SPC [7], DeepC [8], DEFACTO [9], and Streams C [10]) for reconfigurable computing structures by directly using high-level languages such as C and Java. Most of these technologies translate the high-level language into behavioral RTL-HDL description according to the target reconfigurable computing structure and then use high-level synthesis or logic synthesis tools to complete back-end processing of compilation.

The main feature of the compiler of a coarse-grained reconfigurable computing processor is that it is specified for target structure precisely. Almost all research and business projects choose to design individual integrated development tools covering the whole process from programming paradigm to placement and routing for their reconfigurable computing processors. Typical examples include the compilation frameworks of such hardware architectures as PipeRench, RaPiD, Xputer, ADRES, REMUS [11], TRIPS [12], and XPP.

(1) The following takes an DIL compiler oriented to the PipeRench structure as an example [13]. DIL can be regarded as an intermediate representation language similar to high-level language and it is used to describe the pipelined combinational logic circuit. DIL language has syntaxes and operators similar to those of C language, and it uses fixed-point variables which can be defined by the user or derived by the compiler. Each operator used in the DIL program has been declared in the library of the module generator. The compiler expands all modules and functions and fully expands the loops appearing in the application to generate a straight-line single-assignment program, so that a hierarchical acyclic DFG can be constructed. Each node of the DFG represents operation, I/O port, and register. After a global DFG is generated, the compiler can perform some optimizations, including traditional compiler optimization, register reduction, and interconnection simplification. Placement and routing is implemented by using deterministic linear greedy algorithm, which can be implemented based on list scheduling.
(2) DRESC is a C compiler [14] designed for the ADRES system. This compiler uses the basic structure of the IMPACT compiler as the front end. The partition, however, is used to identify the compute-intensive loops in the program. The compiler also uses an innovative modulo scheduling algorithm which is integrated with such processes as placement, routing, and scheduling and is used to map the pipelined loops to the reconfigurable array. One interesting thing about this compilation process is that it can compile a family of ADRES-like structures by modifying the architectural description file.
(3) XPP-VC maps C codes to the XPP architecture [15]. This compiler is based on the framework of the SUIF compiler and it uses new mapping technologies and the pipeline vectorization method mentioned in SPC. The computation is mapped to the ALU and data-driven operator, and the compiler will perform temporal partition for the program as long as the resources input which is required by the program exceeds the resource amount which a single configuration can provide.

In addition, some research on compilers focuses on the structures that tightly couple the main processor with a reconfigurable computing processor. Some of the typical representatives are DySER [16] (based on LLVM [17]), CHIMAERA-C [18], Garpcc [19], and NAPA-C [20].

(1) The programming language used in NAPA-C is composed of the subsets of C language and the extended syntax for specified bit width and computation task, and it is suitable for the NAPA reconfigurable processors formed by coupling of the microprocessor and FPGA. It uses the MARGE tool to generate datapaths. MARGE uses a parametric macroblock library which contains such components as arithmetic logic unit, counter, filter, and SRAM. As for the internal structures of some of these components, the pre-layout and pre-arrangement of wires have been completed so as to shorten the compilation time of the compiler. The Modgen tool is used to generate the macroblocks and the logic structure of the macroblocks can be described by using a format similar to C codes. The MARGE tool can be used to transform the codes input to RTL models and finally map it to the FPGA array in the NAPA structure to complete the corresponding computation.
(2) The Garp-C compiler is a standard C compiler. It is oriented to the Garp structure and is also called garpcc. It is a compiler designed by the researchers of University of California, Berkeley. It uses the concept of superblock to extract the core loops in the codes which are suitable for reconfigurable hardware and optimizes the branch statement "if-then-else" as speculative execution. The basic blocks of the superblock are then transformed to generate a DFG data graph. Thus, the parallelism among operators can be fully explored. Garp-C is also integrated with the software pipeline technology in the traditional compilation design to improve the performance of the very long instruction word processor. Garp-C generates the corresponding structure description by using a pre-defined module, and completes the final mapping and configuration generation by using the Gama tool.

To sum up, a reconfigurable computing processor brings a huge challenge to software engineers. This is because its computation model oriented to hardware architecture forces the programmer of the reconfigurable computing processor to play a role in the design of the hardware architecture. A compilation technology is to bridge the gap between the hardware designer and the software programmer, and to automatically map the algorithm applications described with high-level languages to the reconfigurable computing processor. Some of the above compilation frameworks are affected by collaborative design of hardware and software and the hardware synthesis field, and some other compilation frameworks are to design back-end synthesis tools oriented to the hardware architecture of reconfigurable computing processors for traditional compilers.

4.2 Compilation Methods of a Reconfigurable Cryptographic Processors

The fundamental difference between the compiler of the reconfigurable processor and the compiler of the traditional general-purpose processor is that they are completely different in target hardware architecture. This requests the programming languages, programming paradigms, code optimization methods, etc., used by it to be redesigned and solve the problems such as partitioning and scheduling of intermediate expressions, mapping on the array structure. Compared with the general reconfigurable processor, the compiler design of the reconfigurable cryptographic processor will be different and has its special focuses due to the special application feature of the cryptographic-specific field. The key features of the application in the field of cryptography include computing-intensive code with massive loops, the multilevel parallelism that can be explored, dedicated function operators, and fine-grained bit-level operations. The specific analysis of different types of cryptographic algorithms can refer to Chap. 2. Based on these key features, compilers of reconfigurable cryptographic processor should take appropriate measures to improve the efficiency of their compilation results. The following will discuss the key compilation techniques from the perspective of code transformation optimization methods and mapping method of data flow graph.

4.2.1 Code Transformation and Optimization

As shown in Fig. 4.1, after the compiler front end transforms the language-specific application into an intermediate representation, the compiler back end can generally synthesize the intermediate expression directly into configuration information of reconfigurable processor, but the efficiency will be poor without optimization process of compiler middle end. Since code transformation and optimization are important steps in the compiler middle end, it can optimize application codes based on the hardware architecture of the reconfigurable processor, thus shortening execution time and reducing hardware resources. In addition, the compiler should also perform basic architecture-independent code transformation on code, which is mature compilation technique and will not be discussed in detail here.

The following part discusses the techniques for code transformation and optimization from the perspective of bit-level, instruction-level, loop-level, data-level, and function-level.

1. Bit-Level Transformation

The bit-level code transformation technology is mainly used for fine-grained reconfigurable processors such as FPGAs, but also for mix-grained architectures that include fine-grained reconfigurable unit. It mainly includes the methods such as bit narrowing, bit optimization, and floating- to fix-point transformation.

In many applications, the range and precision of data types are often redundant, meaning that the data bit width required to store the actual value is always less than the bit width defined by the structure. Although this problem does not affect performance on general-purpose processors or coarse-grained reconfigurable computing architectures, this issue is important for fine-grained and mix-grained reconfigurable computing architectures. With the change of bit width, the delay and resource usage of fine-grained reconfigurable unit are different. Reducing the bit width of redundant data defined by programming language can make fine-grained reconfigurable processor more efficient. The standard C language does not support custom data bit widths, but many extended programming languages supports this function, such as NAPA-C, Stream-C, and DIL. The bit narrowing technology can not only analyze the redundant bit width statically, but also dynamically analyze the bit width according to the result of real-time operation, thus obtaining greater benefits.

The bit-level optimization technology also improves the application with more bit-level operations (masking, shifting, concatenating, etc.) by saving hardware resources and reducing execution time. The bit-level optimization is usually implemented by using a binary decision diagram (BDD) in logical synthesis [21]. However, this method will occupy a lot of time and storage space, which is not very efficient for large-scale examples because even a small DFG in a program can produce a large logic diagram. Therefore, many compilers choose other DFG-based bit-level data flow methods, such as BitValue [22]. The DIL compiler is a compiler that takes into account bit-level optimization. Its programming language facilitates bit-level optimizations such as the writing for c = (ab) [7: 0] (“.” represents joining, b represents 8-bit wide) can easily be simplified to c = b [7: 0], which can be directly optimized to lower 8-bit hardware connection implementation between b and c. This is much more convenient than c = (0xff) & ((a ≪ 8) | (0xff & b)) described in standard C language, and much closer to the hardware description language. Further bit optimization needs to determine if there are any redundancies in the hardware implementation of some statements, such as expressions of (ans & 0x8000) == 0x8000 can be simplified to single-bit signal decision. It should be noted that some bit-level analysis and optimization will be discovered until the loop has been expanded.

The floating-point number is a widely used data type in most fields of computing. However, the cost for floating-point arithmetic calculations, including hardware resources and the number of execution cycles, is also very high. Therefore, in the implementation of data signal processing algorithms, we often choose to use the fixed-point number, and implement arithmetic operations of floating-point numbers by using integer arithmetic unit and corresponding scaling factor (usually take an integer power of 2, so binary shift can be easily implemented). The advantage of using fixed-point numbers is to save resources while ensuring the required accuracy. However, the transformation process is a time-consuming task. Therefore, the automatic transformation from floating point to fixed point is also an important technology of compilation of reconfigurable processor, which is studied extensively. It should be pointed out that there is almost no non-integer operation in the cryptographic algorithm, so the technique is less significant for the reconfigurable cryptographic processor.

2. Instruction-Level Transformation

Instruction-level transformation is mainly targeting larger granularity code optimization. The goal is the same as bit-level optimization, which is to reduce the hardware resources and execution cycles required for the target operation. The specific method can be divided into algebraic algorithm simplification and circuit-level optimization. Algebraic simplification, a mature optimization method that is independent hardware structure, can reduce the strength of algebraic operations and transform unsupported operations, including tree height reduction (THR) [21], common subexpressions elimination (CSE), constant folding, constant propagation and strength reduction techniques. The circuit-level optimization method is mainly to simplify the hardware circuit implementation for specific operators. For example, for $2^N + x$ operations, a general adder can be replaced by incrementer and concatenation operation, which is close to datapath design of the reconfigurable processor. The following makes brief introduction to several commonly used instruction-level transformation technologies.

① Tree height reduction. Under the premise of not changing the arithmetic function, it can achieve higher degree of parallelism by reordering the operations, reducing the height of the operation tree. Figure 4.2 is an example. In the most ideal condition, the height of the tree can be reduced from $O(n)$ to $O(\log n)$ by tree height reduction operation. When the operator types are the same, the tree height reduction operation will be easy to perform; when the operator types are different, the tree height reduction can be used to find the best combination of operator. However, the tree height reduction usually requires more hardware resources to complete the parallel execution of the operators. With limited hardware computing resources, tree height reduction may even result in worse performance.

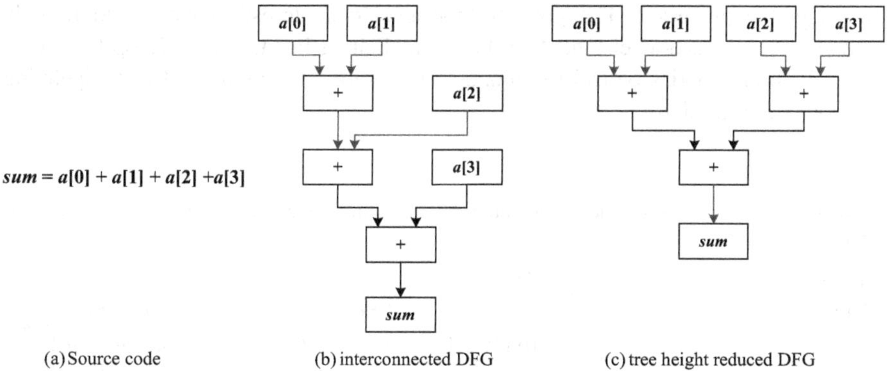

Fig. 4.2 Tree height reduction

② The operation strength reduction can transform a type of operation into a combination of computationally less expensive operations [23]. For example, multiplication and division operations can be transformed into a series of shift and addition and subtraction operations. Therefore, the operation strength reduction usually can reduce the hardware area and delay needed to achieve the computation. For multiplication operations, an efficient hardware-oriented approach is often used, i.e., canonical signed digit (CSD). On this basis, multiplication can be performed by using only a minimal number of add/subtract devices if common subexpression elimination is applied. Table 4.4 shows three cases of computational intensity reduction calculations. It can be seen that the number of additions and subtractions can be significantly reduced by using CSD as well as common subexpression elimination combinations. The operation strength reduction can also be used to process floating-point operations, and have already been used in practice, such as the DeepC compiler.

③ Local and global code motion. Motion can be divided into hoisting and sinking [23]. The size of the code can be reduced through placing the repeated operations of the same expression or subexpression on the common path. During the compilation process of hardware, code motion may make two operations independent of each other, thus achieving parallel execution. It offers the possibility for other optimization techniques, as shown in Fig. 4.3. In this example, the code hoisting and tree height reduction transformation add an adder to the operation, resulting in shortening the length of the critical path and optimizing the performance.

3. Loop-Level Code Transformation

The loop-level transformation method optimizes the loop structure in the algorithm according to the target hardware structure in order to obtain the maximum parallelism. In recent decades, circular optimization has been extensively and deeply studied on other processors (such as CPU, GPU), and on the reconfigurable processor, it has also received a lot of attention, because loop is one of the most important object codes of reconfigurable processor. The goal of loop-level code transformation is to find the instruction-level, data-level, and loop-level parallelism in the loops. Loop-level code transformation includes typical techniques such as loop unrolling, software pipelining, and polyhedral models.

Table 4.4 Contrast cases where the integers are multiplied by constants to reduce the computational intensity

Operation	$231 \times A$		
Method	Binary	CSD	CSD+CSE
	011100111	100-10100-1	Pattern: 100-1
Hardware resources (shifter has not been considered)	Five adders	Two subtractors and one adder	One adder and one subtractor

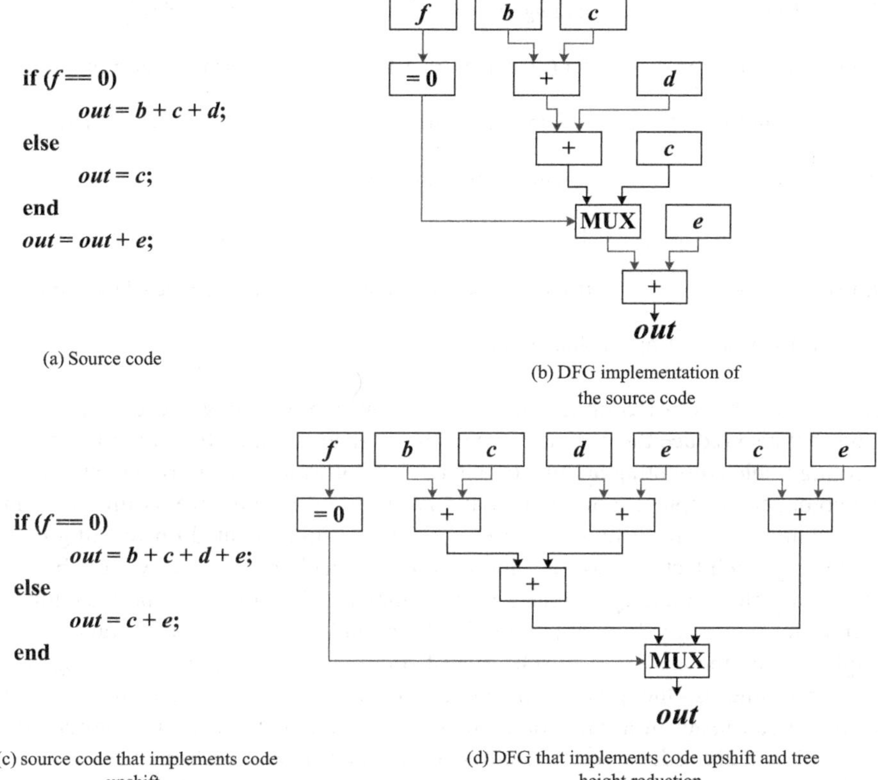

Fig. 4.3 Code upshift

The loop unrolling, a simple and effective loop code transformation technique [24, 25], is widely used in compilers of various processors and also applicable to reconfigurable computing compilers. It can reduce the loop control code through copying the code of loop body, such as iteration count and loop condition judgment, as shown in Fig. 4.4. The benefits of loop unrolling mainly come from two aspects:

① Reduction of control code can improve the execution performance of loop.
② The number of operators that can be simultaneously scheduled by the compiler will increase, helping to find out more parallelism in algorithm (including loop-level parallelism).

The cost of loop unrolling is an increase in the amount of codes and configuration information. For unbounded loops, it is impossible to realize complete loop unrolling. When performing some loop unrolling, we should consider the trade-off between the cost of configuration information and performance benefits because the

```
int img[10];
for (i=0; i<10; i++){
    sum=sum+img[i];
}
```

(a) Source code

```
int img[10];
for (i=0; i<5; i++){
    sum=sum+img[2i];
    sum=sum+img[2i+1];
}
```

(b) partial loop unrolling

```
int img[10];
sum=sum+img[0];
sum=sum+img[1];
...
sum=sum+img[9];
```

(c) Complete loop unrolling

Fig. 4.4 Examples of loop unrolling technique

increase of amount of configuration information may not only exceed the storage limit, but also reduce the performance of the storage system. One of the features of reconfigurable computing architecture is its rich hardware resources, which requires the compiler to explore more computational parallelism to realize its full potential. Meanwhile, the control flow is more costly when implemented on reconfigurable computing architecture, so the loop unrolling technique is very suitable for reconfigurable computing architecture. In summary, the loop unrolling technique is mainly for the relatively simple bounded loop, and will explore the parallelism by implementing the analysis and schedule of some codes, so the higher the degree of loop unrolling is, the greater the benefit is. However, a high degree of loop unrolling can result in a dramatic increase in the amount of code or configuration and will actually reduce your system's storage and overall performance.

The introduction of software pipelining technology can solve the problems about loop unrolling technology. Software pipelining is a widely used method in compilation techniques [26], which refers to the organization of loop-level pipelining in a way similar to instruction-level pipelining. Therefore, it is generally thought that software pipelining is not a code transformation method, but a description of the goal of loop code optimization. The basic starting point is that the data dependencies between loop iterations are not always tight, and the initiation interval (II) between loop iterations (the interval between the beginning of two adjacent loop iterations) can often be shortened. That is to say, the next iteration will start before all the calculations have not been completed in the last iteration, which will reduce the overall execution time as shown in Fig. 4.5. To some extent, software pipelining is also a method of out-of-order execution, and the execution order of multiple loop iterations will overlap. The out-of-order execution technology of processors is achieved by the dynamic scheduling of hardware architecture, while software pipelining is implemented through static analysis of compiler (or manually). Software pipelining has been implemented on many instruction set structure processors, such as Intel IA-64. Meanwhile, it also has been applied to reconfigurable processors, such as ADRES. The advantage of software pipelining technology is to make full use of loop-level parallelism, which can improve the performance of the system with little increase in the amount of code and configuration information, and can optimize unbounded loop to some extent.

The implementation of software pipelining techniques includes kernel identification and modulo scheduling. The former approach is to expand the loop to a certain degree, and then identify the fixed kernel structure of kernel phase as shown in Fig. 4.5 through the searching algorithm. It requires particularly high requirement for recognition algorithms. The representative algorithms include perfect pipelining and Petri Net. The latter approach is to rearrange the body of the DFG graph of the loop body, thus achieving the optimization goal through minimizing the loop initiation interval. Therefore, modulo scheduling is a more direct and effective method, which is currently used by the mainstream software pipelining technology. For reconfigurable structure, modulo scheduling is firstly applied in the DRESC compiler. DRESC uses the modulo routing resources graph (MRRG) to model the system behavior and placement and routing resources of the reconfigurable computing array in temporal extension, and then solve the optimal value of operator scheduling, placement and routing. In order to avoid falling into the locally optimal solution, DRESC uses the simulated annealing method, but it takes much longer time in the scenario where the loop body is large. In order to solve the problem of long simulated annealing time for DRESC, Park proposed the method of modulo graph embedding [27]: using the graph theory technique of graph embedding under the condition of modulo constraint, searching according to the cost function that takes position, dependency and interconnection into the consideration and embedding the loop body into the reconfigurable array's temporally extended structure. The method takes operator nodes as the center, first arranges the

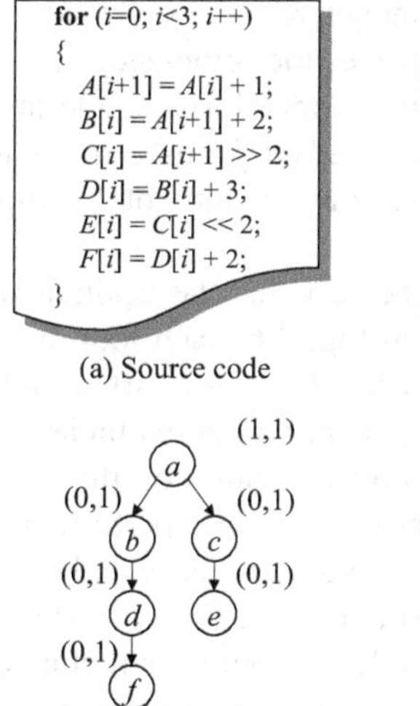

(a) Source code

(b) DFG with (diff, min) information, diff represents the number of iterations between the loop-carried dependence, and min represents the execution time of the source operator

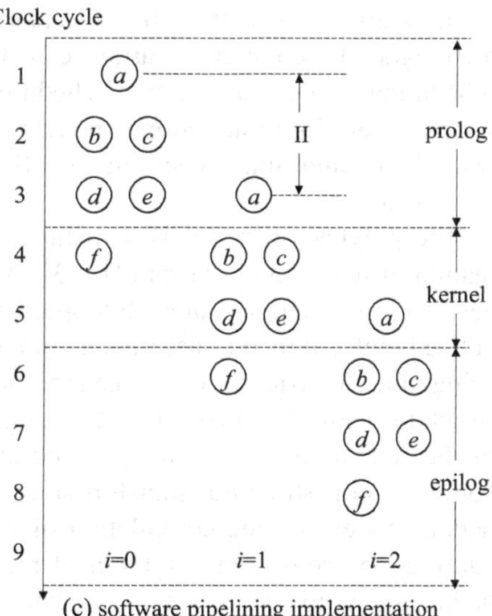

(c) software pipelining implementation

Fig. 4.5 Example of software pipelining principle

operators and then routes them. When the operators have more dependence, it may take longer time and even cause routing failure. In order to solve this problem, Park further proposed the edge-centric modulo scheduling approach [28]. The basic flow of this method is to firstly arrange the dependencies in the data flow graph, and then arrange the operator placement. The search process of routing also has a cost function design to reduce the probability of routing failure. The above two approaches solve the problem of placement and routing efficiency, while the EPIMap method can further improve the performance of modulo scheduling [29], and shorten the loop initiation interval. The main measure is to recomputation, that is, to copy the operators with serious conflicts and reduce operator conflicts by using redundant hardware resources, so that the final placement and routing results can have higher performance. This method models the mapping problem as a surjection subgraph on a reconfigurable array with temporal expansion and designs a heuristic algorithm based on the largest common subgraph method to solve this problem. However, the EPIMap method does not model and use the register resources in the reconfigurable computing array. Hamzeh further proposed the modulo scheduling method for REGIMap [30]. With the reasonable modeling of register resources, recalculation, interconnection, and sharing can be implemented flexibly. The problem of register allocation is modeled as the maximal clique problem for DFG and temporally extended reconfigurable arrays. Finally, this method uses the array's internal register resources to improve performance after mapping. In short, the modulo scheduling technology is the mainstream code transformation method for software pipelining, which requires reasonably mapping reconfigurable computing and loop structures and taking full account of the feature of the hardware structure. In particular, the abundant resources provided by the reconfigurable computing array can help improve the efficiency of modulo scheduling. In general, modulo scheduling is mainly applied to a single-layer loop structure, and for nested loops, modulo scheduling needs to be used in conjunction with loop unrolling, resulting in the increase of amount of configuration information.

The polyhedral model is a mathematical framework for the optimal transformation of nested loop codes [31–33]. As shown in Fig. 4.6, each loop instance of nested loop or iteration of each loop body can be considered as a virtual polyhedral lattice point and then transforms them into an equivalent but more efficient form by using affine transformation or generic non-affine transformation on this polyhedral structure with the goal of optimizing loop code execution performance. The mathematic foundation of the polyhedral model is mature in theory and suitable for complex loop structural transformation. The literature [34] first introduces polyhedral model into the compilation of reconfigurable computing structures, which analyzes the constraint condition of reconfigurable structures, then uses the polyhedral model to optimize the parallelism of the loops and then perform mapping, which proves the effectiveness of the proposed method. Subsequent researches focus on different reconfigurable computing structures, such as one-dimensional array and two-dimensional array, and systematically model and study the effect of polyhedral model [35, 36].

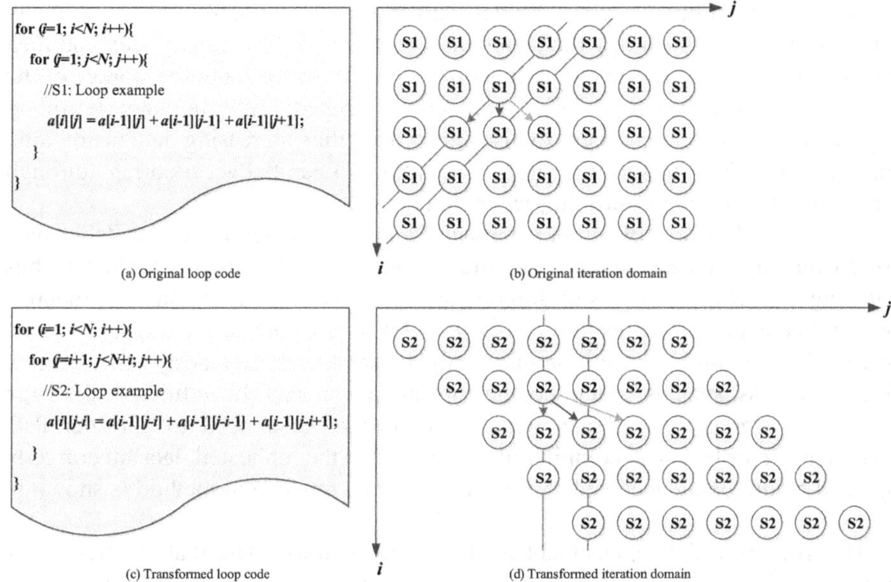

Fig. 4.6 Example of polyhedral model

In summary, the code transformation optimization for loops on the reconfigurable processor has attracted a lot of domestic and foreign research institutions to conduct extensive and in-depth research in the past decade. This is because the loop structure is usually the most time-consuming part in the algorithm. A classic description is that 10% of the code takes 90% of the time, so code optimization for the loop structure can create the most noticeable performance improvement. It should be noted that the above loop-level code transformation method is first used in general-purpose processors. When we apply it to the reconfigurable computing structure, it is no longer independent of the hardware structure, and the mapping efficiency of transformed code on different reconfigurable computing structures is the emerging problem. If insufficient consideration is given to the structural feature of the hardware, the execution efficiency of the code may be significantly affected; otherwise, if hardware structural feature is fully utilized, the efficiency of the code transformation may also have room for improvement.

4. Data-Level Code Transformation

The flexibility of reconfigurable computing structure, especially in terms of the configuration and organization of storage system, makes data-level transformation methods, such as data packets, data replication, scalar replacement, particularly suitable for these structures. Many other transformation methods, in particular, loop-level transformation methods, can be combined with data-level transformation methods to optimize the computation of many variable arrays manipulated s by affine indexing functions in signal and image processing.

The data distribution method divides a given program variables array into many different subarrays, each of which contains some data of the original array and then allocates the data on each subarray to a different memory unit, but the storage of the original data still remains. This approach, combined with the loop unrolling approach, allows parallel access to distributed data, thus increasing data bandwidth. Fig. 4.7 shows how to split the target array into two parallel access arrays through data grouping and loop unrolling technology.

The goal of data replication approach is also to increase the available data bandwidth, but its measure is to copy the data into multiple subarray variables, thus enabling concurrent accesses to different storage modules that the data are mapped onto. Of course, data replication consumes a lot of storage space compared to data distribution, so this approach applies only to small constant arrays that require frequent access, such as system coefficients and parameters. In addition to the large amount of storage space, data replication methods also need to consider data consistency problem. The compiler must ensure that the replicated data are correctly updated before being accessed. An example of data replication method is shown in Fig. 4.8.

The goal of scalar replacement method is to optimize data that are frequently used and have a long life cycle by caching it in scalar variables. For general-purpose processors, the compiler maps these data into internal register files to improve performance. For reconfigurable processors, these data will be stored in the distributed registers or internal memory to avoid frequent access to external storage and improve computing performance and data bandwidth. An example of scalar replacement method is shown in Fig. 4.9. Scalar replacement methods also need to consider data consistency, especially for variables that are read and written repeatedly in the loop. After the scalar replacement is over, the data in the scalar need to be updated back into the original array, thus preventing subsequent operations from using the wrong data.

```
int img[n][n];

for (i=0; i<n; i++){

   for (j=0;  j<n;  j++){

      ...=img[i][j];

   }

}
```

(a) Source code

```
int img_1[n][n/2], img_2[n][n/2];

for (i=0;i<n; i++){

   for (j=0;  j<n;  j+=2){

      ...=img_1[i][j/2];

      ...=img_2[i][j/2];

   }

}
```

(b) Code after data are distributed

Fig. 4.7 Example of data grouping method

```
int img[n][n];
for (i=0;i<n; i++){
    for (j=0;  j<n; j++){
        ...=img[i][j];
    }
}
```

(a) Source code

```
int img_1[n][n],img_2[n][n];
for (i=0; i<n; i++){
    for (j=0;  j<n;  j+=2){
        ...=img_1[i][j];
        ...=img_2[i][j+1];
    }
}
```

(b) Code after data are copied

Fig. 4.8 Example of data replication method

```
int img[n][4], para[4];
for (i=0; i<n; i++){
    for (j=0;j<4;j++){
        ...=img[i][j]*para[j];
    }
}
```

(a) Source code

```
int img[n][4], para[4];
P0=para[0]; P1=para[1]; P2=para[2]; P3=para[3];
for (i=0; i<n; i++){
    ...=img[i][0]*P0;
    ...=img[i][1]*P1;
    ...=img[i][2]*P2;
    ...=img[i][3]*P3;
}
```

(b) Code after scalars are replaced

Fig. 4.9 Example of scalar replacement method

5. Function-Level Code Transformation

The function-level code transformation method includes function inline and function outline, which is complementary and should be selected according to compiling constraints and optimization goals.

Function inline can explore more instruction-level parallelism through expanding the function in the location where the function is called. Function inline actually eliminates the explicit sharing of resources and repeatedly implements the function when calling it, and trades for higher computational parallelism with hardware

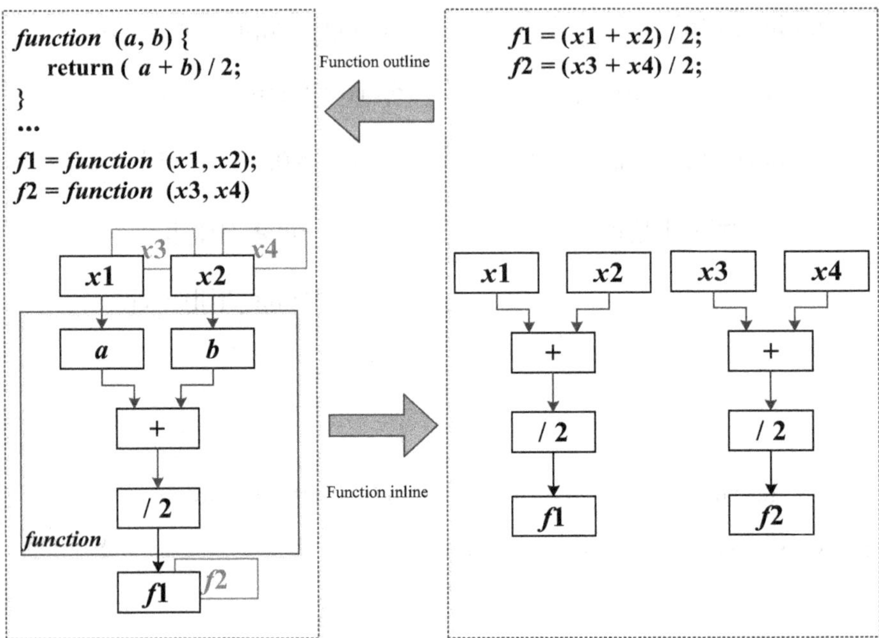

Fig. 4.10 Example of function outline and function inline method

resources, s shown in Fig. 4.10. In fact, the purpose of the function inline technique is to discover function-level parallelism and task-level parallelism in the program. In contrast to the function inline, the goal of function outline is to sacrifice the chance of developing task-level parallelism on the hardware and to reduce the hardware resources through time-multiplexing mechanism. Figure 4.10 also shows an example of function outline method.

4.2.2 IR Partition and Mapping

In addition to code transformation and optimization methods, another important step of middle-end processing of compiler is the partition and mapping of IR, i.e., how IR is implemented on target reconfigurable processor architecture. Because the current reconfigurable processor architecture often includes multiple reconfigurable array structures and memories, and the size of the array structure may also be greatly different, for example, the hardware size of the FPGA is very large, but the size of dynamic reconfiguration processor is small, IR needs to generate a group of subtasks that are suitable for the target hardware structure according to the temporal and spatial partition. The computation of the subtasks needs to be mapped to the computing resources of each reconfigurable array, and variables in the subtasks also should be mapped to different levels of memory system.

1. IR Partition

The IR partition is divided into temporal partition and spatial partition. The former partitions the IR into subtasks of reconfigurable processors at different times and realizes time multiplexing through dynamic reconfiguration. The latter partitions the IR into subtasks on the different arrays of the reconfigurable processor and implements multitask concurrent execution through spatial parallelism. An example of the two partitions is shown in Fig. 4.11. The PE function of the spatial partition does not change in the program execution while the PE function of the temporal partition changes with time during the program execution. Temporal partition and spatial partition are not separate, but often used together.

(1) Spatial partition

Spatial partition divides the IR of a task into multiple subtasks, and each subtask corresponds to an independent reconfigurable computing array. The premise of using spatial partition is that a single task cannot be mapped to a reconfigurable array, and the reconfigurable computing system consists of multiple reconfigurable computing arrays that are interconnected (cache and interconnection line, etc.). In spatial partition, data dependencies exist between different subtasks, which are implemented through shared storage or interconnection line. However, special attention should be paid to the number of interconnection ports line and storage access conflicts.

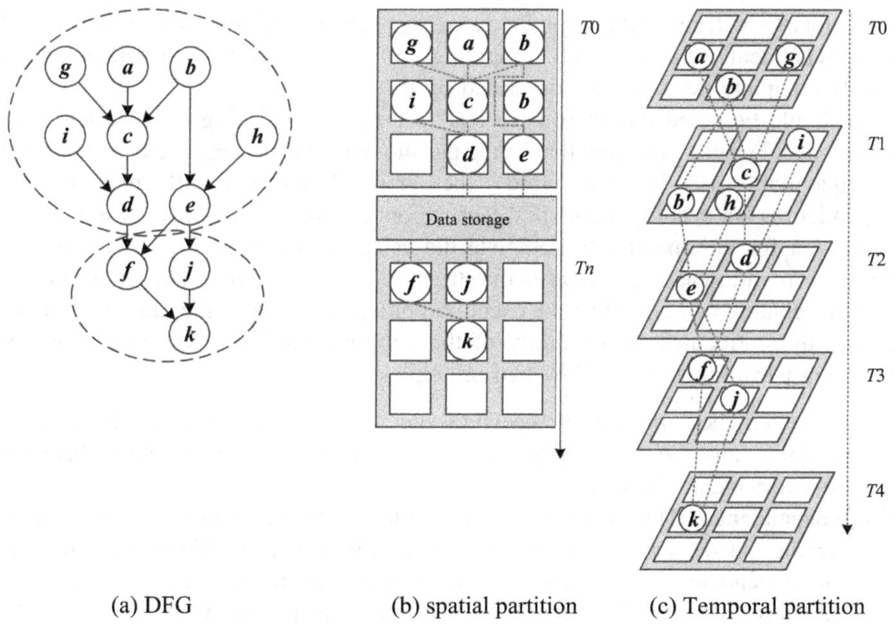

Fig. 4.11 IR division method for reconfigurable processor

Spatial partition is mainly used for reconfigurable processors with multiple array structures. Peterson proposed a spatial partition method targeting multiple FPGAs through simulated annealing techniques [37]; Lakshmikanthan used the Fiduccia–Mattheyses method [38]. The CHAMPION compiling environment also uses spatial partition method [39]. The algorithm and application are divided into modules and then consider the module's IO port as a point of partition only, which preserves module-level information. In the scenario where multiple reconfigurable arrays access the same memory, different operations that access different data sets are divided into different reconfigurable arrays so that they can be executed in parallel.

(2) Temporal partition

Temporal partition will divide the IR of a task into multiple subtasks, and each subtask corresponds to single function configuration of reconfigurable processor. The premise of using temporal partition is that a single task cannot be mapped to one configuration of reconfigurable computing system. Currently, dynamically reconfigurable computing systems all support changing their hardware function with fast and dynamic configuration switching. When reconfigurable computing system executes large tasks that exceed their hardware computing resources, tasks are often divided into a series of subtasks, which are scheduled and executed successively on hardware through multiple configurations, making it possible to configure and execute the same hardware repeatedly. Temporal partition can greatly improve the resource utilization of the temporal-extended hardware system. Figure 4.11b shows a temporal partition of a task, which divides a larger DFG into five subtasks (configurations) whose scales meet the hardware constraints. These five subtasks will reuse the same hardware resource in a time-sharing manner, and then dynamically configure the computing arrays to complete their functions and finally complete the functions described by the overall tasks.

It should be noted that there is a clear feature distinguishing between temporal partition and spatial partition; that is, the partitioned subtasks are strictly executed in a sequential order. In other words, the temporal partition will create control dependences between operators in different configurations. The operator at time T1 must wait for the operator to complete the calculation at time T0. Therefore, the temporal partition actually changes the structure of the task DFG to a certain extent. But this change will not affect the execution efficiency most of the time. In order to ensure the correctness and validity of the temporal partition, we argue that any temporal partition should follow two principles:

(1) Integrity. The functions of the DFG and subtask sequences before and after partition must be completely equivalent and original dependencies between operators cannot be changed.
(2) Independence. The subtask sequence after partition should not have interlocking issues as shown in Fig. 4.12. Operator 3 in Configuration 1 is data-dependent on Operator 2 in Configuration 2, which in turn is data-dependent on Operator 1 in Configuration 1, then Operator 2 will be control-dependent on Operator 3 if Configuration 1 is in front of Configuration

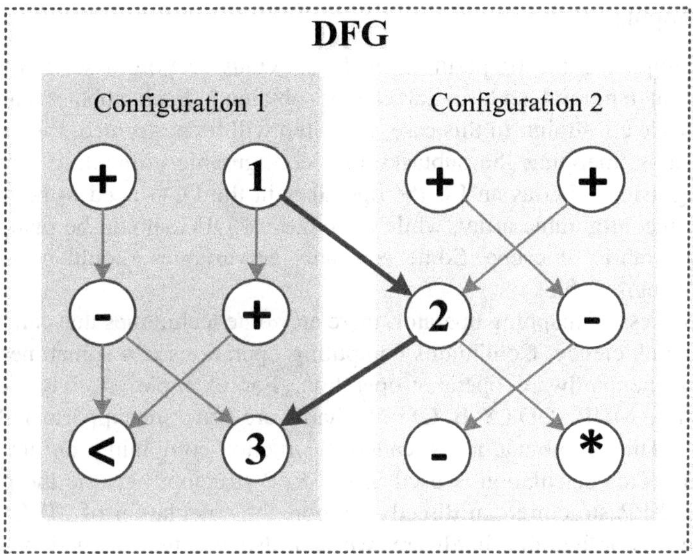

Fig. 4.12 Deadlock partition

2, resulting in the formation of interdependencies between operators 2 and 3, and Operator 1 will be control-dependent on Operator 2 if Configuration 2 is in front of Configuration 1, resulting in the formation of interdependencies between operators 1 and 2. Thus, it will result in a deadlock of two configurations, and we should pay special attention when performing temporal partition.

The effect of temporal partition largely affects the performance of the execution. The modulo scheduling method is a code optimization method based on temporal partition. Different modulo scheduling methods can form different results of temporal partition (software pipelining), and their performances will be greatly different. In general, the method of temporal partition is a multiobjective comprehensive optimization problem, and the execution time and communication should be emphasized. When performing temporal partition on target tasks, we recommend that the execution time should be shortened, which requires to minimize the number of configurations and maximize the utilization of resource. Meanwhile, it should minimize the communication between configurations, which can reduce the cost of storage access.

2. IR Mapping

After completing the IR partition in the previous section, a series of subtasks distributed in temporal and spatial can be obtained. Each subtask satisfies the hardware scale constraint. In this case, next step will be performed: the mapping of the IR, that is, mapping the subtasks to reconfigurable computing array. When considering using DFG as an IR, the operators in the DFG need to be mapped to PEs in the reconfigurable array, while the edges of DFG should be mapped to the interconnect fabric or cache. Some constants or variables should be mapped to registers or register files.

In the process of mapping operator, there are some techniques that can be used to improve the efficiency. Continuous computing operations can sometimes be combined into one hardware operator operation. For example, A × B + C can be combined into MULADD (A, B, C) if the hardware structure supports multiply-add operations. This combination is called instruction-combining during software compilation. This calculation is used in the XPP structure because the processing unit of the XPP structure can directly support the calculation of MULADD. For fine-grained reconfigurable hardware, we can also use the instruction-combining method if the combined instruction has better performance than the separate independent operations. Resource sharing technique can improve the hardware utilization of the mapping results. NAPA-C is a compiler that uses resource sharing techniques.

The register mapping stores the scalar into the RPU's registers and maps the mixed data types, such as arrays, to memory. Thus, the scalar does not need to be accessed from memory and can be used directly on the RPU. Generally, not all variables will need register, some variables can be achieved with the interconnection. As long as the number of variables in each temporal partitioned program exceeds the number of registers in the reconfigurable architecture, the register's assignment algorithm can be used to schedule.

4.3 Compilation Examples of a Reconfigurable Cryptographic Processor

The following part lists examples of implementing the symmetric cryptographic algorithm, the hash algorithm, and the public key cipher algorithm. The code transformation and optimization method, IR partition and mapping methods in the previous section, are all reflected in these implementation examples. In particular, the mapping examples in Sects. 4.3.1 and 4.3.2 are based on the Anole structure in Chap. 5. The public key cryptography mapping example in Sect. 4.3.3 is based on another reconfigurable computing architecture processor designed for general-purpose applications by our team.

4.3.1 Implementation Examples of Symmetric Cryptographic Algorithm

The following details the implementation of AES128 symmetric cryptographic algorithm. By code transformation and optimization, AES's MixColumns computation form will transform to

$$
\begin{aligned}
C00 &= M00 \oplus M11 \oplus B11 \oplus B22 \oplus B33\\
C01 &= M01 \oplus M12 \oplus B12 \oplus B23 \oplus B30\\
C02 &= M02 \oplus M13 \oplus B13 \oplus B20 \oplus B31\\
C03 &= M03 \oplus M10 \oplus B10 \oplus B21 \oplus B32\\
C10 &= B00 \oplus M11 \oplus M22 \oplus B22 \oplus B33\\
C11 &= B01 \oplus M12 \oplus M23 \oplus B23 \oplus B30\\
C12 &= B02 \oplus M13 \oplus M20 \oplus B20 \oplus B31\\
C13 &= B03 \oplus M10 \oplus M21 \oplus B21 \oplus B32\\
C20 &= B00 \oplus B11 \oplus M22 \oplus M33 \oplus B33\\
C21 &= B01 \oplus B12 \oplus M23 \oplus M30 \oplus B30\\
C22 &= B02 \oplus B13 \oplus M20 \oplus M31 \oplus B31\\
C23 &= B03 \oplus B10 \oplus M21 \oplus M32 \oplus B32\\
C30 &= M00 \oplus B00 \oplus M11 \oplus B22 \oplus M33\\
C31 &= M01 \oplus B11 \oplus M11 \oplus B23 \oplus M30\\
C32 &= M02 \oplus B01 \oplus B12 \oplus B20 \oplus M31\\
C33 &= M03 \oplus B02 \oplus B13 \oplus B21 \oplus M32
\end{aligned}
$$

where Bxx is TLU (P $\oplus$ WK) [31:24]; Mxx is TLU (P $\oplus$ WK) [23:16]. TLU is mainly used to realize the operation of SBox lookup table. P represents plaintext (or output of last round operation), WK represents the subkey.

In the optimized AES processing flow, the only difference among the round functions of the first round to the N − 1 round is that the plaintext input of the first round corresponds to the output of last round operation in the second round to the N − 1 round.

The design implementation of the round function in the reconfigurable array is shown in Fig. 4.13. Each round operation requires two rows for implementation: The first row of PE units can implement AddRoundKey, SubByte, and partial MixColumns operations, while the second row of PE unit can implement the remaining part of the MixColumns operation (XOR). Connection row (CR) is used to perform ShiftRows and byte-selection functions. The first row of the reconfigurable array will perform TLU operation on input plaintext P0 [127:0] and WK0 [127:0], each PE receives operand of two-word length (one-word plaintext P0, one-word subkey WK0, PE, the remaining four inputs can be ignored) and outputs four words. Take PE0 as an example, input P [127:96] and WK0 [127:96] and output {B00, M00, D00, E00}, {B01, M01, D01, E01}, M02, D02, E02}, {B03, M03, D03, E03} after passing PE(TLU) unit, where the finite field multiplication data stored by the Dxx and Exx arrays are used for decryption operations and are not required for encryption. CR0 receives the output data of four PEs. After selection and combining, it sends five words to each PE in the next row. Taking PE4 as an example, the output data bits are {M00, M01, M02, M03}, {M11, M12, M13, M10}, {B11, B12, B13, B10}, {B22, B23, B20, B21}, {B33, B30, B31, B32}.

The implementation of final round's round function on reconfigurable array design and is shown in Fig. 4.14.

In Fig. 4.14, the input data T9 [127:0] are the result of the previous round operation. WK9 [127:0] is the subkey. The nineteenth row mainly implements the AddRoundKey and SubByte operations. Take PE72 as an example, data T9 [127:96] and WK9 [127:96] are received, and after the TLU operation, output {B00, B01, B02, B03}. The rest is don't-care data. After CR18 receives these data, it outputs the data after being rearranged by the ShiftRows, and XX is don't-care data. The twentieth row receives the data output of CR18, and subkey WK10 [127:0]. After the key addition, the ciphertext is outputted.

Therefore, the AES algorithm with 128-bit key strength requires ten rounds to complete the encryption operation, and the entire array is designed to be 20 rows. All round operations can be expanded. After the pipeline operation, one block result can be obtained for each cycle. Therefore, the performance of the AES implementation algorithm is B = operating frequency × 128 bit = 300 MHz × 128 bit = 38 400 Mb/s; that is, the chip works at the frequency of 300 MHz, and the performance is about 38 Gb/s for the AES algorithm.

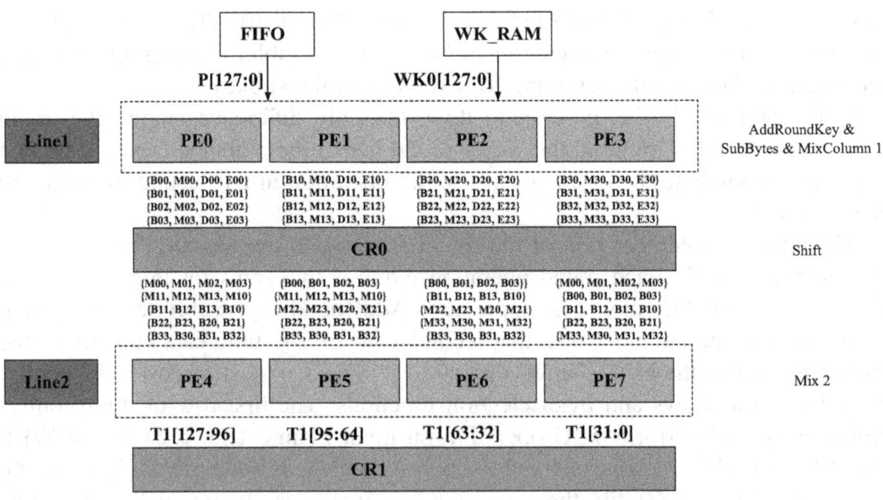

Fig. 4.13 AES round function implements first through N − 1th round

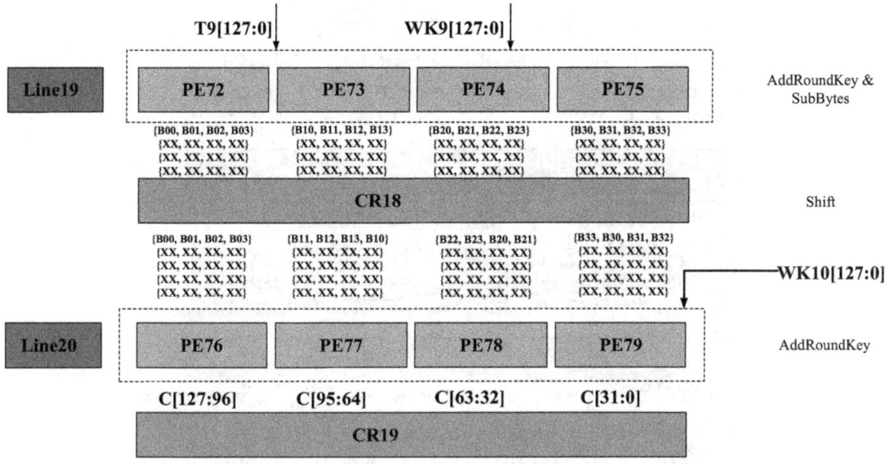

Fig. 4.14 AES round function implements the Nth round

4.3.2 Examples of Hash Algorithm Implementation

This section takes SHA256 (Fig. 4.15) as an example to illustrate the hash algorithm implementation. The SHA256 algorithm performs a series of operations on input bit stream of arbitrary length $L(L < 2^{64})$, resulting in a fixed 256-bit message digest. The basic idea is to divide the original message into N 512-bit data blocks. The hash initial value $H(0)$ will generate $H(1)$ when passing the first message block, and $H(2)$ is generated when $H(1)$ passes second message block, and finally generate H(N), whose eight 32-bit values are concatenated to form a 256-bit message digest.

The original message is divided into N 512-bit message blocks. Each message block is divided into 16 32-bit words, which are identified as M(i) 0, M(i)1, M(i)2, …, M(i)15, and these N message blocks are sequentially processed. It contains W(t) generation and the calculation of the hash value. The following will perform optimization and implementation on these two operations.

For t is 0–15, $W(t)$ is equal to $M(i)$ t, and from 16 to 63, $W(t)$ is:

$$
\begin{aligned}
W(t) &= \mathrm{SSIG1}(W(\mathrm{t}-2)) + W(t-7) + \mathrm{SSIG0}(W(t-15)) + W(t-16) \\
&= (\mathrm{ROTR} \wedge 17(W(t-2))) \oplus \mathrm{ROTR} \wedge 19(W(t-2)) \oplus \mathrm{SHR} \wedge 10(W(t-2)), \\
&\quad + W(t-7) + (\mathrm{ROTR}) \wedge 7(W(t-15)) \oplus \mathrm{ROTR} \wedge 18(\mathrm{W(t-15)}) \\
&\quad \oplus \mathrm{SHR} \wedge 3(W(t-15)), + W(t-16)
\end{aligned}
$$

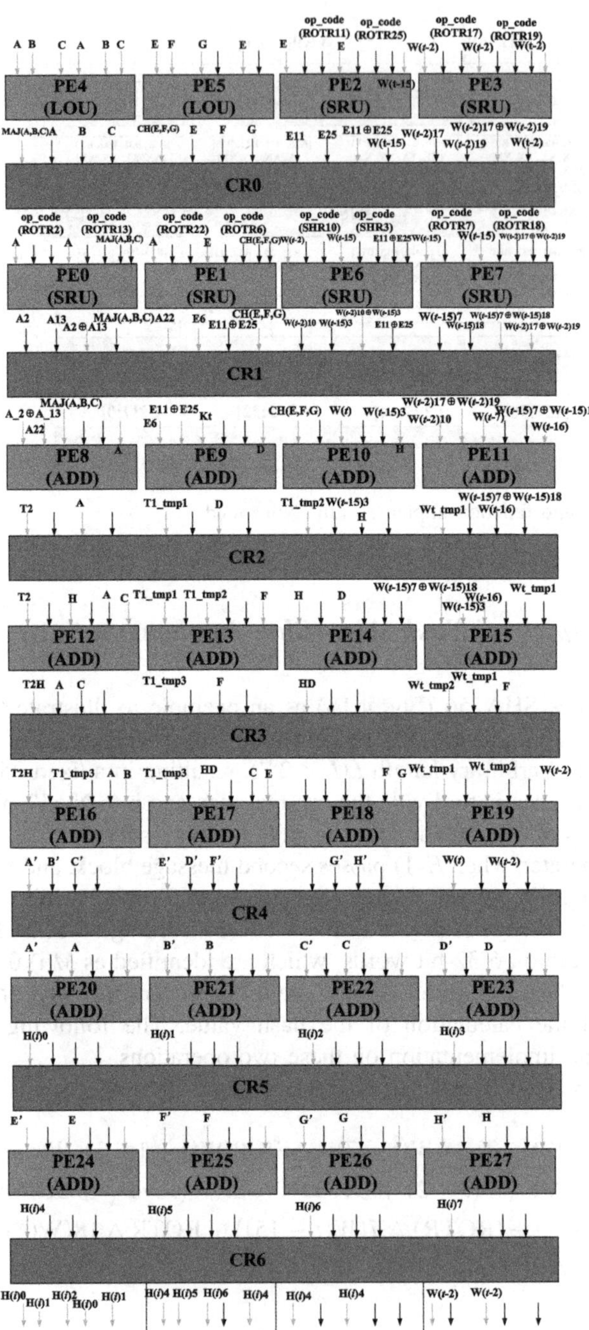

Fig. 4.15 SHA256 implementation plan

There are six shift operators in the *W*(*t*) expression. A single PE unit can perform two shift operations, so three PEs are needed to complete the shift operation. The first PE is set to complete the shift operation of ROTR∧17 (*W*(*t* − 2)) and ROTR∧19(*W*(*t* − 2)), and the following XOR operation, we can obtain the operand ROTR∧17(*W*(*t* − 2)) ⊕ ROTR∧19(*W*(*t* − 2)) (tentatively written as Wt_tmp1); the second PE performs the shifting operation of SHR∧10 (*W*(*t* − 2)) and SHR∧3 (*W* (*t* − 15)); while the third PE completes the shift operation of ROTR∧7 (*W* (*t* − 15)) and ROTR∧18 (*W* (*t* − 15))), and the following XOR operation, we can obtain the operand ROTR∧(*t* − 15)) ⊕ ROTR∧18 (*W* (*t* − 15)) (tentatively written as Wt_tmp2).

$$\begin{aligned}H(i)0 &= a' + H(i-1)0 \\ &= T1 + T2 + H(i-1)0 \\ &= (h + \mathrm{BSIG1}(e) + \mathrm{CH}(e,f,g) + K(t) + W(t)) + (\mathrm{BSIG0}(a) \\ &\quad + \mathrm{MAJ}(a,b,c)) + H(i-1)0 \\ &= (H(i-1)7 + \mathrm{BSIG1}(H(i-1)4) + \mathrm{CH}(H(i-1)4,\ H(i-1)5, \\ &\quad H(i-1)6) + K(t) + W(t)) \\ &\quad + (\mathrm{BSIG0}(H(i-1)0) + \mathrm{MAJ}(H(i-1)0),\ H(i-1)1,\ H(i-1)2)) \\ &\quad + H(i-1)0\end{aligned}$$

$$\begin{aligned}H(i)4 &= e' + H(i-1)4 = (d + T1) + H(i-1)4 \\ &= (d + (h + \mathrm{BSIG1}(e) + \mathrm{CH}(e,f,g) + K(t) + W(t))) + H(i-1)4 \\ &= (H(i-1)3 + (H(i-1)7 + \mathrm{BSIG1}(H(i-1)4) + \mathrm{CH}(H(i-1)4, \\ &\quad H(i-1)5,\ H(i-1)6) \\ &\quad + K(t) + W(t)) + H(i-1)4\end{aligned}$$

$$\begin{aligned}H(i)1 &= b' + H(i-1)1 \\ &= a + H(i-1)1 \\ &= H(i-1)0 + H(i-1)1\end{aligned}$$

$$\begin{aligned}H(i)2 &= c' + H(i-1)2 \\ &= b + H(i-1)2 \\ &= H(i-1)1 + H(i-1)2\end{aligned}$$

$$
\begin{aligned}
H(i)3 &= d' + H(i-1)3 \\
&= c + H(i-1)3 \\
&= H(i-1)2 + H(i-1)3 \\
H(i)5 &= f' + H(i-1)5 \\
&= E + H(i-1)5 \\
&= H(i-1)4 + H(i-1)5 \\
H(i)6 &= g' + H(i-1)6 \\
&= f + H(i-1)6 \\
&= H(i-1)5 + H(i-1)6 \\
H(i)7 &= h' + H(i-1)7 \\
&= g + H(i-1)7 \\
&= H(i-1)6 + H(i-1)7
\end{aligned}
$$

4.3.3 Examples of the Public-Key Cipher Algorithm Implementation

The speed of public key cipher algorithm depends on the efficiency of modular exponentiation and modular multiplication. Therefore, it is necessary to design the code transformation of Montgomery's modular multiplication algorithm when implementing public key cryptography. Meanwhile, it is necessary to optimize the design of modular exponentiation based on the implementation of modular multiplication algorithm, making them more suitable for mapping on reconfigurable processors.

1. Code Optimization of Large Number Modular Multiplication

The RSA public key cipher algorithm relies on the grate number operations. At present, the mainstream RSA algorithms are all based on the 1024-bit grate number operation. This is beyond the computational power of the current computer. The solution is to process the large numbers as an array, and the elements of the array are from every bit of each large number. If the machine can process m-bit numbers, then the grate number is expressed in the form of $k = 2^{\mathrm{m}}$, and zero is filled into the higher order bits, so that the binary large number is transformed into base-k, the value range for each bit is $0 \sim K$, then the number of bits multiplied by the block is limited to m bits, which can be processed by the processor. Any great integer operation can eventually be decomposed into operations between numbers Therefore, a 1024-bit big number becomes an unsigned integer array containing 1024/m (rounded-up) elements. Thus, the large number A can be conveniently expressed in mathematical expressions; that is, $A = \sum_{i=0}^{n} (A_i k^i)$, $0 \leq A_i \leq k$, where Ai is used to record the ith element of the A array. According to the implementation law of multiplication, we can see that it is very difficult for the algorithm implementation because each multiplication requires one judgment and multiple additions.

The most common implementation method of large number modular multiplication is the Montgomery algorithm. The basic idea is to transform the common modular multiplication in an integer ring to the modular multiplication in the Montgomery residual system and to simplify the modular n to a modular for any integer r. When r is chosen to be a power of 2, the modular operation is simplified to a simple shift operation, so as to achieve the purpose of rapid calculation of modular multiplication. The common modular multiplication in an integer ring can be transformed to a modular multiplication in the Montgomery residual system by the following definition.

Definition 1 Let us define n to be a modulus, $n > 0$, $r > 0$ and $(r, n) = 1$. If a is an integer and $0 \leq a \leq n$, then the remaining classes of n are $a' = a * r \bmod n$. The set $\{a' \,|\, 0 \leq a' \leq n, a' = a * r \bmod \mathrm{n}\}$ can be proved to be a complete residual system, called then residual system. Each element in this set is mapped to c $[0, n-1]$ element one to one.

Definition 2 Assuming that a, b, and c are integers in the residual system of n, the Montgomery modular multiplication (denoted by *) may be defined as $c' = a' * b' * r^{-1} \bmod n$, where r^{-1} is the inverse of n of r module.

The current widely used the Montgomery improved algorithm is IFIOS, an improved version of FIOS algorithm. The high- and low-order bits selections are mainly implemented by the right shift and AND operation. The algorithm is shown in Fig. 4.16, where w is the machine-processing word length. Its speed of execution has greater advantages, and algorithm flow is shown in Fig. 4.17. R, C represents carry, S represents bit summary, and T is used to store the intermediate results, where R0, R1 store carry. The data in the process of computation are highly dependent, and the current operation of each step will provide the computing data in the next step.

2. Mapping of Large Number Modular Multiplication Algorithm

Suppose both the loop boundaries i and j are fixed. If these three parameters should be generated at runtime, the generating process can be executed on the main controller. The algorithm will be expanded from the most inner loop, and transform into the DFG form based on the execution order of each statement Since C code is serial, the resulting DFG will be in a pattern that a sequential series of operators. Meanwhile, since there is no multiplication and addition operation in the current PE operator, it can only be split into the form of one multiplication and one addition. Initially, pending data A and B are stored in shared storage. Each multiplication operator reads A [i] and B [j], respectively, from the shared memory and then gives the result of the multiplication of two to the addition operator. The addition operator is responsible for accumulating the results of the multiplications.

Fig. 4.16 IFIOS algorithm

```
IFIOS algorithm
Input: a, b, n, n0;
Output: t=a*b;
//Computational process
for (i=0; i<=s−1; i++) {
   k=t[0]+a[0]*b[i];
   R0=k>>16; S=k&0x0000ffff;
   m=S*n0 % w;
   k=S+m*n[0];
   R1=k>>16; S=k&0x0000ffff;
   for (j=1; j<s−1; j++) {
      k=t[j]+a[j]*b[i]+R0;
         R0=k>>16; S=k&0x0000ffff;
      k=S+m*n[j]+R1;
         R1=k>>16; S=k&0x0000ffff;
      t[j−1]=S;
   }
   (C,S)=R0+R1;
   C=k>>16; S=k&0x0000ffff;
   t[s+1]=t[s+1]+C;
   (C,S)=t[s]+S;
   C=k>>16; S=k&0x0000ffff;
   t[s−1]=S;
   t[s]=t[S+1]+C;
   t[s+1]=0;
}
if (t≥n) t=t−n;
return t;
```

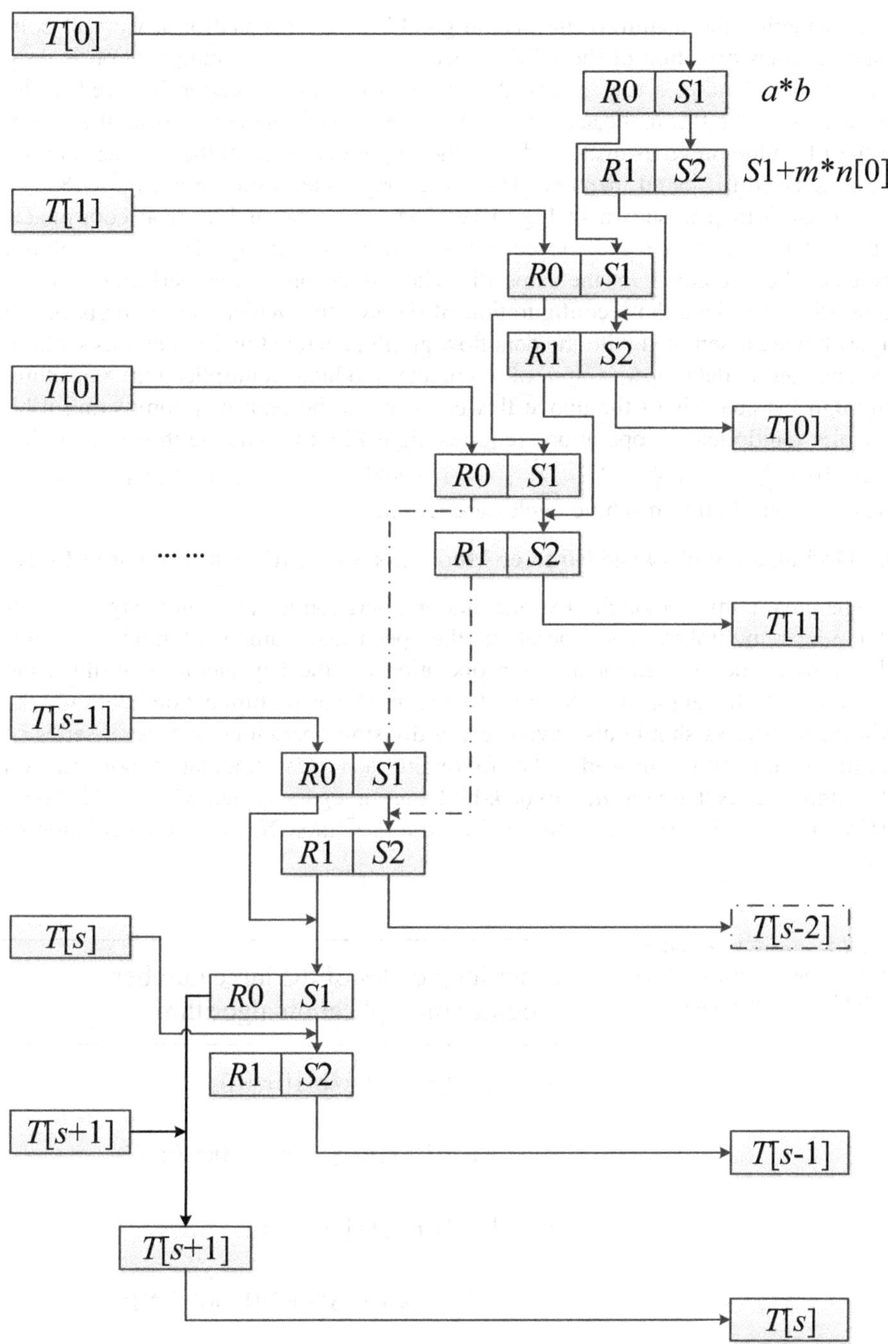

Fig. 4.17 Large number modular multiplication improved the Montgomery algorithm flowchart

Considering the feature of the reconfigurable processor platform, we will continue the transformation of the DFG: Since the result of the multiplication is only used by the PE after a few cycles, the result of the multiplication is saved in the register file of the PE; in the accumulation process, one input comes from the output of the PE in the previous cycle and the other input comes from the register file; the result is saved to shared memory. The inner loop code is shown in Fig. 4.18.

The DFG map is shown in Fig. 4.19. The horizontal and vertical coordinates represent the PE space and the execution time, respectively. The same column indicates the operations on the same PE. The similar operations performed by the same PE can reduce the reconfiguration of the system. When various parts of the algorithm are assembled with the data flow graph generated in the previous section, we can get a data flow graph of complete modular multiplication algorithm. Through the analysis of the above flowchart, it can be seen that completing n-bit modular multiplication operations requires eight PEs to complete the computation in $\frac{n}{16}\left(10+\frac{9n}{16}+5\right)+\frac{n}{16}+1=\frac{9n^2}{256}+n+1$ machine cycles, and when n = 1024, a total of about 38,000 machine cycles are required.

3. Optimization of Large Number Modular Exponentiation Algorithm Codes

The great number modular exponentiation operation is based on a large number of modular multiplications. Therefore, the speed and number of modular multiplications in modular exponentiation operation are the key factors that affect the operation of the algorithm. Similar to the modular multiplication method, the arithmetic process should also avoid using division operations and use a series of additions and shifts instead. The following analyzes modular exponentiation algorithm. Let us define *N*, *a*, *b* to be *k*-bit binary integers, typical *k*'s are 512, 1024, 2048, etc. Modular exponentiation refers to c = a^b mod N, i.e., the calculation of c value.

Fig. 4.18 Inner loop codes of the large number modular multiplication algorithm

Inner loop codes of the large number modular multiplication algorithm

```
k=t[j]+a[j]*b[i]+R0;
R0=k>>16; S=k&0x0000ffff;
k=S+m*n[j]+R1;
R1=k>>16; S=k&0x0000ffff;
t[j-1]=S;
```

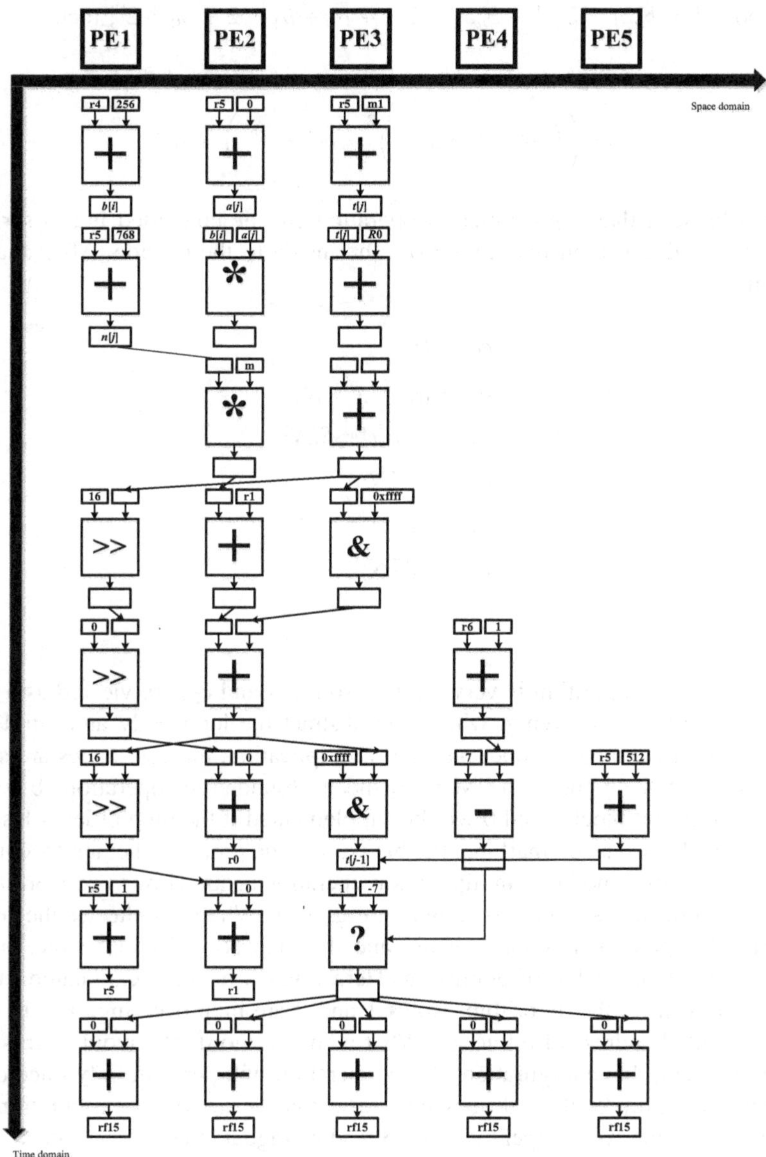

Fig. 4.19 DFG mapping implementation of modular multiplication

Suppose $b = b_{n-1} * 2^{n-1} + b_{n-2} * 2^{n-2} + \cdots + b_1 * 2 + b_0 * 2^0$, then

$$\begin{aligned} a^b &= a^{b_{n-1} * 2^{n-1}} * a^{b_{n-2} * 2^{n-2}} * \cdots * a^{b_1 * 2} * a^{b_0 * 2^0} \\ &= \left(\left(\left(a^{b_{n-1}}\right)^2 * a^{b_{n-2}}\right)^2 * \cdots * a^{b_1}\right) * a^{b_0 * 2^0} \end{aligned}$$

It can be seen that exponentiation operation can be converted into a series of square and multiplication iterations from the inside to the outside. After analysis, we define

$$\begin{aligned} c_0 &= 1; \\ c_1 &= \left(a^{b_{n-1}}\right)^2 \% N; \\ c_2 &= c_1 * a^{b_{n-2}} \% N; \\ c_3 &= c_2^2 \% N; \\ c_4 &= c_3 * a^{b_{n-3}} \% N; \\ c_5 &= c_4^2 \% N; \\ &\vdots \end{aligned}$$

Of course, this algorithm is very regular, so c_{2n-1} and c_{2n} are viewed as a set of iterative body when resolving, which can abstract the loop body and build algorithm based on the large number modular multiplication, the core codes are shown in Fig. 4.20. Among them, ┌ M/32 ┐ means rounded-up operation, b will be divided into 32-bit blocks, and 0 will be supplemented if the high order is less than 32 bits. By the value of marking bit bi, we can determine whether to continue iteration or perform modular multiplication operation again. How to convert it to an easy-to-implement description is shown in Fig. 4.21. Where g indicates the address of storing the processing data, flag bit, and the loop control count table, and [g] indicates the stored value of address g. Here gives a detailed explanation of the special treatment in the algorithm. Rows 1 and 2 are preprocessing. The first row stores the initial address of b into g3. What is more, in order to avoid confusion of data assignment, the configuration of this operation will perform only once during the entire algorithm cycle and does not participate in the loop configuration. The meaning of adding row 3 operations is to read storage address value whose output value can be used as indirect address by other PEs in the next cycle. The fourth row operation is modular operation of large number.

4. Mapping of the Large Number Modular Exponentiation Algorithm

The method of mapping the large number modular exponentiation on the reconfigurable cryptographic processor is shown in the following. According to the number of iterations of the outer loop, we will set the number of the top loops of PE, and the inner loop can be implemented through checking the conditions. In the

Core codes of the large number modular exponentiation algorithm

```
for (i=0; i<= ⌈M/32⌉; i++) {
    for (j=1; j<32; j++) {
            bi=b[i]≪1;
            c=c*c mod n;
            if (bi==0)
              continue;
            else{
              c=a*c mod n;
              continue;
                 }
}
```

Fig. 4.20 Core codes of the large number modular exponentiation algorithm

Optimization of core codes of the large number modular exponentiation algorithm

```
        g3=b's initial address;
    L0: tag=0; g5=[g3];
    L1: G1=t; g2=t; counter=31;
    L2: g1*g2=c;
        t=c;
        if (tag>0) {tag=0; goto L3;}
        bi=g5≫31; g5=g5≪1;
        if ( bi=0) goto L3;
        g1=a; g2=t; tag=1; goto L2;
    L3: counter=counter-1;
        if(counter≥0) goto L1;
        g3=g3+1;
```

Fig. 4.21 Optimization of core codes of the large number modular exponentiation algorithm

Fig. 4.22 DFG mapping implementation of the modular exponentiation algorithm

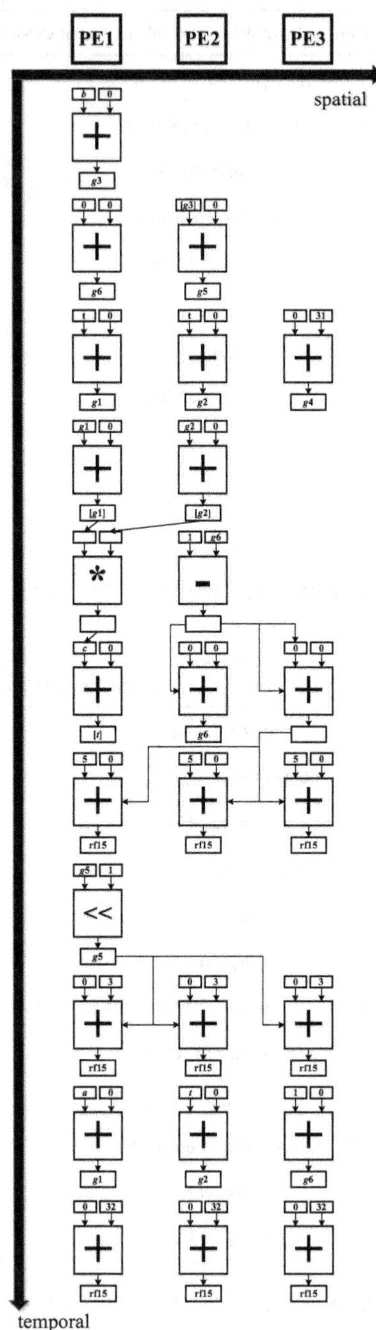

designed modular exponentiation algorithm, the innermost loop of the algorithm is fully expanded. The DFG map of inner loop of modular exponentiation algorithm from row 1 to row 12 is shown in Fig. 4.22. It can be seen from the analysis that nine PEs are required to complete n-bit modular exponentiation operation, and $\frac{9}{128}n^3 + \frac{17}{8}n^2 + \frac{177}{16}n$ machine cycles are needed in the worst case (i.e., all bits in the exponential index are 1); in average, $\frac{27}{512}n^3 + \frac{51}{32}n^2 + \frac{145}{16}n$ machine cycles are required to complete the operation. When n = 1024, 58,303,552 machine cycles are required on the average.

This chapter mainly introduces the compilation method of reconfigurable cryptographic processors. According to the features of the cryptographic algorithm, we focus on the key steps such as the code transformation and the IR mapping and introduce how to get the fine-grained-level, instruction (coarse-grained)-level, loop-level and data-level parallel methods through code transformation and get the task-level parallel method through the partition and mapping of IR, thus improving the compilation efficiency of reconfigurable cryptographic processors. At last, we give examples of the implementation of the cryptographic algorithm and use it an example of the application of these compilation methods. However, the compilers of reconfigurable processors still have many unsatisfactory aspects, hindering the widespread application and further development of reconfigurable computing technologies. For example, though the current reconfigurable computing compiler is based on high-level programming language, it still cannot avoid the help of manual assistance. Without manual assistance, the compiler will provide a poorer effect. This causes an obvious difference between the convenience of reconfigurable computing compiling process and that of software compiling. In addition, reconfigurable computing compilers are more domain-specific and perform worse in general-purpose computing because it relies on library files or artificial code optimization. Therefore, there are three development directions for reconfigurable cryptographic processor compiler:

① Fully automated compiling framework that can accelerate the software development of the reconfigurable cryptographic processor.
② Stronger field adaptability that efficiently supports different types of or new cryptographic algorithms.
③ Improve the programming flexibility and efficiency based on the standard high-level programming language.

References

1. Cardoso JMP, Diniz PC et al (2010) Compiling for reconfigurable computing: a survey. ACM Comput Surv 42(4):1–65
2. Li Z, Liu L (2017) Aggressive pipelining of irregular applications on reconfigurable hardware. In: International symposium on computer architecture

3. Mahlke SA, Lin DC, Chen WY et al (1993) Effective compiler support for predicated execution using the hyperblock. In: International symposium on microarchitecture, pp 45–54
4. Girkar M, Polychronopoulos CD (1992) Automatic extraction of functional parallelism from ordinary programs. IEEE Trans Parallel Distrib Syst 3(2):166–178
5. Galloway D (1995) The transmogrifier C hardware description language and compiler for FPGAs. In: IEEE symposium on FPGAs for custom computing machines, p 136
6. Agarwal L, Wazlowski M, Ghosh S (1994) An asynchronous approach to efficient execution of programs on adaptive architectures utilizing FPGAs. In: IEEE workshop on FPGAs for custom computing machines. IEEE, pp 101–110
7. Weinhaudt M, Luk W (2002) Memory access optimisation for reconfigurable systems. IEE Proc Comput Digit Tech 148(3):105–112
8. Babb J, Rinard M, Moritz CA et al (1999) Parallelizing applications into silicon. In: IEEE symposium on field-programmable custom computing machines, pp 70–80
9. Bondalapati K, Prasanna VK (1999) Dynamic precision management for loop computations on reconfigurable architectures. In: IEEE symposium on field-programmable custom computing machines, pp 249–258
10. Gokhale M, Stone JM, Arnold JG et al (2000) Stream-oriented FPGA computing in the streams-C high level language. In: IEEE symposium on field-programmable custom computing machines, pp 49–56
11. Yin C, Yin S, Liu L et al (2009) Compiler framework for reconfigurable computing system. In: International conference on communications, circuits and systems, pp 991–995
12. Smith AL (2009) Explicit data graph compilation. The University of Texas at Austin doctoral dissertation, Austin
13. Budiu M, Goldstein SC (1999) Fast compilation for pipelined reconfigurable fabrics. In: ACM/SIGDA international symposium on field programmable gate arrays, pp 195–205
14. Mei B, Vernalde S, Verkest D et al (2003) ADRES: an architecture with tightly coupled VLIW processor and coarse-grained reconfigurable matrix. In: International conference on field-programmable logic and applications, pp 61–70
15. Baumgarte V, Ehlers G, May F et al (2003) PACT XPP-a self-reconfigurable data processing architecture. J Supercomput 26(2):167–184
16. Benson J, Cofell R, Frericks C et al (2012) Design, integration and implementation of the DySER hardware accelerator into OpenSPARC. In: IEEE international symposium on high-performance computer architecture, pp 1–12
17. Lattner C, Adve VS (2004) LLVM: a compilation framework for lifelong program analysis and transformation. In: International symposium on code generation and optimization, pp 75–86
18. Ye ZA, Shenoy N, Baneijee P (2000) A C compiler for a processor with a reconfigurable functional unit. In: ACM/SIGDA international symposium on field programmable gate arrays, pp 95–100
19. Callahan TJ, Hauser JR, Wawrzynek J (2000) The Garp architecture and C compiler. Computer 33(4):62–69
20. Gokhale M, Gomersall D (1997) High level compilation for fine grained FPGAs. In: IEEE symposium on field-programmable custom computing machines, pp 165–173
21. Micheli GD (1994) Synthesis and optimization of digital circuits. McGraw-Hill, New York
22. Budiu M, Sakr M, Walker K et al (2000) Bitvalue inference: detecting and exploiting narrow bitwidth computations. In: International Euro-Par conference on parallel processing, pp 969–979
23. Muchnick SS (1997) Advanced compiler design and implementation. Morgan Kaufmann, San Francisco
24. Dongarra JJ, Hinds AR (2010) Unrolling loops in fortran. Softw Pract Exp 9(3):219–226
25. Hartenstein RW, Kress R (1995) A datapath synthesis system for the reconfigurable datapath architecture. In: Asia and South Pacific design automation conference, pp 479–484
26. Lam MS (1988) Software pipelining: an effective scheduling technique for VLIW machines. ACM Sigplan Not 23(7):318–328

27. Hamzeh M, Shrivastava A, Vrudhula S (2012) EPIMap: using epimorphism to map applications on CGRAs. In: Design automation conference, pp 1280–1287
28. Park H, Fan K, Mahlke S et al (2008) Edge-centric modulo scheduling for coarse-grained reconfigurable architectures. In: International conference on parallel architectures and compilation techniques, pp 166–176
29. Park H, Fan K, Kudlur M et al (2006) Modulo graph embedding: mapping applications onto coarse-grained reconfigurable architectures. In: International conference on compilers, architecture and synthesis for embedded systems, pp 136–146
30. Hamzeh M, Shrivastava A, Vrudhula S (2013) REGIMap: register-aware application mapping on coarse-grained reconfigurable architectures. In: Design automation conference, p 18
31. Bastoul C (2004) Code generation in the polyhedral model is easier than you think. In: International conference on parallel architecture and compilation techniques, pp 7–16
32. Bondhugula U, Hartono A, Ramanujam J et al (2008) A practical automatic polyhedral parallelizer and locality optimizer. ACM SIGPLAN Not 43(6):101–113
33. Cohen A, Sigler M, Girbal S et al (2005) Facilitating the search for compositions of program transformations. In: International conference on supercomputing, pp 151–160
34. Hannig F, Dutta H, Teich J (2004) Mapping of regular nested loop programs to coarse-grained reconfigurable arrays-constraints and methodology. In: International parallel and distributed processing symposium, p 148
35. Liu D, Yin S, Liu L et al (2013) Polyhedral model based mapping optimization of loop nests for CGRAs. In: Design automation conference, pp 1–8
36. Liu D, Yin S, Peng Y et al (2015) Optimizing spatial mapping of nested loop for coarse-grained reconfigurable architectures. IEEE Trans Very Large Scale Integr Syst 23(11): 2581–2594
37. Peterson JB, O'Connor RB, Athanas PM (1996) Scheduling and partitioning ANSI-C programs onto multi-FPGA CCM architectures. In: IEEE symposium on FPGAs for custom computing machines, pp 178–187
38. Kum K, Kang J, Sung W (2000) Autoscaler for C: an optimizing floating-point to integer C program converter for fixed-point digital signal processors. IEEE Trans Circuits Syst II: Analog Digit Signal Process 47(9):840–848
39. Ong SW, Kerkiz N, Srijanto B et al (2001) Automatic mapping of multiple applications to multiple adaptive computing systems. In: IEEE symposium on field-programmable custom computing machines, pp 10–20

Chapter 5
Examples of Reconfigurable Cryptographic Processor Design

This chapter describes a reconfigurable cryptographic processor designed by the reconfigurable computing research team at the Institute of Microelectronics, Tsinghua University, and the processor is named Anole. Anole is designed for various symmetric cryptographic algorithms and hash algorithms, and its core structure includes a dynamically and partially reconfigurable processing array and the interconnection between processing elements for the function enhancement. The design optimization goal is to improve the energy and area efficiencies while maintaining flexibility. Three key technologies have been proposed including distributed control network (DCN), concurrent computation and reconfiguration (CCR), and configuration compression and organization (CCO). The basic architecture, key technologies, integrated development tools and chip implementation results of Anole are presented in detail as follows:

5.1 Basic Architecture of the Processor Anole

The Anole uses the generic architecture of reconfigurable computing processor described in Sect. 1.4, as shown in Fig. 5.1. It mainly consists of the reconfigurable computing datapath (RCD) and the reconfigurable computing controller (RCC).

5.1.1 Reconfigurable Computing Datapath

The reconfigurable computing datapath includes the processing element array (PEA), input/output FIFO, register channel (RCH), and general purpose register files (GPRF). Among them, the PEA is the main data processing module, and the RCH is mainly used for data exchange in the single configuration of PEA, while

L. Liu et al., *Reconfigurable Cryptographic Processor*,
https://doi.org/10.1007/978-981-10-8899-5_5

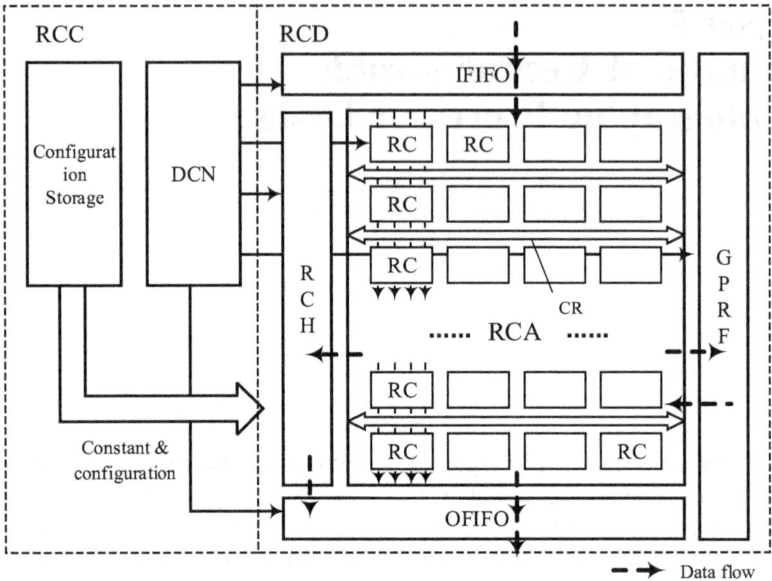

Fig. 5.1 Block diagram of Anole's architecture

GPRF is mainly used for data exchange between the different configurations of PEA.

1. Structural Design of PEA

The PEA consists of a PE array with 4 columns and 32 rows, and the interconnection between PEs. Designed for supporting symmetric cryptographic algorithms, the structural design of PEA can be optimized to some extent and functionally tailored based on the features of symmetric cryptographic algorithms. Table 5.1 lists some of the symmetric cryptographic algorithms and security protocols used for reference in the design process of Anole. It includes block ciphers such as AES, DES, IDEA, Serpent, and Twofish, and stream ciphers such as ZUC and RC4. Table 5.2 summarizes the features of these algorithms, including key length, bit width of operations, granularity and iterations of round function. It should be noted that many hash algorithms use the same operations as those of block ciphers and the same feedback mechanisms as those of stream ciphers, so they are also supported on Anole. The features of a hash algorithm SHA256 is also included in Table 5.2. Based on the result of algorithm analysis, the structure of PEA is designed from the following aspects.

(1) Granularity design of PE's data processing

In order to keep consistent with the granularity of most symmetric cryptographic algorithms, 32 bits are chosen as the datapath width of the PE for data processing.

Table 5.1 Examples of studied algorithm

Algorithm	Brief description
AES	IPSec, SSL, U.S. FIPS PUB 197 by NIST
DES	IPSec, SSL, U.S. FIPS PUB 46 by NIST
IDEA	PSec, SSL, DES alternative
Serpent	AES candidate algorithm
Twofish	OpenPGP, AES candidate algorithm
ZUC	China 4G mobile communication standard
RC4	SSL, TLS

(2) Functional design of the PE

According to the statistical result of operator functions of different algorithms, PE includes five different functional units such as logic unit (LU), arithmetic unit (AU), shift unit (SU), table unit (TU), and permute unit (PU), as shown in Fig. 5.2. Meanwhile, in order to speed up the calculation process, XOR units are added to the inputs and outputs of each functional unit to combine with other operators. This is mainly because XOR operation is a commonly used operator in cryptographic algorithms. In addition, the PE also supports data buffer and bypass. The LU consists of six logic basic cells (LBCs), each of which include AND, OR, NOT, and other logic operations for two inputs. Each LU can be configured to implement more complex logic functions that are composed of these basic logic operations. The AU consists of a 32-bit adder and a 16-bit multiplier. As modulo operations are often accompanied by arithmetic operation, hardware optimization is made in AU for modulo operations, so that those non-2^N modulo operations can be converted into multiplication and addition operations. The SU consists of two 32-bit basic shift units (BSUs). The two BSUs can be combined to implement up to 64-bit rotating shifts and logical shifts. This design is based primarily on shift operations that have a fixed number of bits of shift (typically up to 32 bits) and are determined by algorithms or subkeys. The PU is mainly designed for the permutation operation, including a 64-bit BENES network. The network can provide provable non-blocking transmissions from input to output. The TU includes four 8×8 S-Box units. Multiple 8×8 S-Box units can form a larger S-Box so as to meet the computational demands of widths of output bit greater than 8 bits. Anole's S-Box is implemented by a dedicated memory in hardware, where the S-Box input is the address of the memory and the output is the corresponding data in the memory. The maximum capacity of each memory supports the S-Box with 8-bit input and 32-bit output, and the S-Box's parameters can be switched by configuration contexts. For example, the S-Box in AES is 8-bit input and 8-bit output, and PRESENT uses 4-bit input and 4-bit output. In order to switch algorithms, the S-Box's configuration is changed from the memory with 8-bit address and 8-bit data to 4-bit address and 4-bit data.

Table 5.2 Analysis of algorithm feature

Algorithm			Key length/bit	Operator data width						Granularity/bit	Number of rounds
				S-Box	Permutation	XOR	Shift	Addition and subtraction	Multiplication		
Symmetric cipher	Block cipher	AES	128+	8–8		32/128	R	8	8	8	10–14
		DES	64	6–4	32/48	32/48	R			4	16
		IDEA	128			16	R	16	16	16	8
		Serpent	128 + *	4–4	128	32/128	L/R			4	31
		Twofish	128 + *	4–4		8/32/128	L	8/32	8	4	16
	Stream cipher	ZUC	128	8–8		32	L	32		32	
		RC4	128					8		8	
Other algorithm	Hash function	SHA256	256			32	R			32	64

*128 + stands for 128,192, and 256 bits

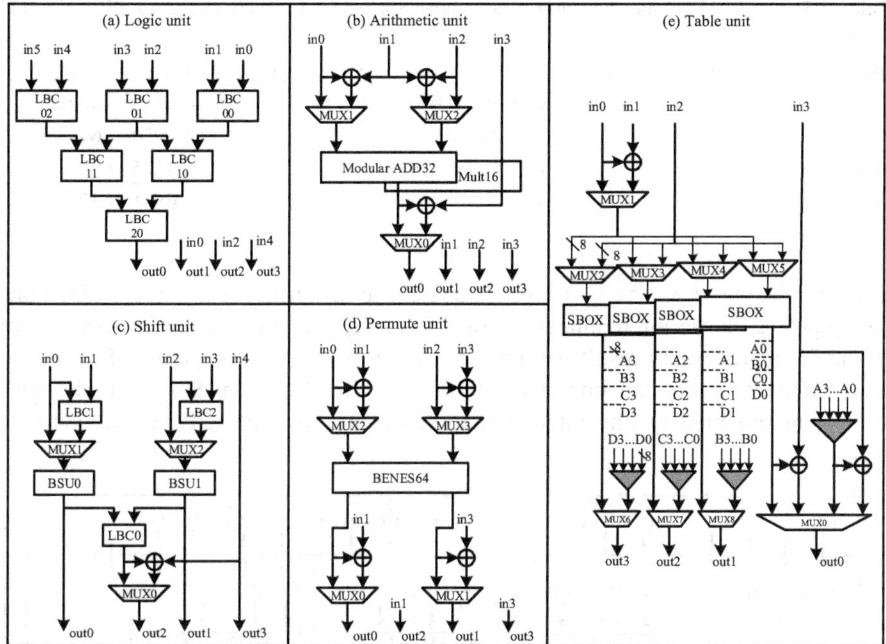

Fig. 5.2 Design block diagram of PE's five functional units

As mentioned above, the functional design of the PE covers most of the functions of the symmetric cryptographic algorithm, but if each PE contains all five kinds of functional units, it will cost enormous area and power even though flexibility is significantly improved. Therefore, PEs are designed heterogeneously based on operator usage frequency and hardware complexity, with each PE containing different functional units. First, each PE contains those frequently used but small-area units including LU, AU, and SU. The multiplier in AU is of high area cost. In order to reduce their area, only a quarter of the PEs contains the arithmetic unit with a multiplier, and the arithmetic units of other PEs have no multipliers. The table unit and the permute unit appear alternately in each PE row as they are less frequently used and have a larger area. The structure of the heterogeneous PE used in Anole is summarized in Table 5.3, and the areas of the heterogeneous PEs are compared. The total area of the heterogeneous PEA is equivalent to 2.29 million logic gates. If the homogeneous design is used, the PEA area is equivalent to 4.0 million logic gates. Therefore, the heterogeneous design can significantly improve the area efficiency.

(3) PEA interconnection design

The interconnected structure of PEA is implemented in the form of the interconnected row between PE rows. According to the structure of different cryptographic algorithms, PEA's interconnected row is designed as a three-layer structure,

Table 5.3 Structural composition and area comparison of Anole's heterogeneous PE (area unit: gates)

RC	AU (846)	AU (multiplier) (2760)	LU (1600)	SU (2670)	PU (1745)	TU (22854)	Total area	Total number
Type1	✓		✓	✓		✓	27,970	64
Type2		✓	✓	✓	✓		8775	32
Type3	✓		✓	✓	✓		6861	32

as shown in Fig. 5.3. The first layer is a one-way mesh network so as to facilitate data pipeline. In the interconnected row, PEA's input can be loaded on any row of PE, and the processing results of any row of PE can also be outputted. The second layer interconnection can implement byte-level shift function. Some simple operations implemented by the interconnection can improve the efficiency. The data

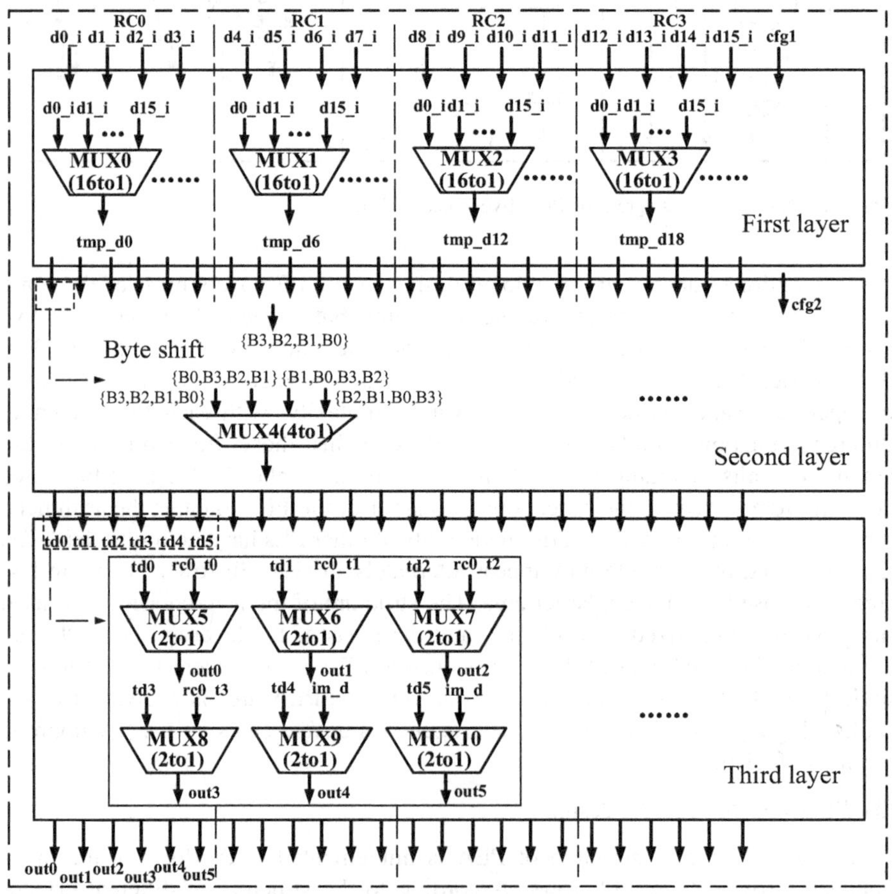

Fig. 5.3 Anole's interconnected row structure

sources for the first two layers of interconnection are from PEs on the previous row. The function of third layer interconnection is to take operands from other cache structures, such as input FIFO and GPRF.

2. Design of Data Cache Structure

Anole's cache structure includes input/output FIFO, RCH, and GPRF. The function of input/output FIFO is to connect the PEA with the external interface, read the data that PEA needs to process from the external interface, and output the results processed by the PEA to the external interface. The input/output FIFO uses 8 128-bit × 256-bit dual-port memories. The function of the RCH is to buffer the data in a single PEA configuration. Its purpose is to rearrange and resort the PEA's calculation results and output them to the output FIFO. The RCH uses 32-bit × 16-bit memories. The function of GPRF is to buffer the PEA data between multiple configurations. It uses 128-bit × 256-bit memories, including 16 reading ports and 16 writing ports.

5.1.2 Design of the Reconfigurable Computing Controller

The reconfigurable computing controller (RCC) can be mainly divided into the configuration module (configuration memory and configuration control module) and the distributed control network (DCN), as shown in Fig. 5.1. The configuration control module mainly initializes the GPRF and S-Box of the PEA, as well as the configuration information memory. During the execution process of Anole, he configuration control module receives and parses the external configuration contexts and configures the DCN and PEA. The DCN controls the PEA data flow, mainly including the read/write control of the I/O FIFO, RCH, and GPRF. It controls the enable signal of the PE in case of congestion, and it controls the configuration context switching when the PEA function needs to be reconfigured.

5.2 Key Technologies of Anole Processors

In the system design, three key technologies are proposed to improve the energy and area efficiencies of Anole.

① The distributed chain network (DCN) introduces multi-thread execution features into reconfigurable structures, which means that each PE can independently perform different tasks according to the enable signals of DCN and perform data-driven computation based on the validity of the input data. This approach can improve the hardware utilization rate and the performance of the array.

② The concurrent computation and reconfiguration (CCR) is used to implement the pipelined computation and configuration process in parallel. This approach can reduce the performance cost of dynamic configuration switching.
③ The configuration compression and organization (CCO) can reduce the size of the configuration contexts and shorten access time by reusing the common configuration contexts in the temporal and spatial domain and adopting a design with a multi-level context cache.

5.2.1 DCN

DCN is the key module of the reconfigurable controller. It controls the PEA's data flow so as to maximize the utilization rate of computing resources in the execution process. As shown in Fig. 5.4, every PE is enabled by a token register. The token registers in the same column can form a token chain which can be used to control the PEs within a block. The DCN contains 16 token chains in total; each token chain can operate independently. One of the important roles of the token chains is to implement branch control. For each token chain, the input FIFO generates an initial token, passes it to the token register corresponding to the PE in the first row, and then successively passes it on to those at lower rows. The tokens determine the lifetime of the PE data processing. A PE does not operate until it receives an enable signal from the corresponding token. As such, the token flow in the token chain actually represents the data flow. Anole's token chain design has the following features:

① The PE token output from the upper row is used by the PE in the lower row.
② The PE in a specific row can use the output token from the PE in the lower row.

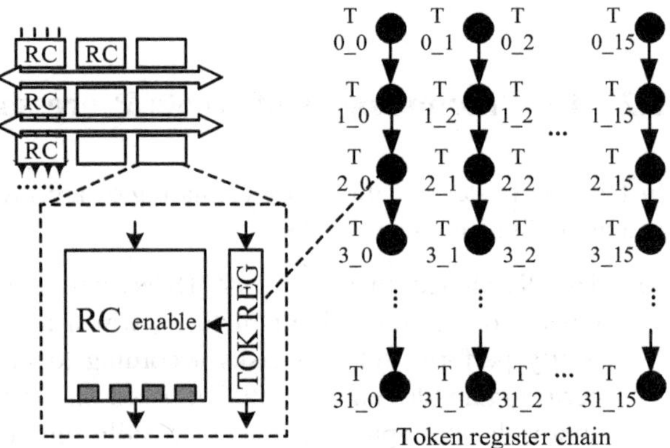

Fig. 5.4 Schematic diagram of DCN

③ The output token of a PE can be used by multiple PEs.
④ The multiple PE output tokens can be used by a PE.

As the length of each token chain is not always the same, the output data of these chains are likely to be out of order when entering output FIFO. Therefore, RCH is specifically designed to adjust the order of the output data.

With the DCN token system, the execution control of the different data flows such as sequence, branch, and loop become feasible.

① The sequential execution is the most common execution mode, and the token flow mapping is relatively simple. If the input FIFO is not empty and the token chain is not locked, the input FIFO sends a block of data and token into the first row of the PE. After the data are processed by the PE in this row, the token and the data will be sent to the PE on the next row. In the meanwhile, next block of data and token will be sent from the upper row. In this way, data and tokens can be pipelined in the PEA.

② An example of the branch execution using c the token chain is shown in Fig. 5.5. When there is a branch in an algorithm, for example there are two cases for output of a PE (numbered as 1_0) (branches shown in a solid line and a dotted line), it is required to determine the operation result of PE (numbered as 1_0) first. If it is the first case, the token is sent to the PE (numbered as 2_0). When the PE (numbered as 2_0) is enabled, the operation is performed and the branches in solid line are executed. For the second case, the token is sent to the

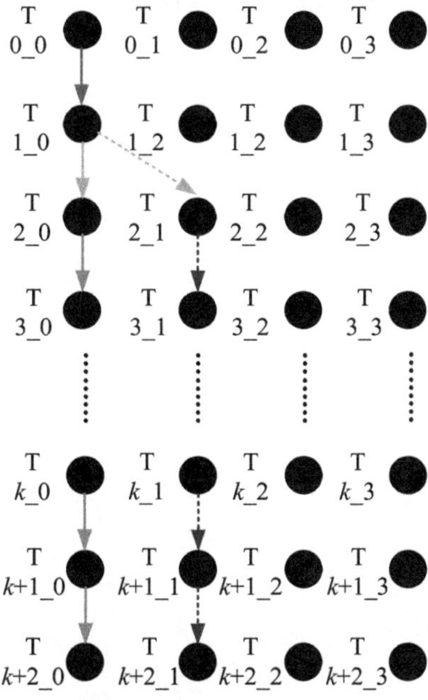

Fig. 5.5 Token chains in branch execution

PE (numbered as 2_1) and the PE (numbered as 2_1) is enabled. The operation is performed, and the branches in dotted line are executed. In both cases, the other branch can only receive an invalid token, and the PE will stay idle.

③ An example of the loop control using the token chain is shown in Fig. 5.6. If there is a loop in an algorithm, for example multiple executions are required for some of the PEs (row number from 2 to k), the token cannot be sent to the PE at the $(k + 1)$th row but to the PE in the second row when the operation of the PE on the kth row is completed. When the condition of exiting the loop is satisfied, the token is then sent to the PE in the $(k + 1)$th row when the operation of the PE in the kth row is completed.

As described above, DCN actually introduces the multi-threading mechanism into the reconfigurable cryptographic processor. Figure 5.7 shows an example of a simple multi-threading mapping. Three token chains have been implemented, each of which corresponds to an independent thread with separate timing configuration in the algorithm. Three groups of PEs in the PEA, corresponding to each thread, operate simultaneously. The input FIFO sends data and the corresponding token to three threads in series. Because Threads 1 and 3 both contain loops, Thread 2 may complete the processing, and then the output result is sent to RCH first. However, the RCH reorganizes the output result of each thread into the output FIFO in the same order as the input FIFO. The multi-threading is implemented on Anole through the DCN design, eliminating the need of designing complex state machine

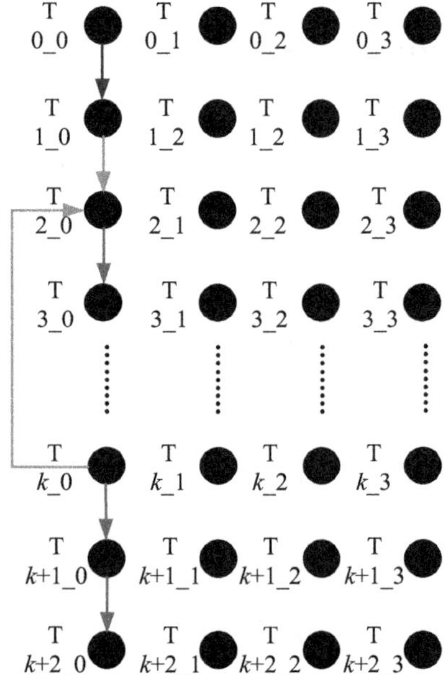

Fig. 5.6 Token chains in loop execution

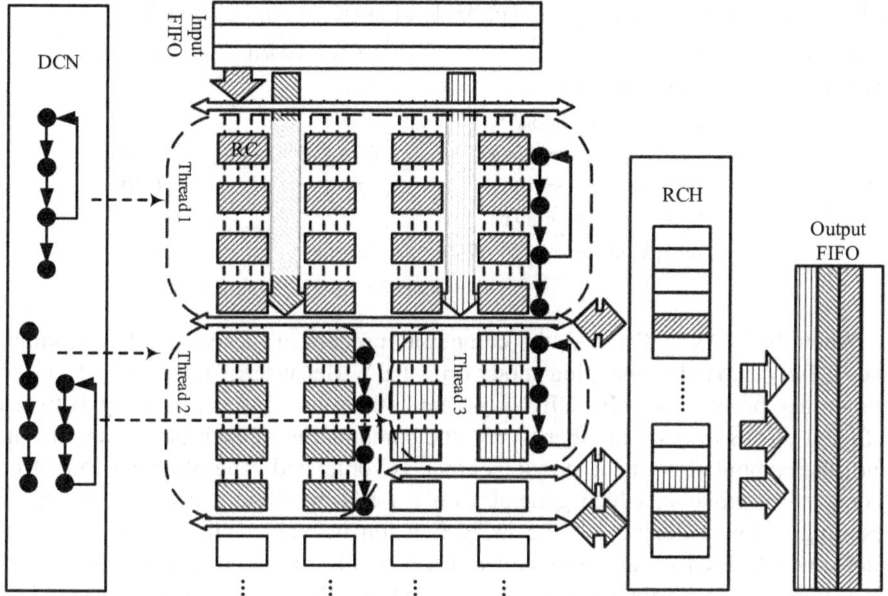

Fig. 5.7 Anole's multithreading application example

controls for each PE. The major advantage of the multi-threading mechanism is the increased utilization of hardware resources, improving parallelism and performance. For previous multithreading technology implemented on reconfigurable computing architecture and ISAP [1–4], multiple subtasks are executed on a single datapath in a time division multiplexing manner. However, for the DCN-based multi-threading, multiple subtasks are executed simultaneously on multiple datapaths by using the space computing advantage of the reconfigurable computing architecture.

To quantitatively demonstrate the effectiveness of the DCN, the energy efficiency of Anole when the DCN is enabled and disabled is compared. The results are shown in Table 5.4. The major advantage of the DCN is to fully utilize the computing resources of the PEA, because the number of idle PEs can be reduced if the algorithm is properly divided into subblocks and they are processed in parallel. The DCN is especially effective for those algorithms that cannot be fully mapped on the PEA or those with low degree of parallelization or data dependency. The MD5 contains four rounds of complicated round functions, which are the typical large-scale algorithm. The AES-CBC is a typical algorithm with feedback. Compared to that when DCN is disabled, the energy efficiency of AES-CBC and MD5 is increased by 8.55 times and 4.57 times, respectively, when DCN is enabled.

Table 5.4 Energy efficiency comparison of Anole when DCN is enabled/disabled

Algorithm	Metric	DCN-disabled	DCN-enabled
AES (CBC)	Performance/(Gb/s)	2.56	40.9
	Power consumption/W	0.38	0.71
	Energy efficiency/(Gb/s/W)	6.74	57.60
MD5	Performance/(Gb/s)	0.64	6.40
	Power consumption/W	0.31	0.68
	Energy efficiency/(Gb/s/W)	2.06	9.41

When the DCN is disabled, the computing process of the AES-CBC algorithm uses a fully expanded mapping mode on a PEA. Because of the feedback in the computing process of AES-CBC algorithm, there is data dependency between different blocks of data. In other words, the computation of each block cannot start until the computation of the previous block is completed. The plaintext cannot be processed in pipeline, which generates only a 128-bit ciphertext every 20 cycles, resulting in low efficiency. If the DCN function is enabled, the thread-level parallelism of AES-CBC can be exploited. Each smallest loop body of the AES-CBC algorithm (i.e., round function) can be mapped to 8 PEs as one thread. These 8 PEs can process a set of data packets completely and independently. This practice is very effective for practical applications, such as network communications, because these applications will process a large number of independent data packets. In this way, the 32-bit × 4-bit PEA can simultaneously support the parallel processing of up to 16 threads, and the data-level parallelism and the thread-level parallelism can be used to improve the performance by 16 times. In summary, the process of Anole processing the AES-CBC can be summarized as follows:

① The keys for 16 data packets are generated in parallel, and then the results are saved in the corresponding registers of the threads.
② The PEA is configured to process 16 separate AES-CBC threads, which will process 16 data packets in a parallel (using the corresponding subkeys).

Figure 5.8 compares the performance of the Anole using DCN with that of the FPGA and other reconfigurable computing architecture by taking AES's two operating modes (AES-CTR without feedback mode and AES-CBC with feedback mode) as an example. FPGA and other reconfigurable computing architectures are better suited for AES-CTR because the pipeline techniques can improve throughput performance in non-feedback modes. However, the pipeline could increase processing latency and decrease performance in feedback modes. DCN allows reconfigurable architectures to improve the performance of the AES with feedback by multiple processing threads in a paralling manner while maintaining the high performance in non-feedback mode. As shown in Fig. 5.8, the performance of Anole in the CTR mode is similar to that of other architecture, but the performance in the CBC mode is 7.99 times of FPGA and 5.39 times of other reconfigurable architectures.

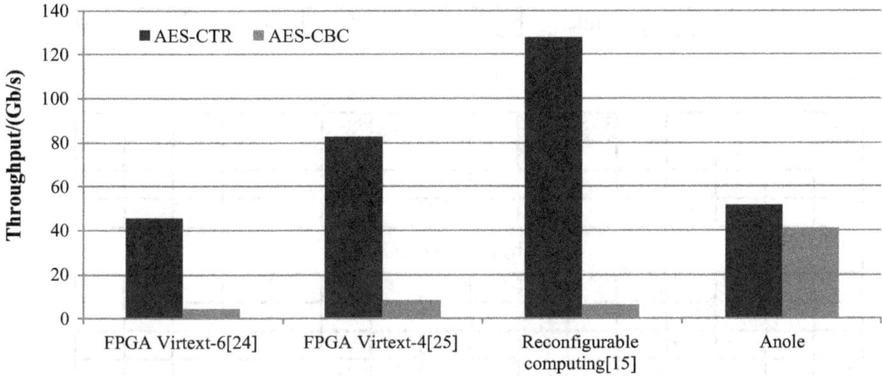

Fig. 5.8 Comparison of the performance of different architectures when the AES algorithm is processed in two modes

5.2.2 Concurrent Computation and Reconfiguration (CCR)

The data-level parallelism in symmetric cryptographic algorithms is very common, so the pipeline design can effectively improve the performance. However, there is a particular problem for the pipeline in reconfigurable cryptographic processors that the switching between different configuration contexts interrupts the pipeline and then affects the pipeline efficiency. The effects of this problem on the tasks with longer pipeline are especially obvious. In general, the configuration and the computing process of the reconfigurable computing architecture are executed exclusively and alternately. Therefore, all the PEs are idle during the configuration switching, as shown in Fig. 5.9a. This results in a large time interval between the operation and the reconfiguration of each PE, thus wasting the hardware resources. Anole uses concurrent computation and reconfiguration (CCR) to parallelly execute the computation and reconfiguration processes of the arrays, reducing the idle time of each PE and improving the executing efficiency of pipeline. Figure 5.10 shows the specific implementation mode of CCR. To reduce the negative effect of the reconfiguration on the performance, the configuration contexts are also treated as special data. The configuration contexts are divided based on the row of PEs, and it is read and distributed on a row-by-row basis, which relieves the bandwidth pressure on the configuration memory. The configuration contexts that are sorted by rows are then distributed to configuration registers. This process is synchronized with data update in the data register, thus forming a parallel synchronous pipeline of data and configuration. Moreover, the switching speed between different threads has also been improved. In short, with the CCR, each PE does not need to wait for the completion of reconfiguration of the whole PEA and can quickly switch to the next task. The PEA can execute two task threads simultaneously, as shown in the fourth cycle of Fig. 5.9b; Task 1 and Task 2 are processed simultaneously in the PEA.

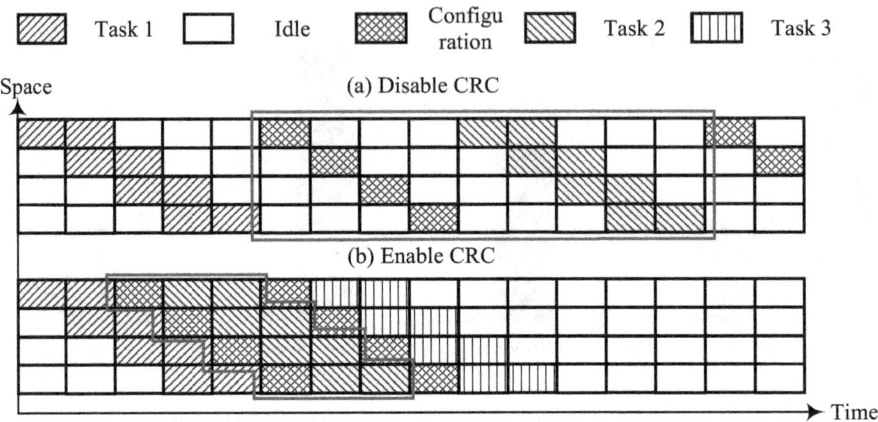

Fig. 5.9 Principle of executing CCR

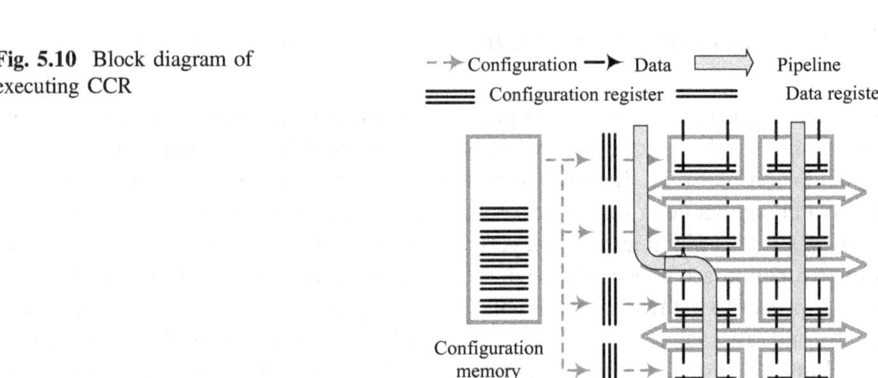

Fig. 5.10 Block diagram of executing CCR

In the reconfigurable computing processors, especially FPGAs, the function of configuring arrays will consume significant time, area, and power, which is considered as the cost of hardware reconfiguration. The number of cycles required for the configuration and computation is usually determined by the algorithm. Under this premise, the main purpose of applying the CCR is to reduce the overall time consumption including both the configuration and computation. For example, for a given application, the computation for both Task 1 and Task 2 takes five units of time, whereas configuration requires 4 units of time, then the entire process consumes 14 units of time before the optimization, as shown in Fig. 5.9a. However, after CCR is utilized, the time consumption is reduced to 8 units of time. Because the configuration and computation overlap, and computation of different tasks can also overlap, the overall time consumed by configuration and computation can be saved.

Table 5.5 Effect of the CCR technology on improving energy efficiency

Algorithm	Metric	CCR disabled	CCR enabled
Serpent	Performance (Gb/s)	2.75	5.49
	Power consumption/W	0.60	0.67
	Energy efficiency/(Gb/s/W)	4.58	8.19
MD5	Performance (Gb/s)	5.12	6.40
	Power consumption/W	0.67	0.68
	Energy efficiency/(Gb/s/W)	7.64	9.41

To quantify the contribution of CCR on improving the energy efficiency, Table 5.5 compares the performance and power consumption of the implementation result with and without CCR. Because the primary contribution of CCR is to reduce the cost of the configuration switching, it is inevitable to switch configuration context when the scale of an algorithm is large enough and such algorithms are likely to benefit more from the CCR. Therefore, Serpent and MD5 are chosen as examples of the block cipher and hash cipher (the size of the stream cipher is usually small), respectively. Serpent and MD5 have 31 and 64 round function iterations, respectively. PEA can accommodate neither algorithm for a single configuration. The results show that energy efficiency of Serpent of MD5 has increased by 179% and 123%, respectively, when CCR is employed. This is achieved by improving the throughput without significantly increasing power consumption. For example, PEA requires at least two sets of configuration contexts when executing MD5. If the CCR is not used, with configuration and computation in series, all PEs and interconnection are determined by the first set of configuration contexts before the algorithm is executed. The PE in the first row and first column does not begin the computation until all PEs in PEA have been configured, and all PEs are reconfigured after the first part of the algorithm is computed. Similarly, the reconfiguration process does not begin until the entire PEA finishes the computation. If CCR is used, with the configuration and computation in parallel, the computation for each PE start immediately after its own configuration is complete without waiting for the PEA to complete reconfiguration. In this way, the total time consumed is significantly reduced and the throughput is significantly increased.

5.2.3 *Configuration Compression and Organization (CCO)*

The above two methods require that the configuration contexts of each PE row can be quickly switched; that is, the configuration switching is completed within one clock cycle. The DCN also requires configuration switch of different parts of the PEA to be synchronized. In order to meet these needs, major efforts in two aspects are made to speed up the configuration process.

① The configuration contexts are organized into the baseline and increments to reduce the amount of the configuration contexts.
② Multiple scenarios of configuration contexts are used to improve the access efficiency of configuration memory.

The switching frequency of different types of configuration contexts is analyzed before optimization. In general, the configuration switching frequencies for opcodes, interconnections, GPRFs, and sequential control are higher than those of S-Box, permutation, RCH, and immediate operands. In addition to the switching frequency, the size ratio is also important. Figure 5.11 summarizes seven sets of configuration contexts (take AES as an example). Similar to other algorithms, opcode and interconnection have the largest amount of configuration contexts, accounting for 60% of the total size. Then, S-Box's configuration contexts account for about 26%, and the remaining configuration contexts are less than 15%. By analyzing switching frequency and size of configuration contexts, the optimization should focus on opcode and interconnection.

Because different PEs, different PE rows, and even different algorithms may have similarities in configuration contexts, configuration contexts can be reorganized and stored in the common subgraphs (baselines) and the incremental form, which can significantly reduce the size of memory for configuration contexts. Subgraphs are abstracted from three levels. The subgraph type (SGT) 1 describes the common part of configuration contexts between PEs. SGT2 describes the common part between PE rows. SGT3 describes the common subgraph between different algorithms. CCR uses SGT2, while SGT1 further reduces the amount of configuration contexts for each row of PEs. Due to the uncertainty of the increments of configuration contexts, two sets of context registers are designed for fast switching, one is for configuration and the other is preloaded at the backstage

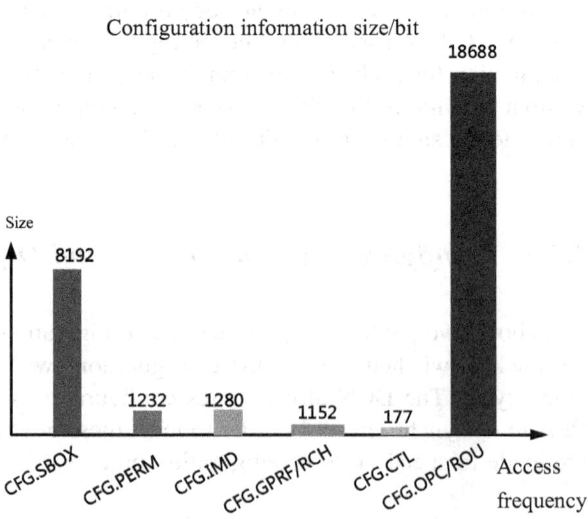

Fig. 5.11 Sizes and use frequencies of different parts of the AES configuration information

Table 5.6 Size of the configuration contexts with and without CCO

Classification	Algorithm	Configuration size/bit	Optimization results	Ratio/%
Block cipher	AES-CTR	5500	264	4.8
Stream cipher	ZUC	20,689	1304	6.3
Hash function	SHA256	10,136	744	7.34

(utilizing SGT3-based contextual relationship). A context cache is designed for SGT1 configuration and each context register has direct access to the cache, which significantly reduces access frequency to configuration memory. When a PE row needs to be configured, the configuration context can be switched immediately within one cycle, which is the basis of the CCR. The context cache is designed with 16 channels, and configuration contexts of different rows can be updated simultaneously via different channels of context cache, which is the basis of DCN.

With the above-mentioned configuration compression and organization (CCO) design, the total amount of configuration information can be significantly reduced. Most algorithms can benefit from the CCOs, so commonly used algorithms for block ciphers, stream ciphers, and hash functions are chosen as benchmarks to analyze the effects of the CCO. Table 5.6 lists the size of configuration context of these algorithms. After utilizing COO, the size of the configuration contexts is reduced to a few Kbs, which is much smaller than that of FPGA (typically Mb level). By comparing the results with and without CCO, the compression ratio of the configuration contexts is less than 10%, which fully proves the effectiveness of CCO. The layout analysis shows that the total area of the configuration memory is 9.2% of the overall area of Anole. The area typically accounts for 20–30% in common reconfigurable processors. As a result, the configuration context compression reduces the size of the configuration memory and reduces the total chip size. To further illustrate the effect of the CCOs, the breakdown of execution time of Anole for three algorithms with and without using CCO is shown in Fig. 5.12. Obviously, the compression of the configuration contexts significantly reduces the configuration time of all the algorithms by over one order of magnitude.

5.3 Integrated Development Tools of Anole

The integrated development tools of Anole are used to generate the configuration contexts with manual aid. The development tool can be used to modify the configuration of PEs, interconnection, tokens, and memory. Each basic unit can be separately configured, providing high flexibility. Meanwhile, the tool can automatically check the configuration to ensure that the configuration meets the rules. The main purpose of the integrated development tools is to manually optimize the configuration contexts and improve the execution efficiency of the configuration. Because of the limitations of the current automated compilation tools, the manual

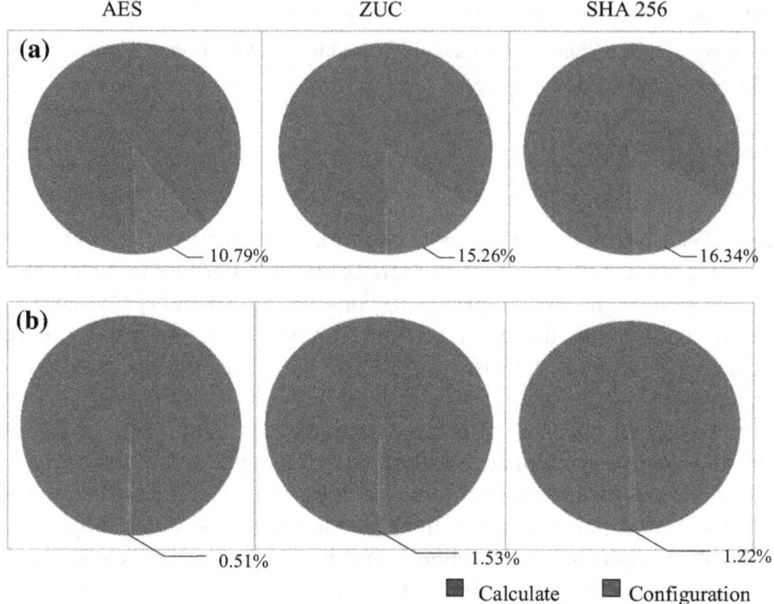

Fig. 5.12 Breakdown of the execution time before and after the CCO is used

optimization process is currently unavoidable for most compilers of reconfigurable computing processors in order to improve the performance of the processors.

5.3.1 Introduction to the Tools

The graphic user interface (GUI) of the integrated development tool of Anole is shown in Fig. 5.13, including the menu, toolbar, project manager, configuration manager, and workspace. The project manager displays the created project file, and the configuration manager is used to configure the function of PEs and interconnection among PEs. The workspace displays the schematic diagram of occupied PEs and simple interconnection. Currently, the tool supports simultaneous configuration of up to 4 × 28 PEs. The basic design flow is as follows.

The file tab in the menu bar can be used to create a new project, specifying the project path and the project name. The filename extension for the corresponding project file is pro.

① After the project is created, use the button on the toolbar to add an RC based on the rule of filling row first and then column. The button can be used to define the position of the added RC. After selecting a RC, the configuration can be made in the configuration window opened in configuration manager.

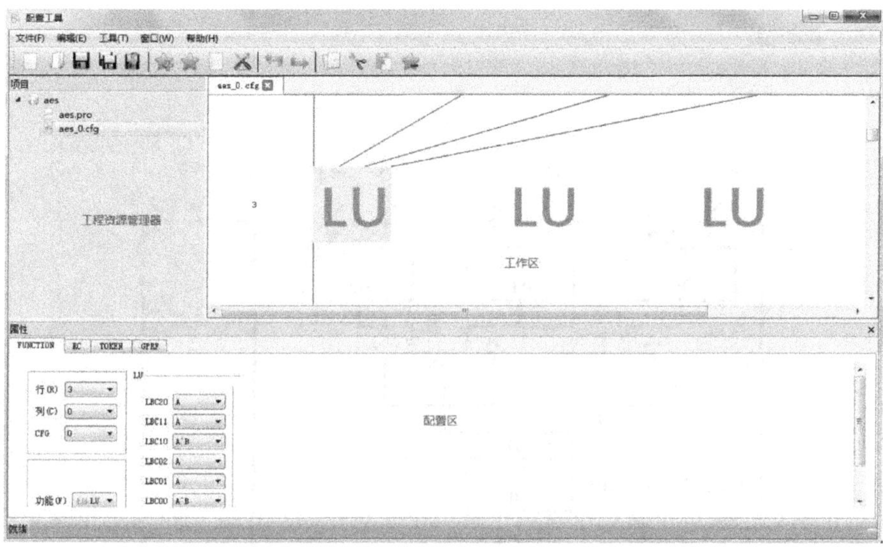

Fig. 5.13 Graphic user interface of Anole's integrated development tools

② To create PEs with the same configuration, the copy function can be used. A prompt will pop-up after copying successfully, and then click "paste" and select the position to paste in the array.

5.3.2 Configuration Method

1. RC Function Configuration

The RCs in even rows can be configured to LU, AU, SU, TU, and the PEs in odd rows can be configured to LU, AU, SU, and PU. The schematic diagram of each functional unit can be found in the drop-down menu in the help bar to assist configuration, as shown in Fig. 5.14. The FUNCTION tab in the configuration manager is used for configuring the function of RCs.

The LU contains six LBC units, and each unit can be configured into eight types of logic operations. One 4-to-1 MUX is used to choose the output.

The AU contains three 2-to-1 MUXs and one ADDER. The ADDER can be configured into four types of addition operations including modulo 2^8 addition, modulo 2^16 addition, modulo 2^32 addition, and modulo 2^32 − 1 addition.

The SU can implement the shift operation and support the simultaneous two-way shift operation in parallel. Each shift operation can be configured into four types of shift (logical left shift, left rotation, logical right shift, right rotation). The number of bits to be shifted can be any number from 0 to 31.

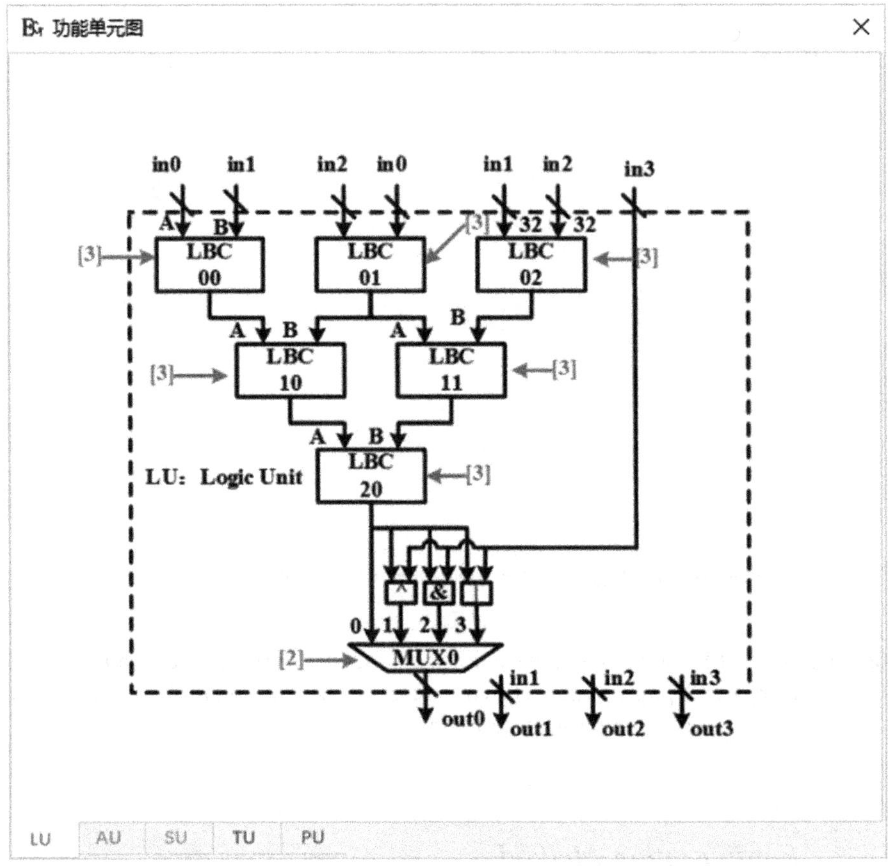

Fig. 5.14 Schamatic diagram of functional units for configuration

The TU can implement the lookup table operation, and only the even rows have TUs. The lookup table source selected by MUX4 is from input0, or the result of exclusive OR of input0 and input1. Then MUX3 further selects from the output of MUX4 or input2. After MUX3, the 32-bit operand data are divided into four lookup table addresses, that is, data [31:24] to S-Box0, data [23:16] to S-Box1, data [15:8] to S-Box2, and data [7:0] to S-Box3 for lookup table operation. Then, four S-Boxes give a 32-bit data, respectively, (S-Box0 outputs {A3, B3, C3, D3}, S-Box1 outputs {A2, B2, C2, D2}, S-Box2 outputs{A1, B1, C1, D1}, S-BOX3 outputs {A0, B0, C0, D0}, A0–A3, B0–B3, C0–C3, and D0–D3 are 1 byte data). MUX1 selects the out0 to be{A3, B3, C3, D3}, or {A3^B3^C3^D3, 0x00, 0x00, 0x00}; selecting out1 to be {A2, B2, C2, D2}, or {A2^B2^C2^D2, 0x00, 0x00, 0x00}; selecting out2 to be {A1, B1, C1, D1}, or {A1^B1^C1^D1, 0x00, 0x00, 0x00}; selecting out3 to be {A0, B0, C0, D0}, or {A0^B0^C0^D0, 0x00, 0x00, 0x00}, where ^ is the exclusive OR operator. MUX2 selects out0 to be {A3, A2, A1, A0} or {A3, B2, C1, D0}. When MUX0 is configured to 0, the output of MUX1, corresponding to out0, is selected. When MUX0 is configured to 1, the XOR result of output of MUX1 and input3 is selected. When MUX0 is configured to 2, the output of MUX2 is selected. When MUX0 is configured to 3, the XORed results of the output of MUX2 and input3 are selected.

The PU, a permutation control unit, is included only in the odd rows. It consists of a BENES64 (64-bit data BENES network, the 352-bit configuration required by the BENES needs to be pre-generated, loaded in the initialization configuration) and 4 MUXs. The data selected by MUX2 and MUX3 are merged into 64-bit data {DAT_MUX2, DAT_MUX3} to be processed by BENES network, and 64-bit data BNDAT are generated by BENES network. MUX0 selects out0 to be BNDAT [63:32] or XORed results of BNDAT [63:32] and input3. MUX1 selects out1, to be BNDAT [31:0] or XORed results of BNDAT [31:0] and input3.

2. RC Datapath Configuration

The RC tab page in the configuration manager is used to configure the datapath of the RCs, including four input sources and four output destinations of the RCs as shown in Fig. 5.15.

As shown in the datapath configuration diagram, an RC has four inputs and four outputs, and its bit width is one word. Each input source can be selected from FIFO, GPRF, RCH, IMD, PRE. FIFO corresponds to the input FIFO of the PEA, GPRF represents memory. RCH is register files, and IMD corresponds to the immediate data. PRE represents the data in the previous row. For every input source, BYTE_SHIFT operation is available. It means the 32-bit input data can be shifted in bytes. When 1 is chosen, it means circular shift by 1 byte. If 2 is selected, two bytes circular shift is made and so forth.

When FIFO is selected, it means that the input of the RC is from the input FIFO of PEA. The bit width of input FIFO is 128 bits, so there are four options for

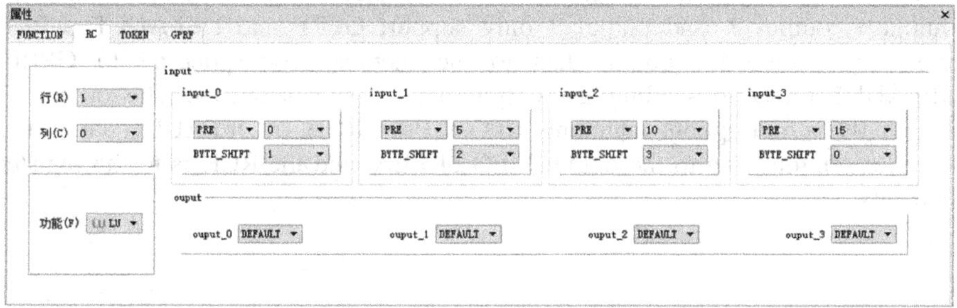

Fig. 5.15 RC datapath configuration diagram

Table 5.7 RCH channel selection

Block no.	Column no.	Channel no.
0	0	Register channel 0
0	1	Register channel 1
0	2	Register channel 2
0	3	Register channel 3
1	0	Register channel 4
1	1	Register channel 5
1	2	Register channel 6
1	3	Register channel 7
2	0	Register channel 8
2	1	Register channel 9
2	2	Register channel 10
2	3	Register channel 11
3	0	Register channel 12
3	1	Register channel 13
3	2	Register channel 14
3	3	Register channel 15

one-word input, represented by 0–3 from the least significant to the most significant word.

When GPRF is selected, the input data of the RC is from the memory. The single-line RC supports up to four-way reading GPRF operations (size: 32 bit 256), and the returned data are 128 bits, that is, four words. Any one of four words can be selected just like FIFO.

For RCH, the data are read from the register channels. There are 16 register channels in the PEA. Data from any one of 16 channels can be selected to be the input of RCs. Refer to Table 5.7 for the specific register channel selection.

Selecting IMD means selecting the immediate data. The IMD contains 128 bits in total, and one of the four words can be selected by choosing the value from 0 to 3.

Selecting PRE means to select from the output data of the previous row. There are 16 outputs from the four RCs on the previous row in total. The input of each RC can be any one of them.

The configuration tool supports the configuration of the output datapath. Output_1, output_2, and output_3 only support GPRF and DEFAULT. GPRF refers to sending the output data to the memory corresponding to GPRF. DEFAULT refers to sending the output data to the next row. The output of Output_0 has four options including FIFO, GPRF, RCH, or DEFAULT. The FIFO refers to sending data to the output FIFO of PEA, and the RCH is to the register channels.

3. TOKEN Timing Sequence Configuration

TOKEN controls whether the RC is working or idle. When TOKEN is enabled, the RC operates. When TOKEN is disabled, the RC remains idle.

When the source of TOKEN is selected as a FIFO, it means that TOKEN is driven by the input FIFO data. When a RC reads data from the FIFO, the TOKEN of this RC is enabled.

When the source of TOKEN is selected as TCH, it means that TOKEN is from token channel. There are 16 token channels in PEA, and 4 token channels in each block.

When the source of TOKEN is selected as FDN, it means that the TOKEN value is taken from the fixed token network. When the corresponding FDN is configured to 0, it means that the value of TOKEN is from the previous line. When it is configured to 1, it means that the special row processing exists. The special row processing is used in the case of multiple loops in a single or multiple rows. When the loop is completed, the TOKEN to next row is enabled.

When the source of TOKEN is selected as TCH|FDN, it means that the TOKEN of the RC is enabled when any TOKEN in TCH or FDN is valid. TCH supports 16 channels, while FDN supports the previous row or the special row processing.

The destination (DEST) of a TOKEN can be FIFO, TCH, or DEFAULT. FIFO corresponds to the output FIFO of PEA, and TCH is the token channel. DEFAULT represents the next row.

4. GPRF Configuration

The GPRF tab is mainly used to configure four memory (MEM) addresses to read from and four memory address to write to an RC. The memory addresses to read form are used to input data, and memory addresses to write to are used to output data. The depth of the MEM for GPRF is 256 (16 rows and 16 columns), and the width is 32 bits.

Any input of a RC can be the data from any address of MEM, so there are both 16 options for the row address and the column address. Two reading modes are supported including the default mode and the increase mode. When the default mode is chosen, the data are read from the corresponding address; selecting the increase mode means reading from the configured starting address, in which the read increase step needs to be configured.

Any output of an RC can be written to any address of the MEM. There are also both 16 options for the row address and the column address, respectively. Like reading from the MEM, two modes, including the default mode and the increase mode, are supported.

5. HEAD Configuration

The HEAD configuration tab is mainly used to configure the overall header information of the algorithm, as shown in Fig. 5.16. The HEAD configuration is divided into FB_SOURCE, NO, RC, OUTPUT, ADNGDN, RCH, S-BOX, FEEDBACK, IMD, and ADN.

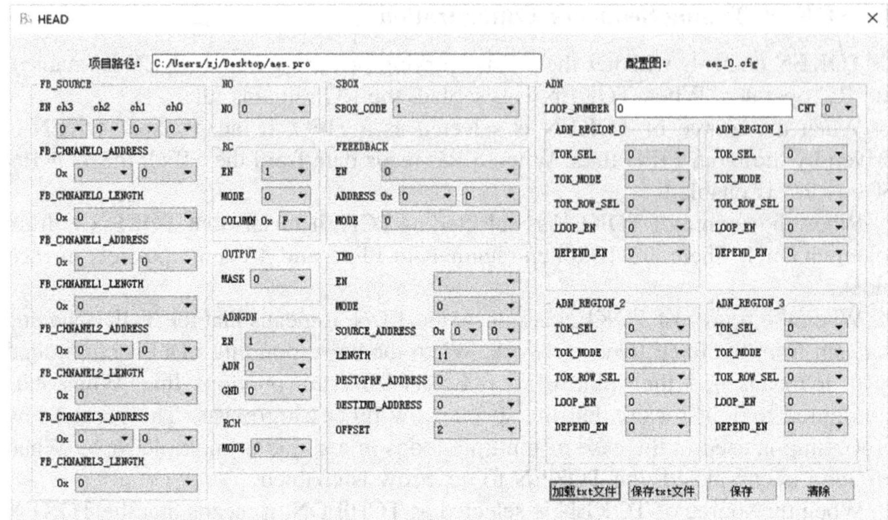

Fig. 5.16 HEAD configuration diagram

NO is the number for each algorithm. This number can be any integer from 0 to 7.

OUTPUT MASK refers to the effective output data width, and its value can be any integer from 0 to 2, which represents that the effective output data width of the array is 128 bits, 64 bits, and 32 bits, respectively.

ADNGDN indicates the switching configuration of the algorithm timing sequence and datapath. The EN option indicates that the timing and datapath configuration of the current algorithm parameters need to be switched. The ADN option indicates that the ADN parameter number to be switched is selected (up to four sets of ADN parameter numbers are stored in the chip, numbered from 0 to 3 respectively). The GDN option indicates that the parameter number of the switching datapath is selected (up to four sets of datapath parameter numbers can be stored in the chip, numbered from 0 to 3, respectively).

ADN is the timing sequence configuration of the algorithm; LOOP_NUMBER indicates the number of loops, the configuration value is 0–1024; CNT indicates the number of data that can be received and processed in one burst; ADN_REGION 0–3 correspond to the configuration of Array Blocks 0–3, the array (4 × 28) is divided into four blocks (0–6 rows are block0, 7–13 rows are block1, 14–20 rows are block2, and 21–27 rows are block3). LOOP_EN is enabling signal for loops, and the value can be 0 or 1. If there is a loop operation in block0, it should select 1, otherwise select 0. TOK_MODE indicates loop operation count selection mode. Mode 0 indicates that only a single datum has participated in the loop computation and the RC specified by TOK_SEL should be directly selected. Mode 1 indicates that multiple data have participated in the loop computation and the row specified by the TOK_ROW_SEL shall be selected as the count reference. TOK_SEL indicates the RC specified by the loop count. TOK_ROW_SEL indicates the row specified by the burst count. DEPEND_EN indicates whether there is any dependency between the data when sending multiple data.

RCH MODE represents the register channel mode configuration, and its value can be any integer from 0 to 2. 0 means that 16 RCHs (numbered RCH 0–15) are used independently by each RC in the RCA. 1 means that the RCA array is configured using RCH 15, RCH 0–14 are obtained from the original RCH 1–15, that is, the value of RCH 1 is assigned to RCH 0, …, the value of RCH 15 is assigned to RCH 14. 2 means the RCA array is configured using RCH 0, RCH 1–15 are obtained from the original RCH 0–14, that is, the value of RCH 0 is assigned to RCH 1, …, the value of RCH 14 is assigned to RCH 15.

IMD is the immediate option configuration used by the algorithm. EN means that the immediate data are required during operation. MODE defines the application mode of the immediate data, and the value can be 0 or 1. 0 means that the immediate value is sent to the IMD port of the RCA array, and 1 means the immediate is sent to the GPRF. SOURCE_ADDRESS is the initial address of the immediate data to be read from the memory for IMD. LENGTH indicates the number of IMD data to be used (1 for 128-bit data). DESTGPRF_ADDRESS is the initial address of the GPRF when the immediate data are sent to the GPRF. (The width of IMD memory is 128 bits, the size of the GPRF is 32 bits × 256. Therefore, the storage of 1 IMD datum occupies four GPRF addresses.) DESTIMD_ADDRESS indicates the start row number of the RCA array when the immediate is sent to the IMD port of the RCA.

S-BOX is the algorithm lookup table configuration. There are six options in this configuration option: 0 disable means there is no S-BOX lookup table; 1 means that a single row (four RCs) uses the same S-box and the S-BOX is marked with the number 1 (up to four sets of S-BOX parameters can be stored in the chip and their numbers are 1, 2, 3 and 4 respectively). 2 means that a single row (four RCs) uses the same S-BOX, the S-BOX is marked with the number 2. 3 means that a single row (four RCs) uses the same S-box and the S-BOX is marked with the number 3. 4 means that a single row (four RCs) uses the same S-box and the S-BOX is marked with the number 4. 5 means that a single row (4 RCs), respectively, use four different S-boxes, RC0 uses the S-BOX marked with the number 1, RC1 uses the S-BOX marked with the number 2, RC2 uses the S-BOX marked with the number 3, and RC4 uses the S-BOX marked with the number 4.

FEEDBACK is the algorithm feedback configuration. Feedback means the results of the array computation is returned to the configuration parameter memory rather than being sent to output FIFO. EN indicates there is feedback in the algorithm configuration. ADDRESS indicates the starting memory address for the feedback. MODE represents the feedback mode, and the value can be 0 or 1. 0 means feedback is sent to the memory for IMD0, and 1 means feedback is sent to the memory for IMD1 which is also the IMD memory used by the array.

FB_SOURCE is the feedback source configuration, EN Chn 3–Chn 0 corresponds to the four words of the 128-bit IMD. For the enabling signal for chn 0–3, 0 means being disabled and there is no feedback. 1 means there is feedback in the corresponding channel. FB_CHANNEL0_ADDRESS indicates the starting memory address of channel 0's feedback data in the GPRF; FB_CHANNEL0_LENGTH is the length of the feedback data of Channel 0 in the GPRF. For other channels (Channel 1–Channel 3), the configuration is the same as channel 0.

6. Initialization Configuration

The initialization configuration page, as shown in Fig. 5.17, is mainly used to preload of the constants that the algorithm needs to use. The S-BOX tab is used to preload S-Box lookup table data, which is only available if the configuration of the S-BOX option in the HEAD configuration is enabled first. The IMD tab is used to preload the immediate data. The PEU tab is used to preload the BENES permutation configuration parameters in PUs.

5.3.3 *Demonstration Cases*

This section discusses how to use the integrated development tools for Anole by taking AES encryption algorithm for example. In the AES encryption configuration, ten rounds of round functions need to be mapped onto RCA sequentially. The AES encryption implementation on the RCA can be divided into three steps.

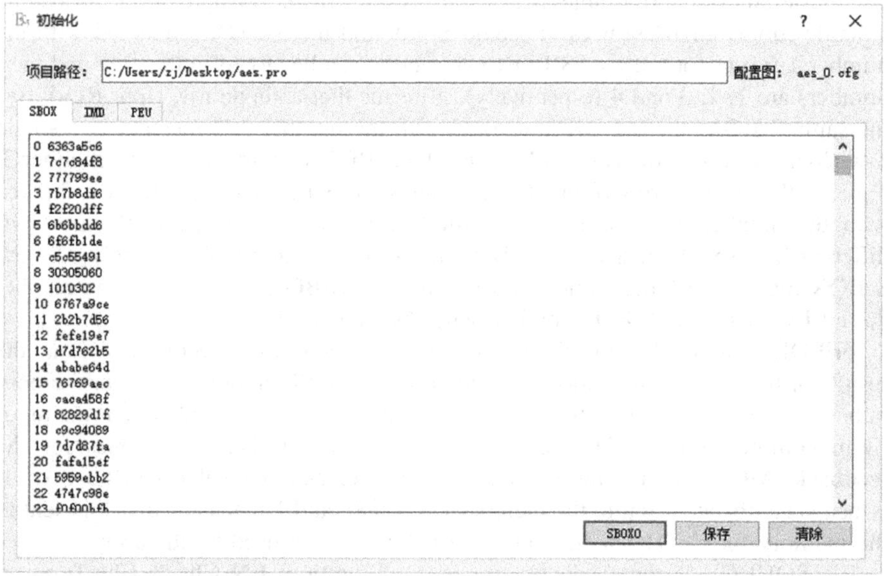

Fig. 5.17 Initial configuration diagram

1. First Round of Round Function

First, the input plaintext is processed by the first round of round function using the two rows of RCs (Row 0 and Row 1). For Row0, the plaintext and the subkeys are XORed, and part of S-Box substitutions and MixComlumn operations are also performed. The plaintext and subkeys are loaded via the input FIFO and the IMD, respectively. As shown in Fig. 5.18, TU is selected by configured to the function required. The specific configuration operation is as follows.

Click RC tab to start the configuration (as shown in Fig. 5.19). First, select the input/output of RCs, where IN_0, IN_1, IN_2 and IN_3 respectively represent 4 inputs, and the input data source can be selected from the drop-down list. The input port IN_0 is chosen for loading the plaintext (from the FIFO). The input port IN_1 is selected for loading subkeys (from the IMD). If the RC output is used for the next row of RC, OUTPUT is selected as default.

The RCs of Row 1 implement shift and XOR operations of four inputs, and the output is the result of the first round of round function. The input data of this row

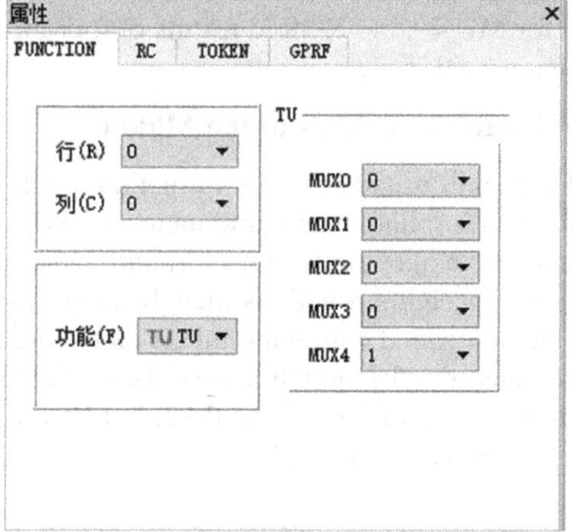

Fig. 5.18 TU function configuration diagram

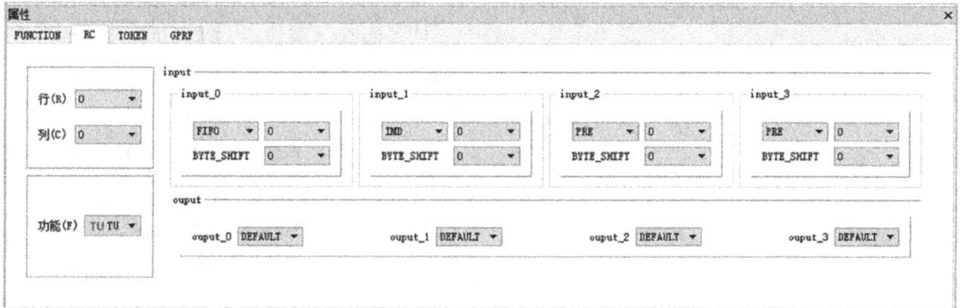

Fig. 5.19 TU configuration diagram

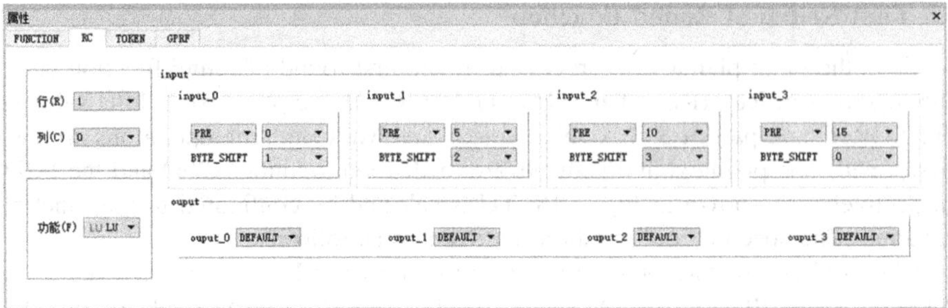

Fig. 5.20 LU configuration diagram

are from the RCs of the previous row, so IN_x is selected as a corresponding port of Preline_RC, as shown in Figs. 5.20 and 5.21. The output is also selected as default and Byte_Shift can be selected as the number of bytes of the input data that have been shifted (for the ShiftRows operation).

To implement the XOR functions, the LU can be used by analyzing the LU operator block diagram; LBC10 and LBC00 are defined as XOR function. After Channel 1 is selected for MUX0, the XORed results of four inputs can be produced through the output channel OUT_1.

2. Eight Rounds of Round Functions in the Middle

The operations of the eight rounds of round functions in the middle are exactly the same as those of the first round. The combination of TU and LU operator can also be used to implement the operations of each round with two rows of RCs. The total number of RC rows of eights rounds is then 16 rows (Row 2–Row 17). The only difference from the first-round configuration is that the input data of the TU are the outputs of the previous round function and subkeys. Therefore, the sources of the two inputs are no longer the FIFO and the IMD, but Preline_RC and IMD. The specific configuration is shown in Fig. 5.22.

Fig. 5.21 LU function configuration diagram

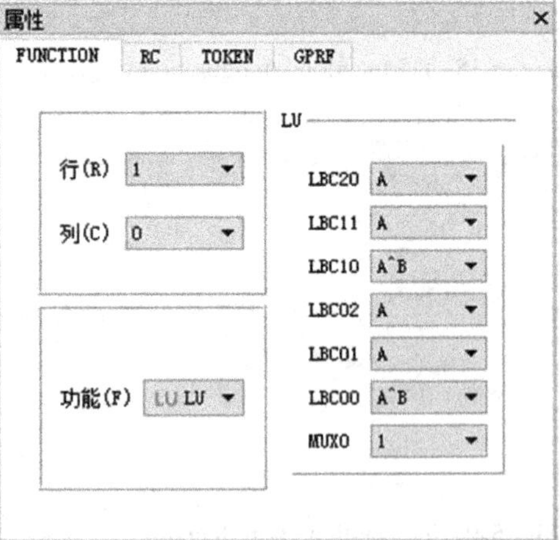

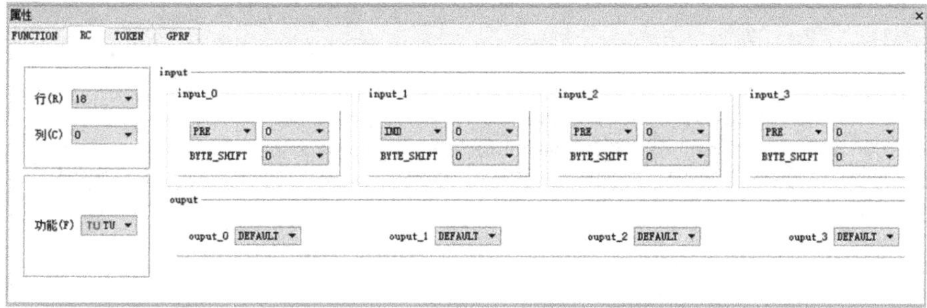

Fig. 5.22 Configuration diagram of the 8 rounds of TU functions in the middle

3. Last Round of Round Functions

Unlike previous rounds, there is no MixColumn operation in the last round. In this round of transformation, the S-Box transformation is implemented first through a row of RCs (Row 18). The RCs in this row use the TU operator, but only 8 bits are valid in the four 32-bit output data of each RC, and the values of the other invalid bits are 0.

The specific operation is shown in Fig. 5.23. The difference from the previous TU operator is that the parameter of MUX1 is set to 1, which can implement the required functions here.

On this basis, an OR operation on the four outputs of each RC of the previous row is performed. The output of Shiftrows operation is produced by merging four 8-bit data into one 32-bit datum. This function can be implemented by one row of RCs (Row 19). For the RCs of this row, the Byteshifts is performed first and the function of LBC10 and LBC00 in the LU operator is selected as 'OR'. The output for MUX0 is chosen as the third channel, as shown in Figs. 5.24 and 5.25.

As the last stop of the entire AES encryption process, the output data of the previous row are XORed with the subkeys to produce the ciphertext using a row of

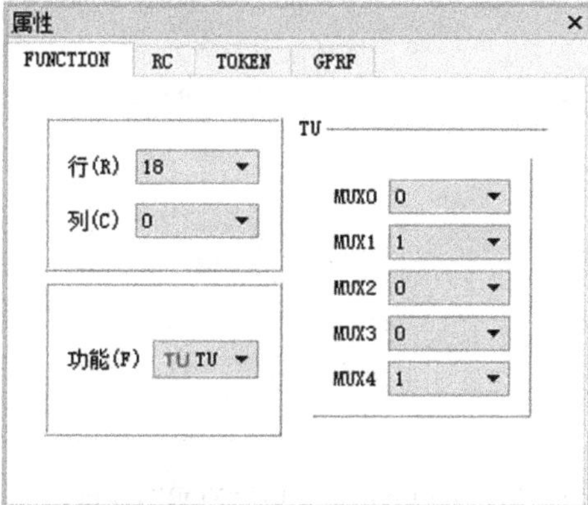

Fig. 5.23 Configuration diagram of the last round of TU functions

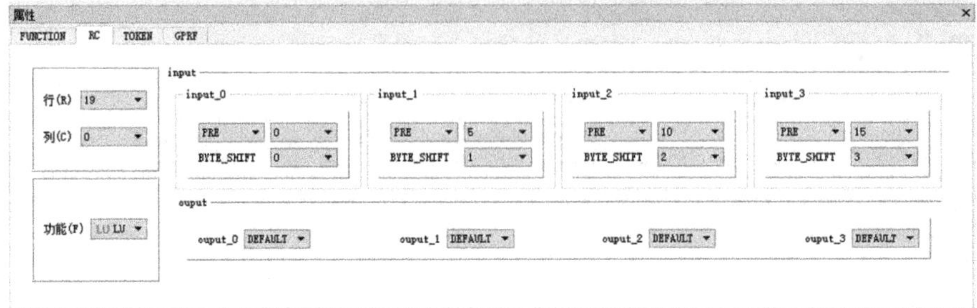

Fig. 5.24 Last round of LU "OR" operation data configuration diagram

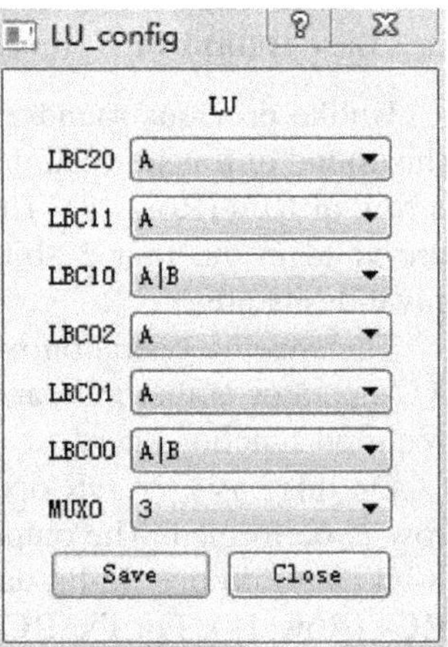

Fig. 5.25 Last round of LU "OR" operation data configuration diagram

RCs (Row 20). LU operators are selected for these RCs and the configuration is relatively simple. As shown in Figs. 5.26 and 5.27, the XOR function is selected for the LBC00.

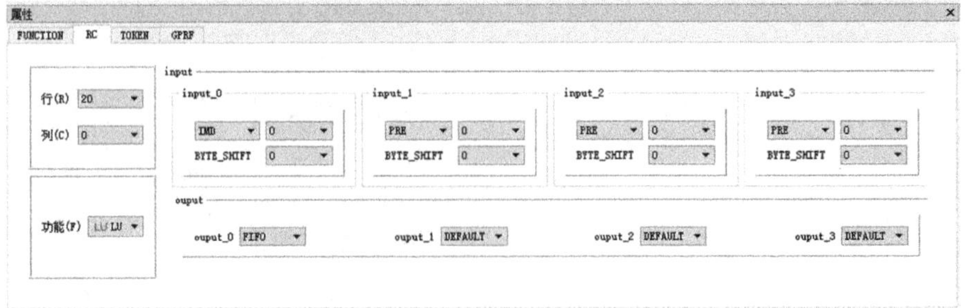

Fig. 5.26 Last round of LU "XOR" operation data configuration diagram

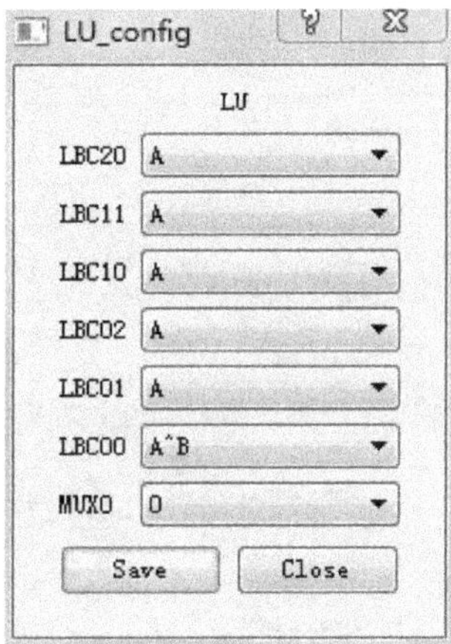

Fig. 5.27 Last round of LU unit "XOR" operation configuration diagram

At this point, the AES encryption algorithm is fully configured. Based on this configuration, the 128-bit plaintext is encoded into 128-bit ciphertext after the array computation. The entire AES encryption algorithm has used 21 (rows) × 4 (columns) of RCs when configuring RPU and 4 columns of RCs work in a parallel. Each column processes 32-bit plaintext encryption, which can greatly ensure the work efficiency of encryption.

4. HEAD Configuration

① Set the AES algorithm number to 0.
② Specify the use of S-Box. A set of lookup table parameters is used in the AES encryption algorithm. Therefore, set the configuration value of S-Box to 1.
③ Specify the use of IMD. In the AES encryption algorithm configuration, the round keys are used as a immediate data. There are 11 subkeys in the AES128, which are directly sent to the IMD ports of the RCs of each row (0, 2, 4, …, 20).
④ In the AES128 encryption algorithm, for the reconfigured RCs, the configuration parameters of RCs need to be switched and the four RCs in each row are used. Therefore, the MASK of the COLUM is set to 0xF.
⑤ Configuration of enabling timing sequence parameters and datapaths. Since there is only one algorithm configuration, the memory addresses of ADN and GDN parameters are both 0.

After the HEAD configuration is completed, the result is shown in Fig. 5.28.

5. Initialization Configuration

Load the S-Box lookup table parameters and the IMD round key parameters, as shown in Figs. 5.29 and 5.30.

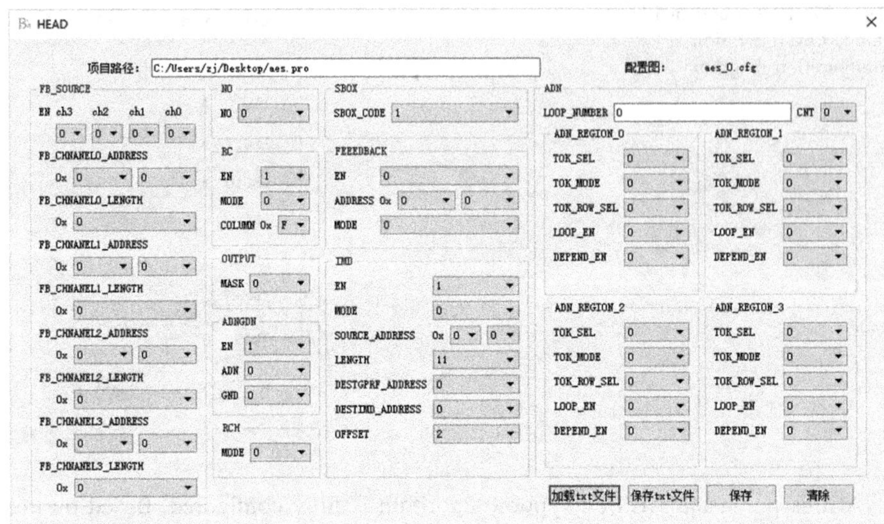

Fig. 5.28 HEAD configuration diagram

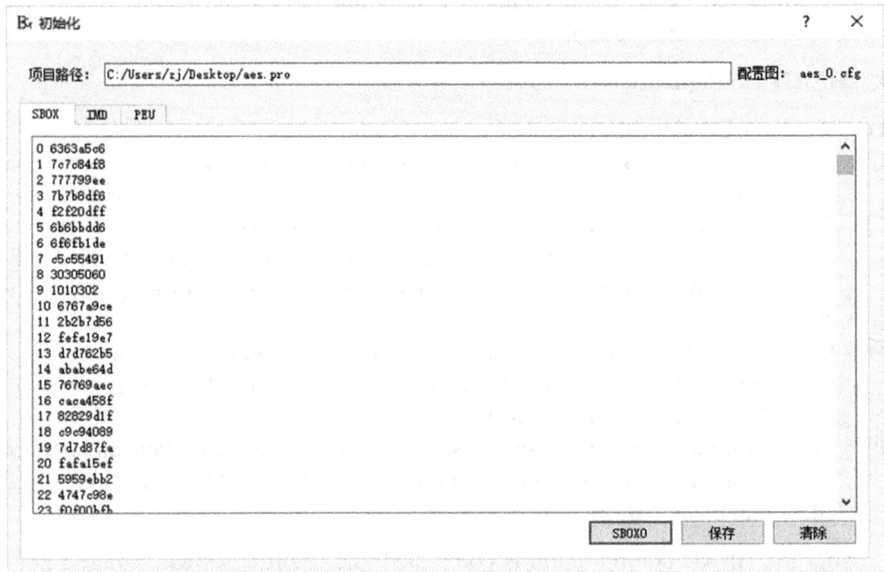

Fig. 5.29 S-Box parameter loading configuration diagram

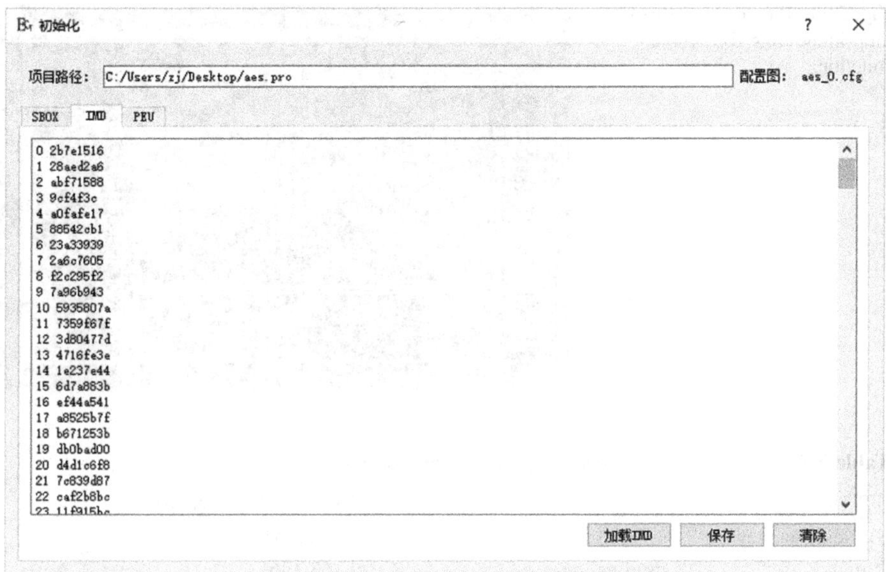

Fig. 5.30 IMD parameter loading configuration diagram

5.4 Analysis of the Implementation Results of the Anole Processor

This section describes the implementation results of the Anole chip in detail and compares them with the latest implementations of the symmetric cryptographic algorithm.

5.4.1 Implementation Results of the Chip

The Anole cryptographic processor is verified on the FPGA, as shown in Table 5.8. The Anole is implemented by adopting the TSMC 65 nm process technology, as shown in Table 5.9. Many symmetric cryptographic algorithms are mapped on the Anole, and the performance is listed in Table 5.10. Unless otherwise described like that for the AES-CBC, the algorithm results of the block cipher are the results in the non-feedback working mode. Figure 5.31 shows a picture of SoC chip integrating Anole as the core processor, and each module is labeled in the figure. RCA, RCC, FIFO, RCH, and GPRF modules have been described in Sect. 5.1.1. The remaining ESRAM, MCU, and SoC fabrics are the peripheral modules of Anole. ESRAM is an on-chip memory, MCU is a micro-control unit, and SoC fabrics refers to the SoC's interface module, including Uart, I2C, GPIO, SDIO, and SWD.

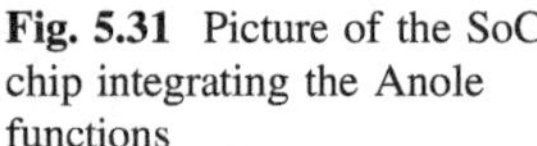

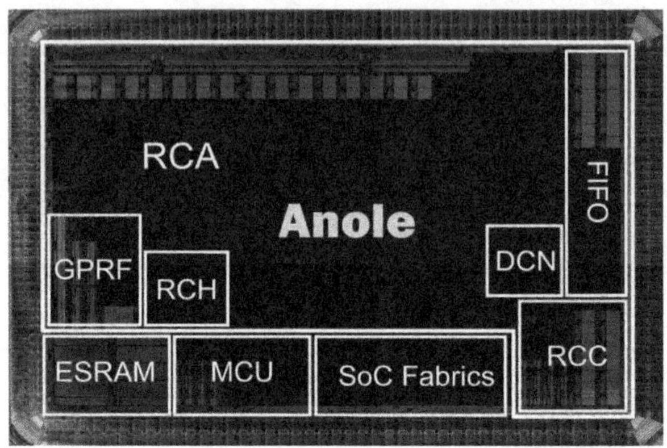

Fig. 5.31 Picture of the SoC chip integrating the Anole functions

Table 5.8 FPGA verification results of the Anole processor

FPGA	Virtex7 XC7VX690T
Basic frequency	108.2 MHz
Number of slice LUTs used as logic	163,136
Number of slice LUTs used as memory	243
Number of slice LUTs used as register	23,221

Table 5.9 Chip parameters of the Anole processor

Process	TSMC65nmGP
Frequency	400 MHz
Gates	1,910,000 equivalent gate
Memory	97.9 kByte
Area	$2.5 \times 3.1\ \text{mm}^2$

5.4.2 Chip Performance Comparison

This section compares the Anole chip parameter with the t state-of-the-art symmetric cipher processors to show the advantages of Anole.

Table 5.11 compares the energy efficiencies of Anole, other dynamically and partially reconfigurable structure, FPGA and GPP (some literature do not provide relevant metrics explicitly). Anole's energy efficiency is significantly higher than those of other structures. Compared with other dynamically and partially reconfigurable processors, Anole's energy efficiency has increased by 4.4–57.6 times. Compared with the FPGA, Anole's energy efficiency has increased by more than 364.8 times. The Anole's advantage in energy efficiency is even more obvious when compared to GPP, which is about nine orders of magnitude higher.

Table 5.12 compares the area efficiency of Anole and other implementations. Compared with the latest dynamically and partially reconfigurable computing

Table 5.10 Comparison of different cryptographic algorithms performance implemented on Anole

Type	Algorithm	Throughput/(Gb/s)
Block cipher	AES	51.2
	SM4	51.2
	Serpent	5.5
	DES	25.6
	Camillia	51.2
	Twofish	25.6
	MISTY1	9.6
	SEED	16.0
	IDEA	11.4
	SHACAL-2	4.0
	AES-CBC	40.9
Stream cipher	ZUC	3.2
	SNOW 3G	6.4
	RC4	1.6
Authentication cipher	AESGCM	22.4
	MORUS-640	2.9
Hash function	SHA256	0.46
	SHA3	0.28
	SM3	0.64
	MD5	6.4

Table 5.11 Horizontal comparisons of Anole's energy efficiency

Type	Algorithm	Reconfigurable structure[a]	FPGA	GPP	Anole
Block cipher	AES	3.79 [5]	5.0 [6]	0.64 [7]	82.6
	Serpent	–	7.21 [8][c]	9.6e−5 [9]	8.19
	DES	0.11 [10][b]	1.46 [11][c]	9.6e−7 [12][d]	55.7
	IDEA	0.61 [13]	0.054 [14]	3.8e−4 [14]	19.7
	AES-CBC	1.03 [15]	1.65 [16]	0.64 [7]	59.3
Stream cipher	ZUC	0.84 [17]	1.10 [18]	2.08e−3[d]	6.67
	SNOW 3G	1.99 [17]	2.41 [18]	1.1e−8 [19]	13.1
Hash function	SHA256	3.2 [10][b]	0.28 [20]	0.032[e]	1.11
	MD5	2.15 [10][b]	0.36 [21]	7.03e−7 [22]	9.41

[a]The reconfigurable structure here refers to the work that claim to be dynamically and dynamically reconfigurable

[b]The power consumption is normalized to the 65 nm process with fixed-voltage scaling)

[c]The FPGA's power consumption is obtained by multiplying the supply voltage by the minimum current (from the datasheet), and its actual value is multiplied by the resource utilization ratio

[d]The power consumption and the area consumption are estimated to be 10 W and 100 mm^2

[e]The experimental platform is a platform with a CPU of Intel core i7-6500u, a memory of 8G, Microsoft Windows 10 Professional operating system, and a compiling environment of Visual Studio 2012-gcc

Table 5.12 Horizontal comparisons of Anole's area efficiency

Type	Algorithm	Reconfigurable computing structure[a]	FPGA	GPP	Anole
Block cipher	AES	0.21 [5]	0.25 [6]	0.15 [7]	6.61
	Serpent	0.069 [23]	0.22 [8][c]	2.98e−5 [9]	0.71
	DES	0.21 [10][b]	0.044 [10]	9.6e−8 [12]	3.30
	IDEA	0.23 [24]	1.08e−3 [14]	3.8e−5 [14]	1.47
	AES-CBC	0.49 [15]	0.083 [16]	0.15 [7]	5.28
Stream cipher	ZUC	0.11 [17]	0.055 [18]	3.12e−4[d]	0.41
	SNOW 3G	0.27 [17]	0.12 [18]	1.1e−8	0.83
Hash function	SHA256	0.11 [10][b]	0.015 [10]	4.85e−3[d]	0.06
	MD5	0.072 [10][b]	0.016 [10]	3.16e−7 [22]	0.83

[a]The reconfigurable structure here refers to the work that claim to be dynamically and dynamically reconfigurable
[b]The power consumption is normalized to the 65 nm process with fixed-voltage scaling
[c]The FPGA area is obtained by multiplying the smallest area (100 mm^2) by the utilization rate of resources
[d]The experimental platform is a platform with a CPU of Intel core i7-6500u, a memory of 8G, Microsoft Windows 10 Professional operating system, and a compiling environment of Visual Studio 2012-gcc

processors, Anole's area efficiency is increased by 30.5 times. Compared with the FPGA, the increasing ratio is 4–1361.1 times. Compared with GPP, the increasing ratio of area efficiency is also the largest, which is ranging from 12.4 times to seven orders of magnitude. Tables 5.10 and 5.11 do not exactly match each other. This is mainly because some of the literature provides incomplete parameters, but this does not affect the validity of the comparison results.

Table 5.13 shows a comprehensive comparison using the same benchmark algorithm, i.e., AES. Compared metrics include performance, power consumption, area consumption, and the efficiencies. All the metrics are normalized to the 65 nm process by using the fixed-voltage scaling because the operating voltages of the products produced by the 90 nm or below process technology have been virtually unchanged. Anole's energy efficiency and area efficiency are 82.6 Mb/s/mW and 6.6 Gb/s/mm2, respectively. Compared with the latest reconfigurable computing architecture, Anole's energy efficiency and area efficiency are increased by 31.2 times and 13.0 times. Compared with FPGA, Anole's energy efficiency is increased by 9.7–36.2 times and area efficiency by 8.1–69.5 times. Compared to GPP,

Table 5.13 Horizontal comparison of Anole's technologically normalized energy efficiency and area efficiency

Metric		Reconfigurable structure[a]			FPGA[b]			GPP		Anole
		[15]	[25]	[26]	[6]	[27]	[28]	[7]	[29]	
Process technology/nm		45	90	250	90	40	90	65	32	65
Performance/(Gb/s)		128	19.1	0.845	25.1	40.8	30.8	1.02	6.30	51.2
Basic frequency/MHz		1000	149	66	196	319	241	1210	3300	400
Performance/basic frequency/(bits/cycle)		128	128	13	128	128	128	1	2	128
Normalized energy efficiency/ (Mb/s/mW)	Power consumption	6.2	10	0.259	5	11.0	5	1.58	129.9	0.62
	Efficiency	14.3	2.6	12.5	7.0	2.3	8.5	0.6	0.024	82.6
	Increased ratio[c]	5.8	31.2	6.6	11.9	36.2	9.7	127.9	3459.5	
Normalized area efficiency/ ($Mb/s/mm^2$)	Area	6.3	100	6.29	100	100	100	6.63	100	7.75
	Efficiency	6.7	0.5	7.6	0.7	0.1	0.8	0.2	0.008	6.6
	Increased ratio[d]	1.0	13.0	0.9	9.9	69.5	8.1	42.9	878.9	–

[a]The reconfigurable structure here refers to work that claim to be dynamically and dynamically reconfigurable
[b]The FPGA here refers to the literature on the specific function implementation in FPGA
[c]The increased ratio here refers to the increased ratio of Anole's energy efficiency compared with that in the literature
[d]The increased ratio here refers to the increased ratio of Anole's area efficiency compared with that in the literature

Anole's energy efficiency and area efficiency are increased by two to three orders of magnitude.

It is important to note that Anole's energy efficiency can reach 30% of the hardwired logic (for a particular algorithm without any change of the functions). The GPP can only reach 0.1%. The normalization performance of Anole and other reconfigurable structures is 128 bits per cycle, and the difference in actual throughput performance is mainly determined by the operating frequency. What's particularly important, Anole's energy efficiency is 5.8 times that of cryptoraptor, while the area efficiency is basically the same. Anole's goal is not to increase the efficiency of the execution of a particular algorithm, but rather to achieve a general improvement across a wide scope of cryptographic algorithm.

References

1. Wallner S (2003) A reconfigurable multi-threaded architecture model. In: Advances in computer systems architecture, Asia-Pacific conference, pp 193–207

2. Zaykov PG, Kuzmanov G, Gaydadjiev G (2009) Reconfigurable multithreading architectures: a survey. In: International conference workshop on embedded computer systems: architectures, modeling and simulation, pp 263–274
3. Watkins MA, Albonesi DH (2010) Dynamically managed multithreaded reconfigurable architectures for chip multiprocessors. In: International conference on parallel architectures and compilation techniques, pp 41–52
4. Zaykov PG, Kuzmanov G (2011) Architectural support for multithreading on reconfigurable hardware. In: International symposium on applied reconfigurable computing, pp 363–374
5. Garcia AP, Berekovic M, Aa TV (2008) Mapping of the AES cryptographic algorithm on a coarse-grain reconfigurable array processor. In: International conference on application-specific systems, architectures and processors, pp 245–250
6. Good T, Benaissa M (2005) AES on FPGA from the fastest to the smallest. In: International workshop on cryptographic hardware and embedded systems, pp 427–440
7. Liu B, Baas BM (2013) Parallel AES encryption engines for many-core processor arrays. IEEE Trans Comput 62(3):536–547
8. Najafi B, Sadeghian B, Zamani MS et al (2004) High speed implementation of serpent algorithm. In: IEEE international conference on microelectronics, pp 718–721
9. Nazlee AM, Hussin FA, Ali NBZ (2009) Serpent encryption algorithm implementation on compute unified device architecture (CUDA). In: IEEE student conference on research and development (SCOReD), pp 164–167
10. Hauser JR (2000) Augmenting a microprocessor with reconfigurable hardware. University of California, Berkeley, California
11. Mcloone M, Mccanny JV (2003) High-performance FPGA implementation of DES using a novel method for implementing the key schedule. IEE Proc Circuits Dev Syst 150(5):373–378
12. Zhou Y, Li Y (2014) The design and implementation of a symmetric encryption algorithm based on DES. In: IEEE international conference on software engineering and service science, pp 517–520
13. Shan W, Shi L, Fu X, et al (2014) A side-channel analysis resistant reconfigurable cryptographic coprocessor supporting multiple block cipher algorithms. In: Design automation conference, pp 1–6
14. Granado JM, Vega MA, Sanchez JM et al (2006) Implementing the IDEA cryptographic algorithm in Virtex-E and Virtex-II FPGAs. In: IEEE Mediterranean electrotechnical conference. IEEE, pp 109–112
15. Sayilar G, Chiou D (2014) Cryptoraptor: high throughput reconfigurable cryptographic processor. In: International conference on computer aided design, pp 155–161
16. Chen D, Shou G, Hu Y et al (2010) Efficient architecture and implementations of AES. In: International conference on advanced computer theory and engineering, pp V6–V295
17. Wang B, Liu L (2015) A flexible and energy-efficient reconfigurable architecture for symmetric cipher processing. In: International symposium on circuits and systems, pp 1182–1185
18. Zhang L, Xia L, Liu Z et al (2012) Evaluating the optimized implementations of Snow3G and ZUC on FPGA. In: IEEE international conference on trust, security and privacy in computing and communications, pp 436–442
19. Jairaj V, Pohjonen J, Shemyak K (2011) High performance implementation of Snow3G algorithm in memory limited environments. In: International conference on new technologies, mobility and security, pp 1–4
20. Li C, Zhou Q, Liu Y et al (2011) Cost-efficient data cryptographic engine based on FPGA. In: IEEE international conference on Ubi-media computing, pp 48–52
21. Jarvinen K, Tommiska M, Skytta J (2005) Hardware implementation analysis of the MD5 hash algorithm. In: IEEE international conference on system sciences, p 298a
22. Wang F, Yang C, Wu Q et al (2012) Constant memory optimizations in MD5 crypt cracking algorithm on GPU-accelerated supercomputer using CUDA. In: International conference on computer science and education, pp 638–642

23. Damaj I, Itani M, Diab H (2006) Serpent cryptography on static and dynamic reconfigurable hardware. In: IEEE international conference on computer systems and applications, pp 680–684
24. Singh H, Lee M, Lu G et al (2000) MorphoSys: an integrated reconfigurable system for data-parallel and computation-intensive applications. IEEE Trans Comput 49(5):465–481
25. Mancillaslopez C, Chakraborty D, Henriquez FR (2010) Reconfigurable hardware implementations of tweak able enciphering schemes. IEEE Trans Comput 59(11):1547–1561
26. Wang M, Su C, Horng C et al (2010) Single- and multi-core configurable AES architectures for flexible security. IEEE Trans Very Large Scale Integr Syst 18(4):541–552
27. Wang Y, Ha Y (2013) FPGA-based 40.9-Gbits/s masked AES with area optimization for storage area network. IEEE Trans Circuits Syst II Express Briefs 60(1):36–40
28. Good T, Benaissa M (2007) Pipelined AES on FPGA with support for feedback modes (in a multi-channel environment). IET Inf Secur 1(1):1–10
29. Nishikawa N, Iwai K, Kurokawa T (2012) High-performance symmetric block ciphers on multicore CPU and GPUs. Int J Netw Comput 2(2):251–268

Chapter 6
Physical Attack Countermeasures for Reconfigurable Cryptographic Processors

The physical attack countermeasures for reconfigurable cryptographic processors are mainly achieved in two ways. One way is to implement all the universal countermeasures to the reconfigurable architecture. Another way is to develop new countermeasures by using the characteristics of the reconfigurable computing. Traditional universal countermeasures do not take full advantage of the characteristics of reconfigurable computing and result in significant performance, area, and power overhead. In addition, new threatening attack methods such as the local electromagnetic attack with the attack precision of gate level, multiple fault attack which can introduce more than one fault in a single execution, and attacks based on ultra-low frequency (kHz level for instance) acoustic or electromagnetic signal continue to emerge. Various existing traditional countermeasures cannot be used to effectively resist these attacks. Compared with the direct application of traditional countermeasures, countermeasures designed based on the characteristics of reconfigurable cryptographic architecture can effectively reduce the performance, area, and power overhead caused by security improvement through resource reuse. What is more, the new countermeasures are expected to resist novel attack methods that have not been effectively overcome. On the one hand, the dynamic and partial reconfiguration feature can be fully exploited to develop countermeasures based on time and spatial randomization. When each execution of the cryptographic algorithm is performed at a different time and circuit region in the array, various precision attacks will not take effect. It just like that when an attacker wants to attack the backdoor of the cryptographic implementation, the randomization method keeps the position of the backdoor changing rapidly, making it difficult for the attacker to attack even when he or she has the key to the backdoor. On the other hand, we can make full use of the structural advantages of reconfigurable processors and fully combine the countermeasure design with the reconfigurable architecture, thus maximizing the advantages of reconfigurable computing. The rich array computing units and interconnection resources on the reconfigurable cryptographic processors can be used to resist physical attacks. With the resource reuse, the consumption caused by countermeasures can be significantly reduced. For example,

L. Liu et al., *Reconfigurable Cryptographic Processor*,
https://doi.org/10.1007/978-981-10-8899-5_6

a physically unclonable function (PUF) can be constructed based on array computing units, and lightweight authentication or security keys can be generated after the basic encryption/decryption operations are performed. The rich interconnection resources on the array can also be fully developed to resist attacks. When various topology attributes of interconnection network changed slightly and randomness was introduced, physical attack countermeasures can be implemented besides the normal data transmission.

6.1 Countermeasures Based on Time and Spatial Randomization

Most of the physical attacks aim at specific processing steps of sensitive data operations. This can be mainly divided into two dimensions: time and space. On the one hand, the attack needs to be for a specified time period, that is, corresponding to one or several clock cycles. On the other hand, because the operation of sensitive data is actually performed only on the corresponding circuit region of the chip, the effective attack is actually aiming at the corresponding valid space area. The reconfigurable architecture has the ability to dynamically and partially reconfigure based on configuration information. This makes it possible for the time and spatial randomization based on reconfiguration. In addition to inserting random delays and dummy operations in the execution process through the configuration control mechanism, dynamic and partial reconfiguration technology can map sensitive operations randomly to different space areas of the array. Under the two-dimensional randomization, the difficulty of attack brought by the randomness will increase in a square relation, and so the physical attack resistance of reconfigurable cryptographic chips can significantly improve. It is worth mentioning that this spatial randomization method has very significant effects on the emerging countermeasures precisely against the attacks on local circuit areas such as local electromagnetic attacks and laser injection fault attacks.

6.1.1 Fault Attack Countermeasure Based on Randomization Technologies

1. Introduction to Fault Attacks

To successfully complete the fault attack, an attacker usually needs to inject certain faults in the specified operation. The following is an example of a bit fault (introducing a 1-bit fault in the algorithm state). For the byte fault, there is a similar analysis [1]. In fact, single-bit faults can be further divided into bit-set (bit-reset) [2] and bit-flip faults. A study of laser fault injection points out that the probability of introducing bit-set and bit-reset faults is much greater than that of introducing bit-flip faults. For a fault injection process, the type of fault will affect the minimum number of injection attempts required for successful attack. For bit-set or bit-reset type faults, the fault occurs only if the target intermediate value is 0 or 1. During a normal attack, the value of an intermediate variable can be considered as random in various operations (since plaintext is random and the key is unknown), and the probability of successful injection of bit-set or bit-reset is 50%, thus causing the wrong output compared to the bit-flipping fault. In the evaluation of the randomized attack countermeasure, the ratio of the number of faults injected before and after randomization is defined as the attack resistance coefficient. This means that the effect of the absolute value of the number of failed injection attempts caused by different fault types will eventually be eliminated.

Take the AES algorithm, for example, in the brief introduction to the fault attack implementation method. Let us define the 128-bit output state of the *r*th round of the AES encryption algorithm as S^r, which is obtained by inputting the 128-bit input state S^{r-1} and the 128-bit round key K^r into the round function, wherein the 128-bit round key K^r is obtained through the key expansion operation of the AES algorithm. The status in the AES algorithm can be thought of as a 4 × 4 matrix. Each element in the matrix is a byte in the GF(2^8) field, which can be expressed by the expression $s^r_{m,n}$, indicating that the byte is in the mth row and nth column of the matrix, where $0 \leq$ m, n ≤ 3. Let us define the relation $S^r_{m,n} \leftarrow \mathrm{SB}(S^r_{m,n})$ as the byte substitution operation and define the relation $S^r_{m,n} \leftarrow S^r_{m,(n+m)\mathrm{mod}4}$ as the row shift operation. In the column obfuscation step, each byte row is treated as a polynomial in the GF (2^8) field, which is multiplied by a fixed polynomial $a(x) = \sum_{0 \leq k \leq 3} a_k x^k = 02 + 01x + 01x^2 + 03x^3$ to obtain the result of the row shift transformation. In the round key adding operation, the status byte and round key will perform XOR operation, which is represented by the relationship $s^r_{m,n} = s^r_{m,n} \oplus k^r_{m,n}$.

Lets define $\tilde{C}$ as a faulty ciphertext. The ciphertext C is the correct ciphertext. One method to introduce a single-bit fault to the AES algorithm is to inject a 1-bit fault in the ninth round of AES encryption operation. Assuming that one byte of the bytes (m, n) is modified before the last round is started, it can be expressed by the relation $\tilde{s}^9_{m,n} = s^9_{m,n} \oplus \varepsilon$, where ε is the introduced fault value and the only nonzero byte of all $\tilde{C} \oplus C$ results will appear in Column [(n − m) mod 4] of Row m, i.e.,

$$(\tilde{C} \oplus C)_{m,(n-m)\mathrm{mod}4} = s^{10}_{m,(n-m)\mathrm{mod}4} \oplus \tilde{s}^{10}_{m,(n-m)\mathrm{mod}4} = \mathrm{SB}(S^9_{m,n}) \oplus \mathrm{SB}(\tilde{s}^9_{m,n}) \quad (6.1)$$

An attacker finds a subset of $s^9_{m,n}$ that satisfies the above equation by guessing the value of the fault ε. By generating a certain amount of faulty ciphertext, an intersection of each subset is found, and the attack is successful when the intersection contains only one element. In this way, an attacker can get a round key of one byte, i.e.,

$$k^{10}_{m,(n-m)\bmod 4} = S^{10}_{m,(n-m)\bmod 4} \oplus SB(s^{9}_{m,n}) \tag{6.2}$$

All of the round keys can be obtained by performing the above operation on the 16 bytes of the round key. After the round key is obtained, the attacker can obtain the key by performing an inverse byte substitution operation.

The traditional fault attack countermeasures mainly use the idea of redundancy and comparison; that is, the fault attack countermeasures use time redundancy and hardware redundancy. It should be pointed out that these redundancy-based countermeasures are based on the assumption of a single-fault attack; that is, an attacker can only successfully inject a fault during the single execution of the algorithm. In recent years, the accuracy of fault injection has been greatly improved. For example, the space precision of laser injection has reached the logic gate level, and the time accuracy has reached the level of nanoseconds [3]. Researchers have realized the threat of double-fault attacks and proved the feasibility of double-fault attacks [4]. Because existing laser injection and electromagnetic injection have reached a very high level both in space and time, laser injection and electromagnetic injection are more suitable for double-fault attack compared to the traditional clock glitches and power glitch injection [5]. Under the condition that an attacker can accurately control the accuracy of fault injection, there is a great possibility that an attacker injects two or more faults into the circuit during a single execution of the algorithm. For the above-mentioned redundancy and comparison-based countermeasures, the comparison process will fail if an attacker can inject two identical faults simultaneously into the normal execution path and the redundant path, and the security of the cryptographic processor will be seriously threatened. Theoretically, the security problem caused by the gradual injection accuracy improvement of the fault attack can be solved by increasing the number of redundant circuits, but this will significantly affect the performance of the cryptographic processor and increase the size of the cryptographic processor, which is not a feasible attack countermeasure obviously.

Another shortcoming of the traditional fault attack countermeasures is that the countermeasures passively detect faults in the circuit after the faults are injected, but not actively prevent injection when the faults are injected. For a cryptographic circuit with a fixed encryption path, it is theoretically possible for the attacker to bypass the fixed error detection mechanism in the circuit after making enough attempts when the accuracy of fault injection is high enough, so that the circuit may output attacker-desired faulty ciphertext. By introducing time and space randomness in the computation process, the success rate of fault injection can be effectively reduced. If the circuit's encryption path is not fixed, it will make it difficult for attackers to successfully inject faults. In addition to the traditional redundancy measures, randomness is introduced to both the main processing path and the redundancy path of the cryptographic processor. The cryptographic processor can improve fault attack resistance capabilities with relatively little performance loss, power consumption, and hardware resource consumption. In order to implement the randomization of time and space, the hardware must have the ability to randomly change the circuit structure and the calculation timing steps during the execution of the algorithms. The dynamic and partial reconfiguration characteristics of the reconfigurable cryptographic processor have the ability to implement the time and spatial randomization. Compared with traditional cryptographic processors, the reconfigurable cryptographic processors can implement the reconfiguration with the minimum performance loss, minimum power consumption, and minimum hardware resource consumption. This randomization-based attack countermeasure is especially important, as for precision attacks based on laser and electromagnetic injection with the ability to introduce multifault attacks.

2. Injection Effort Model

The traditional metric to measure the countermeasure resistance is the fault detection rate, in which only the proportion of detected faults to the total number of successfully injected faults is considered and the attack resistance of the circuit in the fault injection phase is not considered. In order to study the process of fault injection and evaluate the attack resistance of the randomized attack countermeasures adopted in this process, a new metric model for countermeasures is needed. Here, we introduce an injection effort model (IEM) to evaluate the anti-attack ability of the randomized fault attack countermeasures. The IEM model analyzes the fault injection for an attack in detail and divides the actual injection process into the search phase and the continuous injection phase.

In a fault attack, the actual fault injection operation process of the attacker is mainly divided into two phases: the search phase and the continuous injection phase. The search phase is the phase in which an attacker attempts to inject faults into various parts of the circuit and determines whether a desired faulty ciphertext at the output can be obtained by injecting a fault in this part. The continuous injection phase is a phase in which after an attacker finds the part of the circuit that has some specific effects on the output ciphertext, he or she injects a large number of faults into the part of the circuit. This process does not complete until the attacker has obtained sufficient designated faulty ciphertexts. Finally, the attacker uses the resulting faulty ciphertexts to reveal the secret key.

In the injection effort model, the number of fault injections is abstracted to denote the attack effort and represent the difficulty of the attack. As the number of fault injections increases, the total attack time cost increases, and the probability that the attack causes a permanent damage to the entire circuit becomes larger or even directly results in the circuit failure and makes the attack fail.

The computational steps that can produce some faulty ciphertexts of the specified type during an attack are called sensitive points (SPs). The goal of the search phase is to search for the execution time node and the spatial circuit position of the sensitive point. After the first sensitive point is found, the search process is completed. After the attacker obtains the information about the time and spatial position of the sensitive point in the search phase, he or she continues to inject faults into the sensitive point to get enough faulty ciphertexts. Let us, respectively, define the search effort and the continuous injection effort as the numbers of injection faults required in both the search phase and the continuous injection phase. Therefore, the total effort can be defined as

$$EF_{total} = EF_{search} + EF_{inject} \tag{6.3}$$

Before explaining the various efforts in fault attacks, we first introduce the concept of attack elements (AEs). In order to stimulate sensitive points, the attacker needs to search the entire chip during the execution of the algorithm. The common strategy is the strategy of dividing the whole into many pieces and finding one piece after the other (the divide and conquer strategy), i.e., the chip surface and the chip execution time are divided into many equal units, and then search for each unit, and finally get the information on the time and the spatial position of the sensitive point. If a certain area of circuit at a time is abstracted from the circuit and be used as an attack element, then the entire circuit can be abstracted as a set of attack elements at the runtime.

Each attack element can be expressed with (t_i, s_j), where t represents the time, s represents the space position, $1 \leq i \leq \varphi_t$, $1 \leq j \leq \varphi_s$, where φ_t and φ_s, respectively, represent the total search ranges of time and space, where $\varphi_s = 16$, $\varphi_t = 3$ in Fig. 6.1.

During the search phase, the attacker searches for each attack element. For each attack element, a sufficient number of faults will be injected into the attack element to determine if the attack element is a sensitive point, the number of which is denoted $EF_{s\_s,\ ae}$. The attacker must carefully adjust the time and space parameters

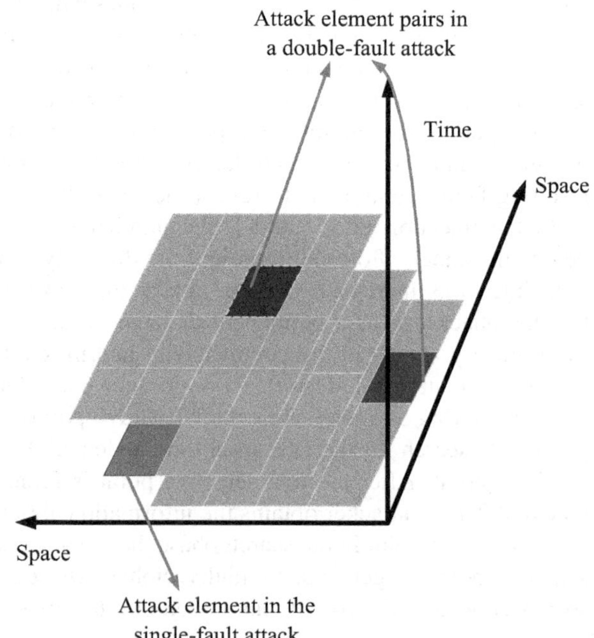

Fig. 6.1 Schematic diagram of attack element model

and observe whether the output is the required ciphertext. The attack element that can obtain the ciphertext at the output is a sensitive point. When an attack element is identified as a sensitive point, the search phase ends and the attack enters the continuous injection phase. In order to obtain available faulty ciphertexts, an attacker must continually inject a certain number of faults into sensitive points based on the time and space information obtained during the search phase, the number of which is expressed as $EF_{i\_s,\ ae}$. This process usually lasts ω times while obtaining ω fault ciphertexts at the output, where ω is a constant and ω is determined by the specific method of analyzing these fault ciphertexts.

Based on the above analysis, search efforts and injection efforts can be described with Eqs. (6.4) and (6.5)

$$\mathrm{EF}_{\mathrm{search}} = \mathrm{EF}_{\mathrm{s\_s,ae}} \times E(N_{\mathrm{search}}) \tag{6.4}$$

$$\mathrm{EF}_{\mathrm{inject}} = \mathrm{EF}_{\mathrm{i\_s,ae}} \times \omega \tag{6.5}$$

N_{search} represents the total number of attack elements searched when the first sensitive point is found, and $E(N_{\mathrm{search}})$ represents the expectation of the total number of attack elements to be searched when the first sensitive point is found.

The above attack model also applies to double-fault attacks. Compared with a single-fault attack, the difference between the models is that a single attack element in a single-fault attack becomes an attack element pair in a double-fault attack. Because in a double-fault attack, an attacker must inject two identical faults simultaneously, each sensitive point is a combination of the same operations. At this point, the attacker's goal is the attack element pair. In a double-fault attack, an attacker needs to search each attack element pair one by one to find a sensitive point for continuous injection. These attack element pairs are

$$((t_{i0,}s_{j0}), (t_{i1,}s_{j1}), \quad i \leq i,\ i1 \leq \phi_t;\ 1 \leq j0,\ j1 \leq \phi_s \tag{6.6}$$

As a supplement, here, we elaborate how to determine whether the attack element pair is a sensitive point in the double-fault attack. Generally, if there is no alarm signal that triggers a redundant error detection mechanism after the fault is injected (the alarm is caused by the difference between the main processing path and the redundant path) and a fault ciphertext with special fault propagation pattern can be obtained at the output, the attack element here is sensitive point. It should be noted that, under certain circumstances, the circuit may output a random number instead of triggering an alarm signal after detecting a fault. Random numbers are often incorrect on every byte compared to the correct ciphertext, so an attacker can still differentiate this failed attack of output random number from successful attack on a sensitive point by using a fault diffusion pattern.

The following will discuss how to select time and space unit dimensions of attack element in the model. The dimension size of time and space unit depends mainly on the specific implementation of the target hardware and encryption algorithm on the circuit. Because of the length of time and size of the injection source (such as the laser injection source) is very small, it is appropriate to select clock cycle and circuit unit (processing unit in a reconfigurable array) as a unit [6]. The guide strategy of selecting time nodes and space units is to increase the unit size as much as possible while ensuring that two corresponding sensitive operations do not appear in the same attack element. On the one hand, the size of large attack element will reduce the number of attack elements that the attacker needs to search; on the other hand, if the selected space and time unit are too large, the attack success rate will be reduced because the large unit size will make the attacker miss sensitive operation in the same attack element. Taking fault attack on the reconfigurable array as an example, assuming that an attacker injects a fault by using laser injection method, a single clock cycle of the reconfigurable array can be selected as a time unit in the attack. In the most intuitive case, each PE and its corresponding interconnection can be selected as the injected space unit. In a reconfigurable array, the number of PEs is huge, and the entire search space for attackers can be huge, especially for attackers who launch a double-fault attack. Because encryption algorithms often have many block-based operations, PEs typically operate in groups for processing data blocks. In this case, PE groups can be used as attack units. This will greatly reduce the number of search space units, so as to assess attackers' injection efforts under the most objective conditions.

The core aspect of the fault attack countermeasure lies in mapping sensitive points onto multiple attack elements (or attack element pairs). It should be noted that the time randomization and the spatial randomization are independent of each other. In a single-fault attack, let us define $\varUpsilon_s$ and $\varUpsilon_t$ as the space randomness and time randomness, respectively. When the space randomness is $\varUpsilon_s$, the sum of all the positions where sensitive points may appear is $\varUpsilon_s$. Assume that the position of the sensitive point at an attack element is marked as (t_0, s_x) in every operation, where x $(0 \le x \le \varUpsilon_s - 1)$ is the label of the space unit. x satisfies a discrete uniform distribution in different operations. For the probability density function $f(x)$, its expression is $f(x) = \frac{1}{\varUpsilon_s}$ in which $x = 0, 1, \ldots, \varUpsilon_s - 1$. If $\varUpsilon_t = 1$ and $\varUpsilon_s = 3$, the sensitive points will be randomly mapped onto the three attack elements (t_O, S_O), (t_O, S_1), and (t_O, S_2), where 0, 1, and 2 represent the labels of the space units. Of course, the above definition can also be applied to time randomization. In the case of a double-fault attack, let us define $\gamma_{s,0}$, $\gamma_{t,0}$, $\gamma_{s,1}$ and $\gamma_{t,1}$ as the randomness of the two corresponding operations in the sensitive points as we did in the above single-fault attack. In order to reduce the impact of the absolute quantities of $EF_{s\_s,ae}$, $EF_{i\_s,ae}$ ae and ω, the ratio between the attack efforts before and after randomization is defined as the attack countermeasure coefficient. An apostrophe is added to the upper right corner of a randomized variable to distinguish the variable before the randomization from the variable after the randomization.

The attack resistance coefficient of the search phase and that of the continuous injection phase is

$$R_{\text{search}} = \frac{\text{EF}'_{\text{search}}}{\text{EF}_{\text{search}}} \tag{6.7}$$

$$R_{\text{inject}} = \text{EF}'_{\text{inject}} \Big/ \text{EF}_{\text{inject}} \tag{6.8}$$

Meanwhile, let us define the unit effort gain, that is

$$A_{\text{s\_s, ae}} = \text{EF}'_{\text{s\_s, ae}} \Big/ \text{EF}_{\text{s\_s, ae}} \tag{6.9}$$

$$A_{\text{i\_s, ae}} = \text{EF}'_{\text{i−s, ae}} \Big/ \text{EF}_{\text{i−s, ae}} \tag{6.10}$$

The attack resistance coefficient of the search phase and that of the continuous injection phase is

$$R_{\text{search}} = \frac{E'(N_{\text{search}}) \times A_{\text{s\_s,ae}}}{E(N_{\text{search}})} \tag{6.11}$$

$$R_{\text{inject}} = A_{\text{i\_s,ae}} \tag{6.12}$$

In order to better understand the fault injection model, we first introduce the attack resistance coefficient in a single-fault attack then introduce the attack resistance coefficient in a double-fault attack and finally give a unified formula. For the sake of simplicity, it is assumed that there is only a single sensitive point in the assumed algorithm.

(1) The attack resistance coefficient in the single-fault attack

The expressions for $A_{\text{s\_s, ae}}, A_{\text{i\_s, ae}}, E(N_{\text{search}}), E'(N_{\text{inject}})$ are

$$A_{\text{s\_s,ae}} = \gamma_s \times \gamma_t \tag{6.13}$$

$$A_{\text{i\_s,ae}} = \gamma_s \times \gamma_t \tag{6.14}$$

$$E(N_{\text{search}}) = \frac{\varphi_s \times \varphi_t + 1}{2} \tag{6.15}$$

$$E'(N_{\text{search}}) = \sum_{k=1}^{\varphi'_s \times \varphi'_t - \gamma_s \times \gamma_t + 1} \frac{C(\varphi'_s \times \varphi'_t - k,\, \gamma_s \times \gamma_t - 1)}{C(\varphi'_s \times \varphi'_t,\, \gamma_s \times \gamma_t)} \times k \tag{6.16}$$

The expression of the attack resistance coefficient is

$$R_{\text{search}} = \frac{\left[\sum_{k=1}^{\varphi_s' \times \varphi_t' - \gamma_s \times \gamma_t + 1} \cdot \frac{C(\varphi_s' \times \varphi_t' - k, \gamma_s \times \gamma_t - 1)}{C(\varphi_s' \times \varphi_t', \gamma_s \times \gamma_t)} \times k\right] \times \gamma_s \times \gamma_t}{\frac{\varphi_s \times \varphi_t + 1}{2}} \tag{6.17}$$

$$\mathrm{R}_{inject} = \gamma_{\mathrm{s}} \times \gamma_{\mathrm{t}} \tag{6.18}$$

It can be seen that the probability that a sensitive point appears at the attack element is $1/(\gamma_s \times \gamma_t)$ for each attack element, no matter whether it is in the search phase or in the continuous injection phase. In other words, among all the injections, only the injection of $1/(\gamma_s \times \gamma_t)$ is effective.

(2) The attack resistance coefficient of the double-fault attack

Through the above derivation, we can get the expressions of $A_{\mathrm{s\_s,ae}}$, $A_{\mathrm{i\_s,ae}}$, $E'(N_{\mathrm{search}})$ in a double-fault attack, which are

$$A_{\mathrm{s\_s,ae}} = (\gamma_{s,0} \times \gamma_{t,0}) \times (\gamma_{s,1} \times \gamma_{t,1}) \tag{6.19}$$

$$A_{\mathrm{i\_s,ae}} = (\gamma_{s,0} \times \gamma_{t,0}) \times (\gamma_{s,1} \times \gamma_{t,1}) \tag{6.20}$$

$$E'(N_{\mathrm{search}}) = \sum_{k=1}^{U} \frac{C(C(\varphi_s' \times \varphi_t', 2) - k, (\gamma_{s,0} \times \gamma_{t,0}) \times (\gamma_{s,1} \times \gamma_{t,1}) - 1}{C(C(\varphi_s' \times \varphi_t', 2), (\gamma_{s,0} \times \gamma_{t,0}), (\gamma_{s,1} \times \gamma_{t,1}))}, \tag{6.21}$$

where, the upper limit U of the summation is

$$\mathrm{U} = \mathrm{C}(\varphi_{\mathrm{s}}' \times \varphi_{\mathrm{t}}', 2) - (\gamma_{\mathrm{s},0} \times \gamma_{\mathrm{t},0}) \times (\gamma_{\mathrm{s},1} \times \gamma_{\mathrm{t},1}) + 1 \tag{6.22}$$

The expression of the attack resistance coefficient in the double-fault attack is not shown here because it is too long.

As we did in the above single-fault attack, $A_{s\_s,\,ae}$ and $A_{i\_s,\,ae}$ are proportional to the total randomness. Since $\gamma_{\mathrm{s},\,0}$, $\gamma_{\mathrm{t},\,0}$, $\gamma_{\mathrm{s},\,1}$, $\gamma_{\mathrm{t},\,1}$ are independent of each other, it is considered that the sensitive points are mapped randomly to $(\gamma_{\mathrm{s},\,0} \times \gamma_{\mathrm{t},\,0}) \times (\gamma_{\mathrm{s},\,1} \times \gamma_{\mathrm{t},\,1})$ attack elements.

(3) Unified expression

No matter whether the attack is a single-fault attack or a double-fault attack, the attack resistance of the search phase and the attack resistance of the injection phase can be respectively and uniformly expressed as R_{search} and R_{inject}. Let χ represent the number of all attack elements (attack element pairs) and σ represent the number of all sensitive points, then the following attack resistance coefficient can be obtained, that is

$$R_{\text{search}} = \frac{\left[\sum_{k=1}^{\chi'+1-\sigma'} \frac{C(\chi'-k,\sigma'-1)}{C(\chi',\sigma')} \times k\right] \times A_{s\_s,ae}}{\sum_{k=1}^{\chi+1-\sigma} \frac{\sigma C(\chi-\lambda,\sigma-1)}{C(\chi,\sigma)} \times \lambda} \tag{6.23}$$

$$R_{inject} = A_{i\_s,ae} \tag{6.24}$$

The expressions of $A_{s\_s,ae}$, $A_{i\_s,ae}$, χ, and σ can be derived from the descriptions on the single-fault attack and the double-fault attack.

In the single-fault attack, there are

$$A_{s\_s,ae} = \gamma_s \times \gamma_t \tag{6.25}$$

$$A_{i\_s,ae} = \gamma_s \times \gamma_t \tag{6.26}$$

$$\chi = \gamma_s \times \gamma_t,\ \sigma = 1 \tag{6.27}$$

$$\chi' = \varphi'_s \times \varphi'_t,\ \sigma' = \gamma_s \times \gamma_t \tag{6.28}$$

In the double-fault attack, there are

$$A_{s\_s,ae} = (\gamma_{s,0} \times \gamma_{t,0}) \times (\gamma_{s,1} \times \gamma_{t,1}) \tag{6.29}$$

$$A_{i\_s,ae} = (\gamma_{s,0} \times \gamma_{t,0}) \times (\gamma_{s,1} \times \gamma_{t,1}) \tag{6.30}$$

$$\chi = C(\gamma_s \times \gamma_t, 2),\ \sigma = 1 \tag{6.31}$$

$$\chi' = C(\varphi'_s \times \varphi'_t, 2),\ \sigma' = (\gamma_{s,0} \times \gamma_{t,0}) \times (\gamma_{s,1} \times \gamma_{t,1}) \tag{6.32}$$

The above is based on the situation of a single sensitive point, but the following is based on the situation of a number of sensitive points. Let us define Λ to be the number of sensitive points in the entire algorithm without loss of generality and assume that the randomness of each sensitive point is the same. In this case, the attack resistance in the search phase and the injection phase can be calculated by using the previous expressions as well, but the expressions of σ, σ', $A_{i\_s,\ ae}$ change correspondingly. Unlike $A_{i\_s,\ ae}$, the expression of $A_{s\_s,\ ae}$ does not change because all attack elements or attack element pairs searched before the first leak point is found are not sensitive points. Therefore, the number of sensitive points does not affect the values of $E'_{s\_s,\ ae}$ and $E_{s\_s,\ ae}$.

Next, we will discuss the effect of the randomized fault attack countermeasure on the coincidence of multiple sensitive points. First, we analyze the situation where the sensitive points do not overlap. Sensitive points do not overlap means that each sensitive point will not appear in the same attack element or attack element pair during different executions of the algorithm. In fact, in the cryptographic processor, multiple sensitive points have the probability of showing in the same attack element or attack element pair. According to the previous definition, σ is the total number of attack elements or attack element pairs at which the sensitive points may appear. In the case of multiple sensitive points, σ is Λ times of that in the case of a single sensitive point. In the case of multiple sensitive points, the total number of attack elements or attack elements that are contained in the sensitive points in single-fault attacks and double-fault attacks is expressed as follows.

In a single-fault attack, there are

$$\sigma = \Lambda \tag{6.33}$$

$$\sigma' = \Lambda \times \gamma_s \times \gamma_t \tag{6.34}$$

In a double-fault attack, there are

$$\sigma = \Lambda \tag{6.35}$$

$$\sigma' = \Lambda \times (\gamma_{s,0} \times \gamma_{t,0}) \times (\gamma_{s,1} \times \gamma_{t,1}) \tag{6.36}$$

During the continuous injection phase, the target attack element or attack element pair corresponds to only one sensitive point, so the expression of $A_{i\_s,\ ae}$ is the same as that in the case of a single sensitive point.

Figure 6.2 shows the statuses of multiple sensitive points in an assumed single-fault attack. Figure 6.2a shows the case where two sensitive points SP_1 and SP_2 are mapped to two attack elements $AE(t_2,s_1)$ and $AE(t_3,s_2)$. The spatial randomization case without sensitive point (Fig. 6.2b) overlap, where the spatial randomness $\gamma_s = 3$. In Fig. 6.2, attack elements that may contain sensitive points are labeled, and the corresponding probability is marked in front of the sensitive point. During each execution, two sensitive points are randomly mapped to the attack elements according to the cases shown in Fig. 6.2.

(a)

(t_i,s_j)	t_1	t_2	t_3	t_4
s_1		○		
s_2			□	
s_3				
s_4				

(b)

(t_i,s_j)	t_1	t_2	t_3	t_4
s_1		1/3 ①		
s_2		1/3 ②	1/3 [1]	
s_3		1/3 ③	1/3 [2]	
s_4			1/3 [3]	

(c)

(t_i,s_j)	t_1	t_2	t_3	t_4
s_1		1/4 ①	1/4 ②	
s_2		1/4 ③	1/2 ④[1]	1/4 [2]
s_3			1/4 [3]	1/4 [4]
s_4				

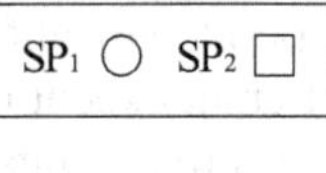

Fig. 6.2 Multiple sensitive points

In the case of no sensitive point overlap, we can obtain the values of σ, σ′, $A_{i\_s, ae}$, $A_{s\_s, ae}$

$$\sigma = 2 \tag{6.37}$$

$$\sigma' = \Lambda \times (\gamma_{s,0} \times \gamma_{t,0}) \times (\gamma_{s,1} \times \gamma_{t,1}) = 2 \times 3 \times 1 = 6 \tag{6.38}$$

$$A_{i\_s,ae} = A_{s\_s,ae} = 3 \times 1 = 3 \tag{6.39}$$

When some sensitive points overlap with each other, the case will become more complicated. σ′ will no longer be the case in which the randomness is simply multiplied by the number of sensitive points. When the sensitive points overlap, $EF'_{i\_s, ae}/EF_{i\_s, ae}$ will be different at each sensitive point. At this moment, Ai_s, ae should be $EF'_{i\_s, ae}/EF_{i\_s, ae}$'s expectation before and after randomization of all sensitive points.

Figure 6.2c shows the sensitive point overlap case, where $Y_s = 2$ and $Y_t = 2$. In this case, σ′ = 7, its value decreases (σ′ = 8 in the case without overlap), compared with the case where the sensitive points do not overlap.

If each of all the six attack elements contains only one sensitive point, we can get $\frac{EF'_{i\_s,ae}}{EF_{i\_s,ae}} = 4$. For the attack element that contains two sensitive points, it means that the number of the sensitive points appearing at that attack element is 2 for every four computations, so we get $\frac{EF'_{i\_s,ae}}{EF_{i\_s,ae}} = 2$. According to the above derivation, in the case of sensitive point overlap, $A_{i\_s, ae}$ is

$$A_{i\_s,ae} = 6 \times \left(\frac{1}{7} \times 4\right) + \left(\frac{1}{7} \times 2\right) = 3.7 \tag{6.40}$$

Because the overlap of sensitive points will lead to the reduction of $A_{i\_s,ae}$ and the decrease of the attack resistance, this situation should be avoided during the specific randomization. However, it is hard to avoid the overlap of sensitive points in the process of using both the time and spatial randomization at the same time. In the overlap case, the designer can make the best choice by analyzing the specific attack countermeasure according to the above analysis process.

The last conclusion to be drawn is that the circuit's attack resistance increases linearly with the increase of the total randomness. Due to the repetitiveness, the increased attack resistance during the continuous injection phase plays a decisive role in improving the overall attack resistance of the circuit. After the search phase is over, the fault injection operation for revealing the key in the continuous injection phase should repeat ω times to effectively reveal it. When multiple keys are reused by turns, the entire continuous injection phase will be further repeated based on different keys. Compared to the continuous injection phase, the enhancement of the attack resistance in the search phase is much slower. Taking $\phi'_s = \phi_s$ and $\phi'_t = \phi_t$ for example, when $\Upsilon_s \times \Upsilon_t$ is increased from 1 to $\Phi_s \times \Phi_t$, the R_{search} attack resistance in the search phase will increase from 1 to $\frac{2 \times \Phi_s \times \Phi_t}{\Phi_s \times \Phi_t + 1}$. It should be pointed out that the above analysis of *IEM* also applies to the analysis of the fault injection attacks with three or more injection faults.

3. Randomized Fault Attack Countermeasures

The above content has shown the specific procedure of using the IEM to analyze the fault attack resistance of the circuit based on the randomness. The following is how to implement the specific randomization with minimal performance loss, power consumption, and area cost. The randomization methods of the fault attack countermeasure mainly include the round-based relocation (RBR), the register pair swap (RPS), and the random delay insertion (RDI).

The RBR and the RDI randomly change the space position and execution time nodes of the sensitive points, thus effectively reducing the probability of successful attacks. The RPS is an enhancement method that improves the attack countermeasure by changing the interconnection between the computing unit and the register. Compared to the RBR, the RPS can be regarded as a relatively randomized operation and can be used in combination with the RBR. It should be noted that, when the above methods are used in the double-fault attack countermeasures, the randomization operation should be performed, respectively, in the main processing path and the redundant path.

(1) Round-based relocation technology

In the RBR, in order to randomly change the positions of sensitive points, the array structure of the reconfigurable cryptographic processor will be reconfigured in each execution process. In order to maximize the resource utilization in the process

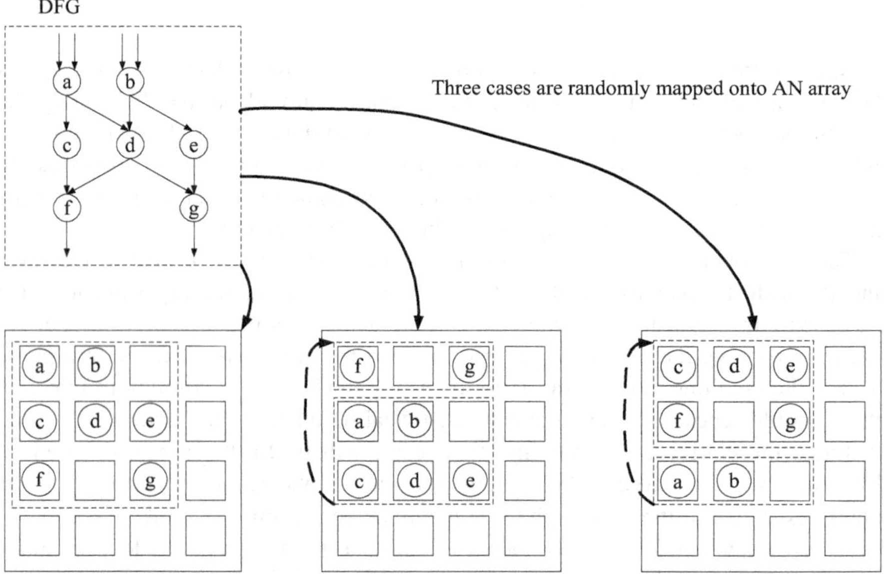

Fig. 6.3 Schematic diagram of the RBR

of randomization, the RBR spatially randomizes each control step. In other words, the randomness can reach m for an algorithm with m stage. The schematic diagram of the RBR is shown in Fig. 6.3.

Assuming that the data flow graph of the round function has three stages, the PEs in the same row in Fig. 6.3 belong to the same stage. When $\Upsilon_{s,0} = 3$, the three mapping modes shown in the figure are randomly adopted. Unlike the fixed mapping mode originally using seven PEs, the RBR method will require two more PEs. In practical applications, the randomness can be less than or greater than three. For example, when $\Upsilon_{s,0} = 4$, the other three PEs in the fourth row will be used. It should be noted that because of the obfuscation feature of iterative encryption, sensitive points mainly exist in the last few rounds of implementation, so relocation technology will be used in the last few rounds. Of course, for greater security, designers can also use the RBR technology in other rounds. The following describes the details of the RBR implementation.

The hardware support diagram of the RBR method is shown in Fig. 6.4. In order to implement the RBR, the control logic in the configuration part needs to be modified accordingly so that it can be driven by a random number. Compared with the non-randomized encryption implementation method, the configuration information of each level in each randomization stage does not need to be repeatedly saved, but the corresponding relation between the configuration information and the actual hardware information needs to be changed.

The main resource consumption of the RBR is produced in the configuration process. In order to reconfigure the round function, the configuration information needs to be taken out of the configuration storage unit and then be used to configure the function. Compared to the entire cipher execution cycle, the additional

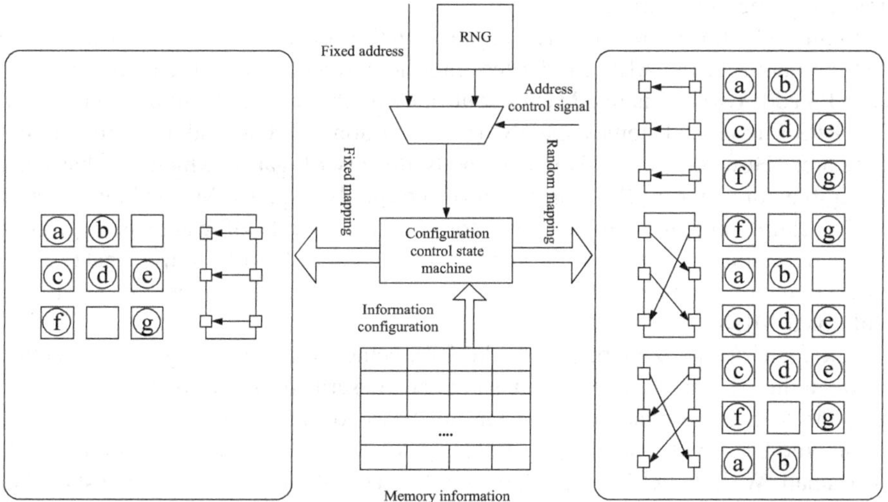

Fig. 6.4 RBR method hardware support diagram

configuration process takes only a few cycles, which is much shorter than the execution cycle of the entire encryption and decryption algorithm (usually several hundred cycles). Therefore, the performance loss and power consumption of the configuration process are very little. When the RBR is adopted, the actual hardware resources consumed increase with the increase of space freedom. Although the total amount of hardware resources cannot be changed after the chip is manufactured, in normal conditions, other free resources may also be used to execute other algorithms or applications from the viewpoint of improving resource utilization. In this case, the discussion of hardware resource utilization will become more meaningful. The RBR selects each computation stage in the array as a space-randomized control step. Compared with the traditional division method based on a sub-function of the algorithm, the reconstruction method with the finer granularity of control can largely reduce the hardware resource consumption of attack countermeasures.

(2) Register pair swap technology

The RPS randomly changes the interconnection relationship between the ALU and the register in each PE so that the outputs of sensitive ALUs (ALUs performing sensitive point operations) in each operation are randomly stored in different registers. The original idea of introducing RPS is that in practical applications, distributed memory units like registers are more susceptible to the threat of fault attacks than traditional combinational logic. After using RPS, the probability of a successful attack after a fault is injected into a target register becomes 1/2, and the space randomness is $\Upsilon_{s,0} = 2$ at this time. By implementing register pair swap technology between multiple PEs, higher space randomness can be implemented. It should be noted that after adding two extra multiplexer units, it must be confirmed that the clock frequency will not be affected; otherwise, a decrease in the clock frequency will result in a reduction in computation speed for the entire reconfigurable cryptographic processor.

Figure 6.5 shows the hardware implementation of the RPS. In order to ensure that the output of the ALU can be fed into the correct PE after the randomization, the RPS adds two alternative MUXs controlled by the same random number to each PE. RPS's overhead comes mainly from additional MUX and random number generator. However, the MUX only adds dozens of gates, which are basically negligible compared to PEs consisting of thousands of gates. Meanwhile, the area of the entire random number generator unit (thousands of gates) is negligible compared to the area of the entire reconfigurable controller (millions of gates). The clock frequency issue should also be confirmed as mentioned before due to the additional MUXs.

The fault attack countermeasure like RPS achieved by changing the connection between the ALU and the register has not been described in any other references. In fact, the RPS is an attack countermeasure based on the structural features of the reconfigurable cryptographic processors. Therefore, its resource consumption is very small. With the RBR adopted, the randomness of the PEs of the same stage has not been developed. The RPS can further improve the randomness between the PEs

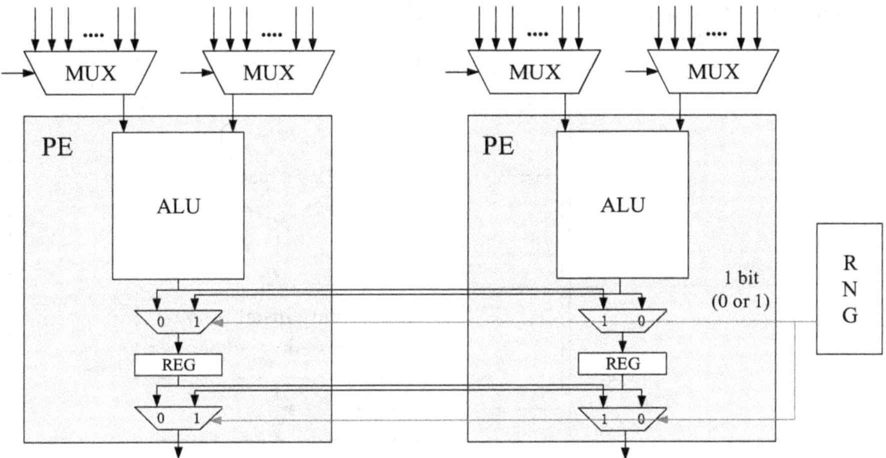

Fig. 6.5 Schematic diagram of the RPS method

of the same stage by changing the connection relationship between the ALU and the registers of the PEs. Because the RPS does not require changing the original configuration information, it does not have any impact on the control part of the RBR technology. The randomness introduced by the RPS and the randomness introduced by the RBR will not overlap. The space randomness of the two methods can be obtained by multiplying their respective random degrees.

(3) Random delay insertion technology

The RDI introduces time randomness by inserting random number of dummy cycles. Figure 6.6 shows a schematic diagram of the random delay technology. Assuming that the sensitive point is located in the Pth cycle of the cryptographic algorithm, the intermediate value after completion of the P − 1 cycles will be fed into the PE to perform the action in the Pth cycle after the random clock cycles have ended. It should be noted that the best operation method of time randomize is to randomly change the order of execution of each operation, so that there will be no performance loss. In order to ensure the correctness of the function of the encryption algorithm, the two operations (f(x) and g(x)) for exchanging the execution order must satisfy f(g(x)) = g(f(x)). In order to ensure the security of the encryption algorithm, the two operations of the encryption algorithm does not always satisfy the function relationship of f(g(x)) = g(f(x)), which means that the orders of most of the operations of the encryption algorithm cannot be exchanged. Therefore, adding dummy cycles is a common attack countermeasure for all the cryptographic algorithms.

Similar to the RBR, some randomization mechanisms need to be added to the circuit control section in order to implement the RDI. The RDI mainly affects the performance of the circuit, in addition to the negligible consumption brought by the random number generator. The dummy cycles increase the execution time of the

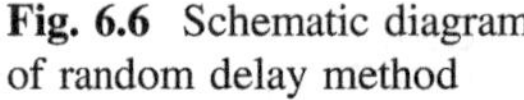

Fig. 6.6 Schematic diagram of random delay method

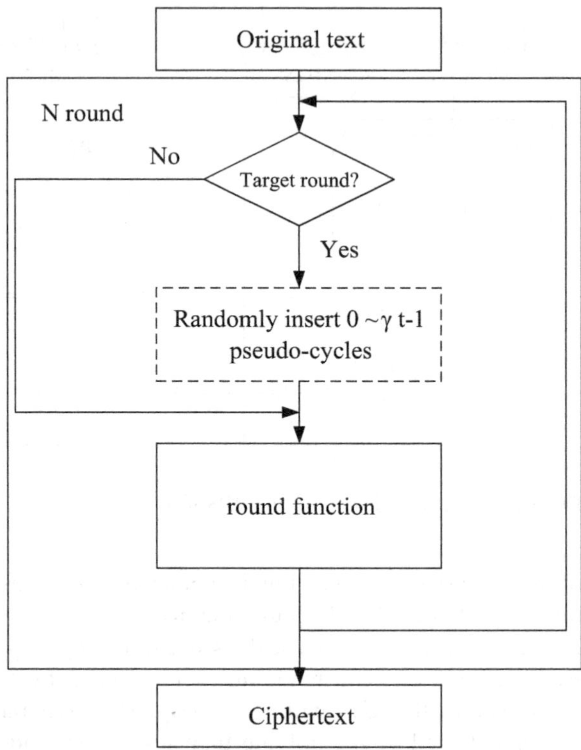

entire algorithm. The total performance reduction is determined by the ratio of time randomness $\Upsilon_{t,0}$ to the total execution cycles.

In the execution of the cryptographic algorithm, the method of inserting the dummy cycles to obfuscate the attacker is widely used in fault attack countermeasures and side-channel attack countermeasures. When using these time-randomized approaches, the most important consideration is how to prevent these cycles from being identified by the basic attack methods such as the simple power attack method. For example, if the hardware resources are idle during the dummy cycles, the power consumption in these cycles will be significantly less than the power consumption of the normal cycles of the algorithm, which makes the dummy cycles easily identified by attackers according to the power consumption curve. In the RDI, the entire reconfigurable cryptographic processor still performs the original operation, in which some of the unrelated data stored in the internal registers instead of the target intermediate value will be sent to the reconfigurable cryptographic processor. When the dummy cycles end, the intermediate value will continue to be sent to the reconfigurable cryptographic processor and perform effective operation.

The attack resistance of the above randomization methods can be evaluated by using the IEM. The performance loss, area consumption, and power consumption introduced by the randomization method can be obtained through the mapping on the actual reconfigurable cryptographic processor platform. Take the AES algorithm against the highly threatening double-fault attack for example. After the attack countermeasure is mapped to the reconfigurable cryptographic processor experiment platform, the fault attack resistance is significantly improved, and the consumption caused by the attack countermeasure is also little. When three kinds of attack countermeasures are used simultaneously, the attack resistance of the reconfigurable cryptographic processor is improved by 2–4 orders of magnitude, the throughput decreases by 5%, the area consumption increases by 35%, and the power consumption increases by 10% [7].

4. Derivation Process of E(Nsearch) and E′(Nsearch) Expressions

First, show the proof of $E'(N_{\text{search}})$. Assume that there are φ attack elements in total. There are $\gamma(\gamma = \gamma_s \times \gamma_t$, and $\gamma_s \times \gamma_t \leq \varphi'_s \times \varphi'_t)$ attack elements that are special because the sensitive points are randomly mapped to these attack elements. Attackers search for attack elements one by one until the first sensitive point is found. N_{search} is the total number of attack elements searched when the first sensitive point is found. $E'(N_{search})$ is the expectation of N_{search}, we can get the following expression, namely

$$E'(N_{\text{search}}) = \sum_{k=1}^{\varphi-\gamma+1} \frac{C(\varphi - k, \gamma - 1)}{C(\varphi, \gamma)} \times k \tag{6.41}$$

The proof of $E'(N_{search})$ expression is given below. The process of searching for a target attack element can be transformed into a ball taking model of probability theory. γ is the total number of special balls that are randomly mapped to φ boxes and each box holds at most one ball. N_{search} is the total number of boxes searched when the first special ball is found. Since attackers do not know which attack element is a sensitive point, the order of the search can be considered as random, and it can be assumed that the positions of the γ special balls are random.

According to the definition of expectations, we can get the following formula, that is

$$E'(N_{\text{search}}) = \sum_{k} P(k) \times k \tag{6.42}$$

where k is all possible values of N_{search}; P(k) is its corresponding probability value.

The proof of probability P(k) is as follows.

When k = 1, there is

$$P(k = 1) = \frac{C(\phi - 1, \gamma - 1)}{C(\phi, \gamma)} \tag{6.43}$$

When k = 2, there is

$$P(k=2) = P(k=2|k \neq 1)P(k \neq 1)$$
$$= \frac{C(\phi-2,\gamma-1)}{C(\phi-1,\gamma)} \times (1 - \frac{C(\phi-1,\gamma-1)}{C(\phi,\gamma)}) \tag{6.44}$$

Because of there is

$$C(\phi,\gamma) = C(\phi-1,\gamma) + C(\phi-1,\gamma-1) \tag{6.45}$$

we can get *P*(*k*) expression, that is

$$P(k) = \frac{C(\phi-k,\gamma-1)}{C(\phi,\gamma)} \tag{6.46}$$

The proof process of $E'(N_{search})$ expression in a double-fault attack process is given below.

The proof in a double-fault attack is similar to the one in a single-fault attack, and the ball taking model of probability theory is also used. There are $\gamma 0$ ($=\gamma_{s,0} \times \gamma_{t,0}$) special red balls and $\gamma 1$ ($=\gamma_{s,1} \times \gamma_{t,1}$) special green balls. These special balls are randomly mapped to φ ($=\varphi'_s \times \varphi'_t$) boxes, and each box can hold a maximum of one ball. During the search, the search process will end after the attacker searched two boxes each time and found a red ball and a green ball in both boxes. Nsearch is the total number of pairs of boxes searched after the first target ball pair is found. From the above theoretical analysis, we can obtain the expression when k = 1, the proof process is as follows.

Since there is

$$\frac{a}{b} = \frac{\frac{(b-1)!}{(a-1)!((b-1)-(a-1))!}}{\frac{b!}{a!(b-a)!}} = \frac{C(b-1,a-1)}{c(b,a)} \tag{6.47}$$

where $a,b \in \{1,2,\ldots\}, b \geq a$.

The expression of *P*(*k*) is

$$P(k=1) = \frac{C(\gamma 0,1)C(\gamma 1,1)}{C(\phi,2)} \tag{6.48}$$

The expression of $E(N_{search})$ can be obtained by using E′(Nsearch) expression, that is

$$E(N_{search}) = \frac{\varphi_s \times \varphi_t + 1}{2} \tag{6.49}$$

6.1.2 Randomization-Based Electromagnetic Attack Countermeasure Technology

1. Introduction to Electromagnetic Attacks

In 1996, after Kocher et al. [8] first proposed side-channel attacks, one after another attacker used power analysis attacks and timing attacks to attack cryptographic processors. However, electromagnetic attacks have not been involved. In 2002, Agrawal et al. [9] first analyzed the characteristics of electromagnetic attacks and pointed out that electromagnetic attacks are more aggressive than power analysis attacks in attack efficiency. Then, the electromagnetic attack gradually attracted the attention of the researchers. A brief introduction to the source of electromagnetic information leakage and the basic knowledge of electromagnetic probes will be given in the following.

Electromagnetic leakage of chips mainly comes from the changing currents in the various modules of the circuits, such as the control units, the I/O interfaces, the computing units. In fact, not only the changing currents in various parts of the chips independently radiate out electromagnetic leakage, but also the electromagnetic fields generated by various parts couple themselves to form coupled electromagnetic leakage. The electromagnetic leakage in electromagnetic attacks mainly comes from the data processing module, so attackers usually lock their targets in the registers, the S-Box, and other circuit modules of the cryptographic processor. By gaining the electromagnetic leakage at a particular position and at a particular time, an attacker can obtain the information that is directly related to the key. The leakage position of electromagnetic information at a particular time is defined as a leakage point. The source of electromagnetic leakage will be analyzed in the following, starting from the underlying structure of the circuit. CMOS cryptographic processors are typically made up of millions of transistors and their interconnection wires. Circuits will leak electromagnetic information when the logic state inside the chip changes. Generally, the logic state changes in the circuit are controlled by the clock. Most cryptographic processors are synchronous circuits. Therefore, it can be considered that the internal logical state change of the cryptographic processor is synchronized with the change of the clock signal. An attacker can determine the time node information of a leakage by determining the specific execution cycle. Attackers can use the small size of the electromagnetic probe to sample the electromagnetic leakage of a specific position. Compared to the power analysis attack in which the attacker can only get the total current changes of the chip and cannot get the leakage information of a specific position, the biggest advantage of the electromagnetic attack is the locality of the sampled information. Two types of electromagnetic leakage will be described in the following. Electromagnetic leakage can be divided into direct leakage and coupled leakage. The source of direct leakage is the transient current in the circuit, while the source of the coupled leakage is the electromagnetic information formed by coupling the electromagnetic information leaked from various parts. Electromagnetic leakage can also be divided into near-field leakage and far-field leakage from the perspective of electromagnetism.

According to the definition of electromagnetism, it is called near-field leakage when the distance between the sampling point and the leakage target is less than $2\pi/\lambda$; otherwise, it is called far-field leakage, where λ is the wavelength of radiation electromagnetic waves. Recent studies show that the closer the distance to the chip surface, the higher the locality of the electromagnetic leakage, the higher the signal-to-noise ratio, and the higher the success rate of attack. In 2013, Sugawara et al. [10] for the first time demonstrated a local electromagnetic attack (LEMA) aiming at gate-level circuits, which completely invalidates the traditional attack countermeasures such as increasing noise and masking, causing great attention of circuit designers. The reason why electromagnetic attacks make all traditional attack countermeasures fail is that the electromagnetic leakage is localized. The reconfigurable cryptographic processors enable time and space randomization through dynamic reconfiguration, thereby defeating the locality of electromagnetic attacks.

The most important tool of electromagnetic attackers used to sample electromagnetic signals—the electromagnetic probe will be described in the following. According to Faraday's Law of Electromagnetic Induction, the changing electromagnetic field can generate induced current in the closed circuit. The attacker will analyze the correlation of the key based on the sampled induced current and then infer the correct key. The commonly used electromagnetic probes include the far-field probe and the near-field probe. As its name implies, the far-field probe mainly measures the far-field radiation, and the near-field probe mainly measures the near-field radiation. Because the near-field radiation has less information noise and higher attack success rate, near-field probes are primarily considered here, and studies of far-field probes can be viewed in other literature [11].

Figure 6.7 shows the physical map of the electromagnetic probe. As for the design of the electromagnetic probe, we may refer to the EMC standard [12]. Heyszl et al. [13] pointed out that the smaller the electromagnetic probe is, the higher the resolution will be. With the higher resolution probe, the attacker is able to sample more accurate electromagnetic leakage information, and the main electromagnetic information comes from the leakage source closest to the probe. The closer to the chip the distance is, the more accurate the electromagnetic information sampled by the attacker is. For example, when the electromagnetic probe is placed about 1 cm above the chip, the sampled electromagnetic information is very similar to the sampled power information in the power analysis attack because the electromagnetic leakage mainly comes from the main leakage sources such as the power wires. As the probe gets closer to the surface of the chip, the probe picks up the information on the specific part of the chip. According to the different position of the probe relative to the chip, the electromagnetic probes can be divided into vertical probes and horizontal probes. Figure 6.8 shows the schematic diagram of vertical and horizontal probes. The vertical probe is placed perpendicular to the surface of the chip and mainly samples the electromagnetic leakage information that is parallel to the chip. The horizontal probe is placed parallel to the chip surface and mainly samples the electromagnetic leakage information perpendicular to the chip surface.

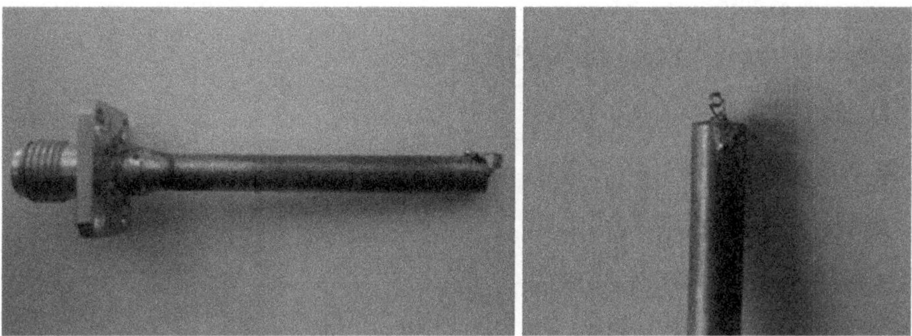

Fig. 6.7 Near-field probe pictures

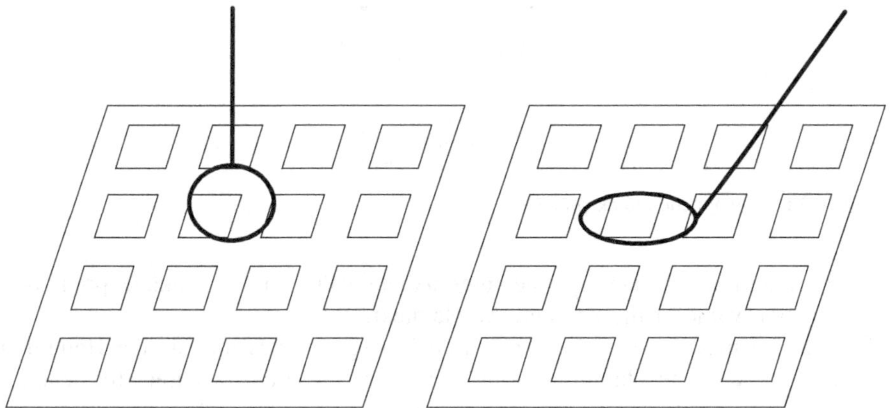

Fig. 6.8 Schematic diagram of a vertical probe and a horizontal probe

The attacker calculates the model output according to the attack model and obtains the electromagnetic leakage information of the chip as shown in Fig. 6.9. After obtaining the results of these two parts, the attacker statistically compares the results of the physical output and the model output to obtain the correct key. The attack process is described in detail as follows.

① Plaintext choice. According to the previous analysis of the electromagnetic leakage source, it shows that the internal state change of the chip will reveal the electromagnetic information. In order to obtain the electromagnetic leakage curve, the attacker must change the plaintext input of the chip to obtain electromagnetic leakage. The attacker samples a certain number of curves at a given leakage point, and the total number of sampling curves is directly proportional to the number of the input plaintext.

② Intermediate value computation. After selecting a certain number of plaintexts, the attacker needs to determine the intermediate value of the leakage point in

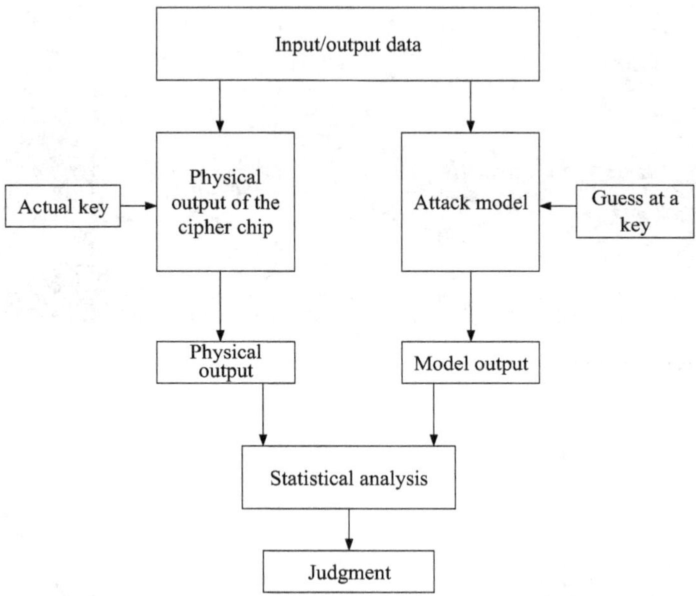

Fig. 6.9 Electromagnetic attack process

the algorithm. Generally, the intermediate value of the leakage point is a function value of the key and the plaintext.

③ Leakage model. The commonly used leakage models include the Hamming distance and the Hamming weight models. The more accurate the leakage model is, the fewer the number of curves needed for subsequent statistical analysis is and the fewer the data processing steps are. A brief introduction of these two leakage models is now described in the following. We first introduce the definitions of the Hamming weight and the Hamming distance. The Hamming weight refers to the number of nonzero elements in a string. For example, if the string X = "0100110", the Hamming weight of the string X is $W_H(X) = 3$. The Hamming distance refers to the number of the different characters between two strings with equal length. Assuming the string X1 = "1011001", the Hamming distance between the strings *X* and *X1* is $D_H(X) = 7$.

Brier et al. [14] proposed a Hamming distance model based on the power consumption of CMOS circuits. The attacker calculates the total number of conversions from "0" to "1" and from "1" to "0" of an intermediate value of the cryptographic processor at a certain period of time and then describes the total power consumption through the total number of conversions. The Hamming distance model can also be used in electromagnetic attacks, which will describe the electromagnetic radiation in a period of time through the total number of state conversions inside the chip.

The Hamming distance model assumes that there is electromagnetic leakage information when the output of the CMOS gate circuit is converted from “0” to “1” or from “1” to “0”, and the gate circuit will not leak electromagnetic information when the output does not change. Due to the similarity between electromagnetic radiation and power consumption, the electromagnetic information radiated by the CMOS gate circuit can also be characterized by the Hamming distance model, that is

$$E = aH_w(Y \oplus Y') + b \tag{6.50}$$

where Y is the state of the output of the CMOS gate circuit before it is converted; Y' is the state after the output of the CMOS gate circuit is converted; E is the electromagnetic energy radiated by the CMOS gate circuit during the conversion period; a is a scaling factor; and b is the noise.

Messerges et al. [15] proposed a Hamming weight model to describe the power consumption of CMOS gate circuits. The Hamming weight model assumes that the power consumption of a CMOS gate circuit in a period of time is directly related to the output state. Similar to the power analysis attack, the electromagnetic energy radiated by the CMOS circuit can also be described with the Hamming weight, that is

$$E = aH_w(Y) + b \tag{6.51}$$

where Y is the state of the CMOS gate circuit.

In the power analysis attack, it is considered that the CMOS gate circuit has the same power consumption when the output converting from “0” to “1” and from “1” to “0”. However, in the electromagnetic attack, since there is a difference in the direction of current flow between the conversions from “0” to “1” and from “1” to “0”, the directions of the induced voltages generated in the electromagnetic probe are opposite when the circuit output converts from “0” to “1” and from “1” to “0”. Attackers can see the induced voltages in two different directions on the oscilloscope.

④ Obtaining the electromagnetic leakage curve. An electromagnetic attack experiment platform is shown in Fig. 6.10. In order to sample the correct information, the attacker first samples the electromagnetic leakage of the entire chip. After obtaining the electromagnetic information of the entire chip, the attacker selects the sampling point with the largest signal-to-noise ratio as the leakage point. According to the previously identified leakage point information, the electromagnetic probe is placed at the leakage point to continuously sample the electromagnetic information at this position, and the sampled result will be displayed on the oscilloscope. The electromagnetic leakage curve is a continuous curve of voltage. Each curve corresponds to the voltage value sampled by the electromagnetic probe at a specific time (the sampling speed of the oscilloscope is far greater than the rate of change of the magnetic field).

⑤ Data preprocessing. Because the acquired leakage curve contains some irrelevant information such as noise, making it more difficult to extract the correct key by using the statistical model later, the data needs to be preprocessed. The purpose of data preprocessing is to reduce the data dimension and extract the main information. The commonly used data preprocessing method is the principal component analysis (PCA).

⑥ Statistical model computation. After getting the preprocessed data, the attacker analyzes the result through a statistical model. The most commonly used statistical method is the correlation coefficient. The attacker performs statistical analysis by calculating the correlation coefficient between the model output and the physical output, and the correlation coefficient of the correct key will be much larger than that of the incorrect key.

2. Electromagnetic Leakage and Probe Sampling Analysis

According to the definition of electromagnetism, the electromagnetic attack process is similar to the antenna transmitting and receiving process. The current wire on the chip acts as the transmitting antenna to radiate the electromagnetic waves that contain the key information, and the electromagnetic probe acts as a receiving antenna to receive the electromagnetic waves. Maxwell's equations are

Fig. 6.10 Electromagnetic attack experiment platform

$$\Delta \cdot D = \rho \tag{6.52}$$

$$\Delta \cdot B = 0 \tag{6.53}$$

$$\Delta \cdot E = J + \frac{\partial B}{\partial t} \tag{6.54}$$

$$\Delta \times H = J + \frac{\partial D}{\partial t} \tag{6.55}$$

where $J = \sigma E$, σ is the conductivity of the medium, E is the electric field strength, $D = \varepsilon E$, ε is the permittivity of the medium, $B = \mu H$, μ is the permeability of the medium, D is the electric flux density, and H is the magnetic field strength.

The Helmholtz equations can be further obtained, i.e.,

$$\Delta^2 E + k^2 E = 0 \tag{6.56}$$

$$\Delta^2 B + k^2 B = 0 \tag{6.57}$$

where $k^2 = \varepsilon_0 \mu_0 \omega^2$.

According to the Helmholtz equations, the electric field strength E and the magnetic field strength H can be obtained

$$E = A(k) \cdot \exp(-jkr) \tag{6.58}$$

$$H = -\frac{1}{j\omega\mu} \Delta \times E = \frac{1}{\omega\mu} k \times A(k) \cdot \exp(-jkr) \tag{6.59}$$

where A is the wave vector of the electromagnetic wave.

The schematic diagram of the space transmission of electromagnetic wave is shown in Fig. 6.11.

The electromagnetic radiation inside the chip can be described by using Maxwell's equations. When solving Maxwell's equations, it yields the strength of the electric and magnetic fields radiated at any position within the chip. According to the analysis in the previous section, the current inside the chip will leak the key information. The wires carrying the key information are similar to the transmitting antenna in electromagnetism, and the radiated electromagnetic waves will propagate in free space. According to the definition in electromagnetism, when the length of the radiating antenna is less than 1/50 of the length of the radiated electromagnetic wave, the radiating antenna can be regarded as an infinite small antenna, and the electromagnetic wave radiated by the antenna can be approximated by the electromagnetic radiation pattern shown in Fig. 6.12.

From the expressions (6.58) and (6.59), we first need to calculate the electromagnetic wave vector before calculating the electromagnetic field. The wave vector of an infinitely small conductor is

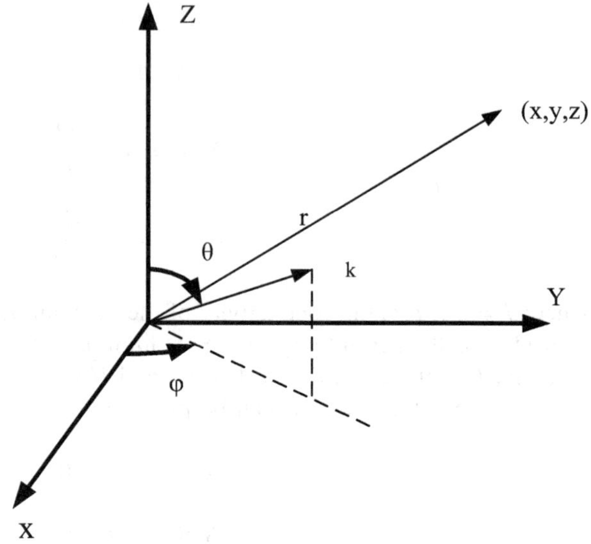

Fig. 6.11 Schematic diagram of electromagnetic wave space radiation

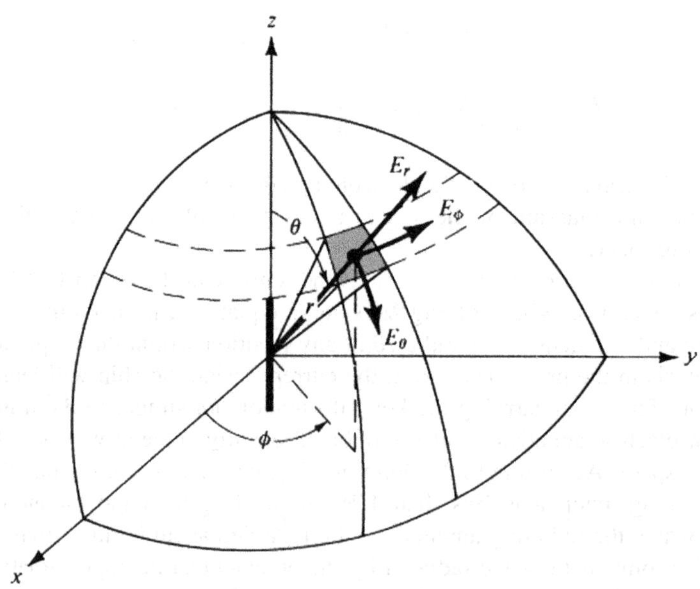

Fig. 6.12 Schematic diagram of infinite small antenna radiation

$$A(x,y,z) = a_z \frac{\mu I_0}{4\pi r} e^{-jkr} \int_{-l/2}^{l/2} \mathrm{d}z' = a_z \frac{\mu I_0 l}{4\pi r} e^{-jkr} \tag{6.60}$$

where r is the distance from the radiating antenna to the radiation point, and its expression is

$$r = \sqrt{\mathrm{x}^2 + \mathrm{y}^2 + \mathrm{z}^2} \tag{6.61}$$

According to the transformation relationship of xyz coordinates and spherical coordinates, we can get

$$\begin{bmatrix} \mathrm{A_r} \\ \mathrm{A}_\theta \\ \mathrm{A}_\varphi \end{bmatrix} = \begin{bmatrix} sin\theta cos\varphi & sin\theta sin\varphi & cos\theta \\ cos\theta cos\varphi & cos\theta sin\varphi & -sin\theta \\ -sin\varphi & cos\varphi & 0 \end{bmatrix} \begin{bmatrix} \mathrm{A_x} \\ \mathrm{A_y} \\ \mathrm{A_z} \end{bmatrix} \tag{6.62}$$

According to the expression (6.62), we can get the expressions of the wave vectors in all directions under the spherical coordinate system as

$$A_r = \frac{\mu I_0 l \mathrm{e}^{-jkr}}{4\pi r} \cos\theta \tag{6.63}$$

$$A_\theta = -\frac{\mu I_0 l \mathrm{e}^{-jkr}}{4\pi r} \sin\theta \tag{6.64}$$

$$A_\varphi = 0 \tag{6.65}$$

According to the expression (6.59), we can get the magnetic field H

$$H = \mathrm{a}_\varphi \frac{1}{\mu r} \left(\frac{\partial}{\partial r} (\mathrm{r}\mathrm{A}_\theta) - \frac{\partial \mathrm{A_r}}{\partial \theta} \right) \tag{6.66}$$

According to the expressions (6.63)–(6.65), we can get the magnetic field expressions in all directions, that is

$$H_r = 0 \tag{6.67}$$

$$H_\theta = 0 \tag{6.68}$$

$$H_\varphi = j \frac{k I_0 \sin\theta}{4\pi r} \left(1 + \frac{1}{jkr}\right) e^{-jkr} \tag{6.69}$$

According to the previous introduction, the electromagnetic leakage studied here is the near-field, so in the near-field area, the expression (6.69) can be simplified as

$$H_{\varphi} \cong \frac{\sim I_0 l}{4\pi r^2} \sin\theta \tag{6.70}$$

Because the size of the radiating antenna inside the chip cannot be ignored compared to the size of the probe, the length of the antenna needs to be considered here. The effect of antenna length on radiation is shown in the Fig. 6.13.

It can be seen from Fig. 6.13 that the final magnetic field strength H can be obtained by integrating each differentiating current element in the antenna

$$H \cong \frac{I_0}{4\pi r} \sin\theta(\cos\alpha - \cos\beta) \tag{6.71}$$

The following part focuses on the characteristics of the electromagnetic probe. The most important characteristic of an electromagnetic probe is its receiving model. Assume the attacker uses a magnetic probe that is a circular probe. The receiving schematic diagram of the electromagnetic probe is shown in Fig. 6.14, where a is the inner diameter of the probe, $a + 2b$ is the outer diameter of the probe, $2b$ is the width of the probe material, and 1 and 2 ends are the open ends of the probe and connected to the oscilloscope system. The sampled induced voltage is finally displayed on the oscilloscope through the oscilloscope system.

The following part gives the specific expression of the receiving mode of the electromagnetic probe. A horizontal probe is placed parallel to the chip surface and mainly samples the electromagnetic information perpendicular to the direction of the chip. The receiving mode is

$$V_{\mathrm{oc}} = \oint Nj\omega\pi\mu_0 H \cos\varphi_i \sin\theta_i \mathrm{d}S \tag{6.72}$$

When substituting the magnetic field expression into Expression (6.72), we can obtain the expression (6.73), that is

$$V_{\mathrm{oc}} = \oint Nj\omega\pi\mu_0 H \cos\varphi_i \sin\theta_i (\cos\alpha - \cos\beta) \mathrm{d}S$$

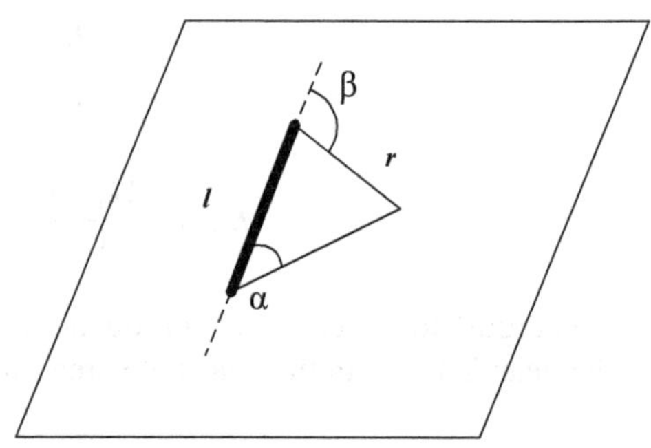

Fig. 6.13 Effect of the antenna length on radiation

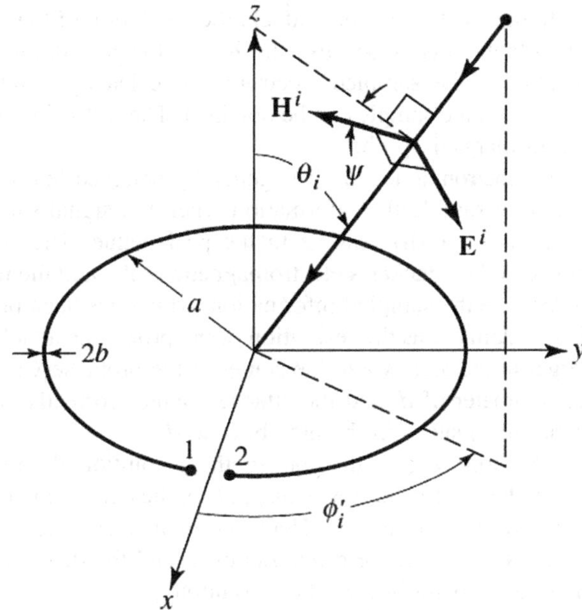

Fig. 6.14 Schematic diagram of the receiving mode of the electromagnetic probe

$$= \begin{cases} Nj\omega\mu_0 \frac{I_0}{4\pi}(\cos\alpha - \cos\beta)\left[4x + 2h\left(\mathrm{a}\tan\left(\frac{a-x}{h}\right) - \mathrm{a}\tan\left(\frac{a+x}{h}\right)\right)\right], x < a \\ Nj\omega\mu_0 \frac{I_0}{4\pi}(\cos\alpha - \cos\beta)\left[4x + 2h\left(\mathrm{a}\tan\left(\frac{x-a}{h}\right) - \mathrm{a}\tan\left(\frac{a+x}{h}\right)\right)\right]\frac{a}{x}, a < x \end{cases} \quad (6.73)$$

where h is the vertical distance from the center of the electromagnetic probe to the chip; x is the distance from the leakage point to the probe center.

The vertical probe is placed perpendicular to the chip surface and mainly samples the electromagnetic information parallel to the direction of the chip. The receiving mode is

$$V_{\mathrm{oc}} = \oint Nj\omega\pi\mu_0 H \sin\varphi_i \sin\theta_i \mathrm{d}s \quad (6.74)$$

where V_{oc} is the voltage when the 1 and 2 ends of the electromagnetic probe are open in Fig. 6.14; N is the number of turns of the electromagnetic probe.

Substituting the magnetic field expression into Expression (6.74), we can obtain the expression (6.75), that is

$$V_{\mathrm{oc}} = \oint Nj\omega\pi\mu_0 \sin\varphi_i \sin\theta_i(\cos\alpha - \cos\beta)\mathrm{d}S$$

$$= Nj\omega\mu_0 \frac{I_0}{4\pi}(\cos\alpha - \cos\beta)\left[4a + 2x\left(\mathrm{a}\tan\left(\frac{h-a}{x}\right) - \mathrm{a}\tan\left(\frac{h+a}{x}\right)\right)\right]\frac{a}{h} \quad (6.75)$$

In the electromagnetic attack, the resolution of the electromagnetic probe is very important. This is because the higher the resolution of the probe, the higher the locality of the sampled electromagnetic leakage, and the lower the noise. Thus, more accurate sample can be obtained. The following describes the resolution of the electromagnetic probe.

In electromagnetism, it is generally believed that an attacker will not be able to correctly sample the information when the signal sampled by the electromagnetic probe drops 6 dB relative to the peak value. Therefore, the farther the distahce between the attacker's electromagnetic probe and the target leakage point, the more inaccurate the sampled information, which has been proved by numerous attackers. Let us define d as the resolution of the probe, the attacker can effectively sample the target in the area where the center of the probe serves as the center of a circle with the diameter of d, and the attacker cannot correctly sample the target electromagnetic information in the area beyond d.

The following will analyze the resolution d through the specific numerical computation. Because N, j, ω, μ_0, I_0 in the expression (6.73) and (6.75) are constant, they are not considered. Here, we mainly consider the influence of the radiation antenna length L, the probe radius a, and the distance h between the probe center and the chip surface on the resolution.

The following will analyze the effect of each variable on the resolution d through the control variable method. Consider the effect of the sampling height h on the resolution of the probe. Let us define the radiation antenna length $L = 10$ μm, the probe radius $a = 10$ μm. For the horizontal probe, the schematic diagram as shown in Fig. 6.15 can be obtained. For the vertical probe, the schematic as shown in Fig. 6.16 can be obtained.

Comparing the two different probe sampling curves, we can see that the voltage curves of the horizontal probe and the vertical probe are different. For the horizontal probe, when the probe center is just above the leakage point, the magnetic flux passing through the probe is always 0. No matter how the current at the leakage point changes, the voltage sampled by the magnetic field probe is always 0. For the vertical probe, when the center of the probe is just above the leakage point, the magnetic line perpendicularly passes through the center of the probe and the maximum voltage is sampled.

The following discusses the resolution of the probe in the vertical direction when the probe is fixed at a horizontal position. It can be seen from Fig. 6.15 that as the height h increases, the voltage value sampled by the probe gradually decreases and the resolution also gradually decreases. When the distance between the probe and the chip surface is h′, the voltage sampled at this moment is exactly 6 dB lower than the peak value. Then, the height h′ can be regarded as the maximum vertical distance when the probe samples. Otherwise, the probe cannot correctly sample the target electromagnetic information. The height h' is

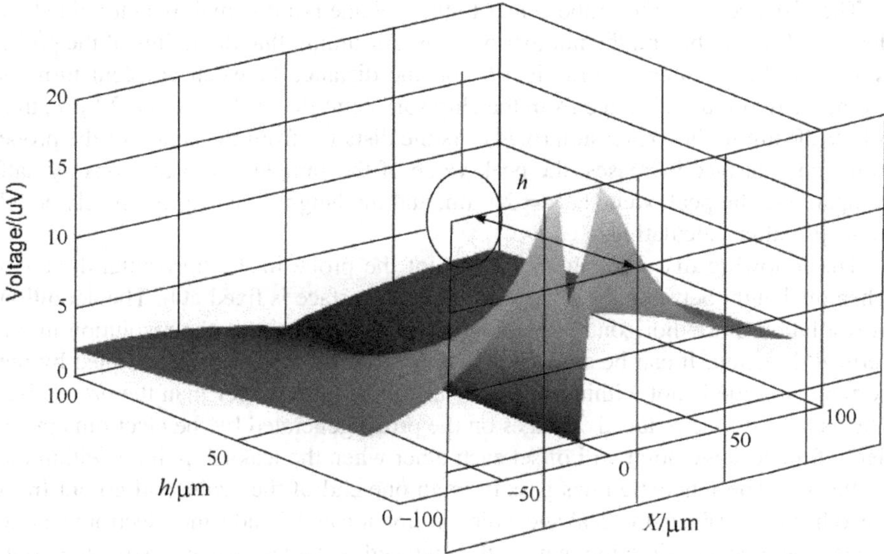

Fig. 6.15 Effect of the sampling height h on the horizontal probe resolution

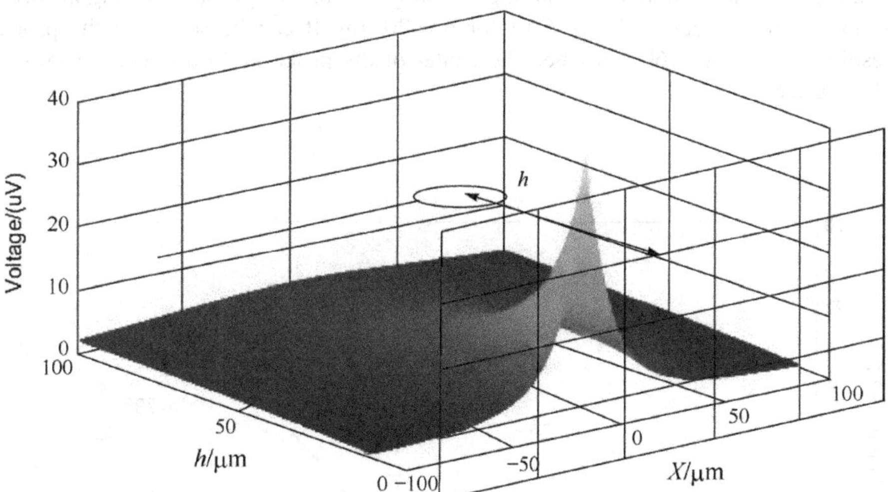

Fig. 6.16 Effect of the sampling height h on the vertical probe resolution

$$20\lg\left(\frac{V(h')}{V(h)}\right) = 6\,\text{dB} \rightarrow \left(\frac{V(h')}{V(h)}\right) = 2 \tag{6.76}$$

where h is the minimum distance between the probe and chip surface.

The distance from the probe center to the leakage point is mainly determined by the size of the probe and the number of turns. Assuming that the radius of the probe is 10 μm, the number of turns is 10, and the distance between adjacent turns is 2 μm, the minimum distance from the chip surface to the probe may be 20 μm, that is $h \approx 20$ μm in the expression (6.74). As the distance from the center of the probe to the chip surface increases, the peak value of the sampled voltage drops by half compared to the peak value at $h = 20$ μm, and the height h' is approximately equal to 40 μm after calculating.

The following discusses the resolution of the probe in the horizontal direction when the height between the probe and the chip surface is fixed at h. The definition of resolution in the horizontal direction is exactly the same as the resolution in the vertical direction. It can be seen from Fig. 6.17 that the peak point sampled by the horizontal probe is not within the probe size but slightly larger than the probe size. This is because the induced voltages on the probe generated by the electromagnetic field of the leakage point will offset each other when the leakage point is within the probe size. The magnetic lines pass through one end of the probe and go out from the other end. When the leakage point is not located inside the electromagnetic probe, there is no situation where the induced voltages cancel each other out. However, as the probe center is further away from the leakage point, the strength of the sampled magnetic field also becomes smaller. Therefore, the sampled peak point of the horizontal probe is slightly larger than the probe size. Figure 6.17 shows a cross section of Fig. 6.15 for $h = 20$ μm. It can be seen that the probe resolution is about 160 μm when the center of the probe is 20 μm away from the chip surface.

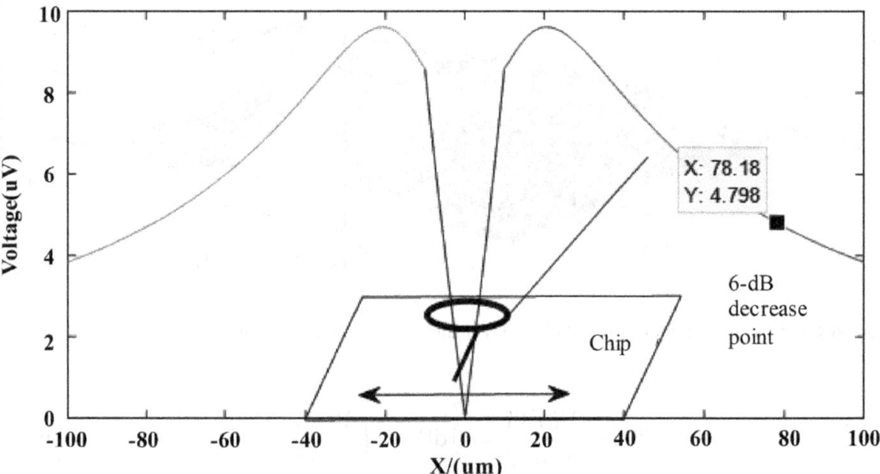

Fig. 6.17 Sectional view of the horizontal probe sampling

For the vertical probe, when the probe is located directly above the leakage antenna, it will obtain the sampled peak value; when the probe is moving in the horizontal direction, and the voltage that is obtained at the distance d from the radiation antenna has dropped 6 dB relative to the peak voltage, we will define d as the resolution of the probe. It can be seen from Figs. 6.15 and 6.16 that when the probe is fixed at a height of h = 20 μm, there is a significant difference between the horizontal valid sampling ranges of the horizontal and vertical probes.

It can be seen from Fig. 6.18 that the resolution of the vertical probe is about 40 μm when the center of the probe is 20 μm away from the chip surface. Therefore, the resolution of the vertical probe is higher than that of the horizontal probe; i.e., the vertical probe can sample more local electromagnetic information at the same probe size. Subsequent discussion mainly focuses on the vertical probe, and we will not make further discussion on the horizontal probe due to its analysis is exactly the same as the vertical probe.

The following analyzes the impact of the probe size on the resolution. We fix the vertical probe just above the leakage point and change the radius of the probe R; then, we analyze the change of probe resolution. Figure 6.19 shows the effect of probe size on probe resolution.

We can see that the larger the probe size, the larger the sampling range. In order to better analyze the effect of the probe size on the resolution, we select the cross section of Fig. 6.19 for analysis. Figure 6.20 shows the cross section of Fig. 6.19 when the probe size is 10 and 20 μm.

It can be seen from Fig. 6.20 that the larger the probe size is, the poorer the resolution is. When the probe radius is 20 μm, the resolution of the probe is about 60 μm. When the probe radius is 10 μm, the resolution of the probe is about 40 μm.

The following analyzes the effect of the length of the radiation antenna on the resolution. The effect of the length of the radiation antenna on the sampling voltage

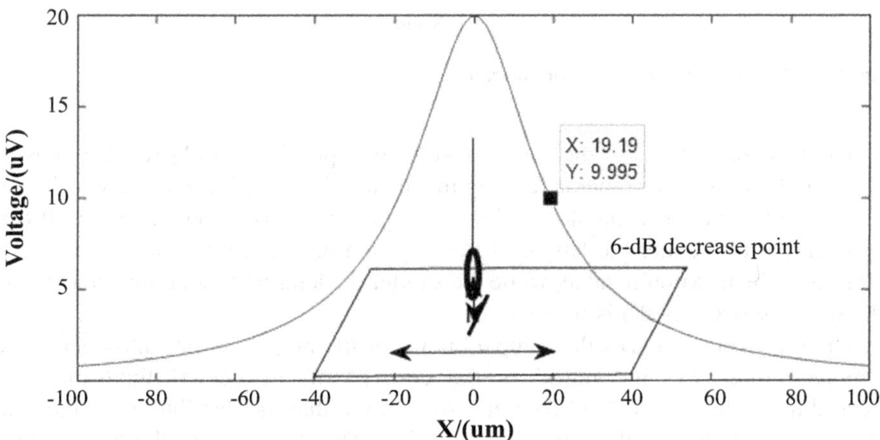

Fig. 6.18 Sectional view of the vertical probe sampling

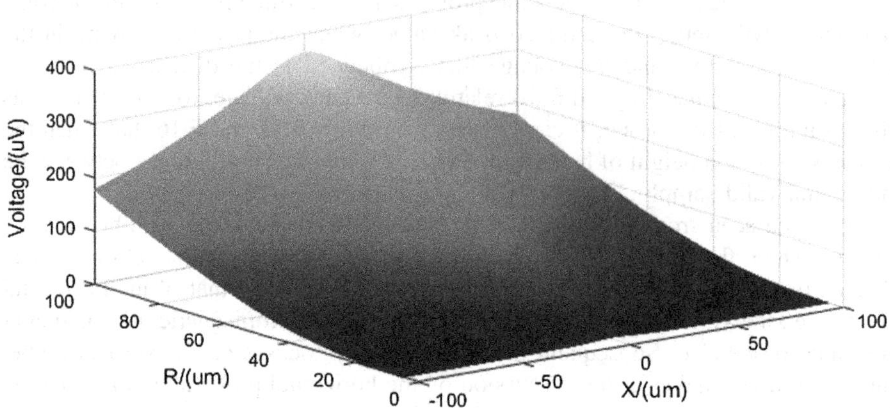

Fig. 6.19 Effect of the probe size on the resolution

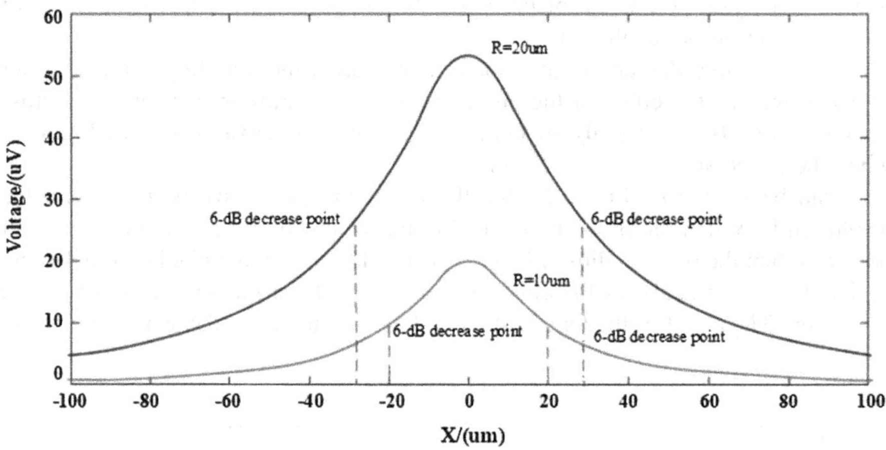

Fig. 6.20 Effect of the probe size on the resolution

is mainly concentrated in the induced voltage expression (6.71) $(\cos\alpha - \cos\beta)$. Figure 6.21 shows the schematic diagram of the change of $(\cos\alpha - \cos\beta)$ in the induced voltage expression along with the length of the radiating antenna L. It can be seen that the longer the length of the radiating antenna, the closer the $(\cos\alpha - \cos\beta)$ is to the maximum value, while the shorter the length of radiating antenna, the closer the $(\cos\alpha - \cos\beta)$ is to zero.

The following analyzes the sampled noise of the probe. Before introducing the sampled noise, we first introduce the concept of the central limit theorem. According to the central limit theorem, we can see that the distribution of the sum of random variables is the normal distribution. The definition of the central limit theorem is as follows: Suppose the random variables X1, X2, …, Xn are

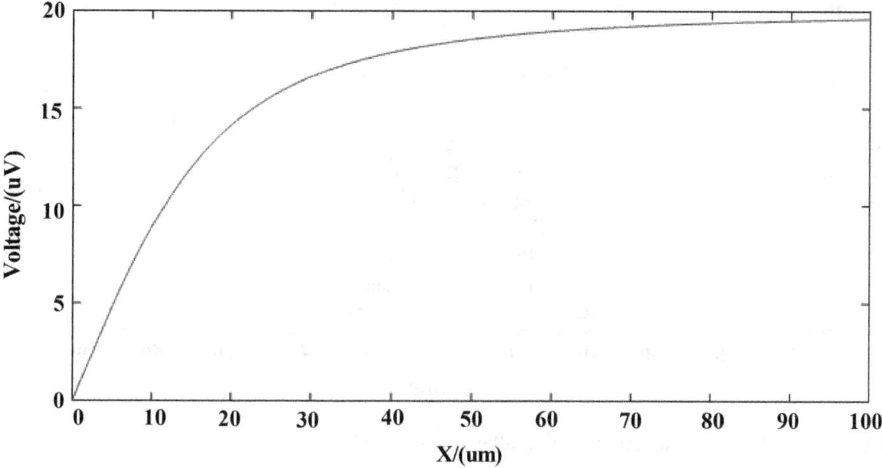

Fig. 6.21 Effect of the radiating antenna length on the resolution

independent of each other and the random variable sequences X_1, X_2, ..., Xn have the same mean value and variance, i.e., EXi = μ, DXi = σ, then the sum of the sequence of random variables $Y_n = X_1 + X_2 + \cdots + X_n$ satisfy $\frac{Y_n - \mathrm{E}Y_n}{\sqrt{D(Y)}} = \frac{Y_n - n\mu}{\sqrt{n}\sigma} = \frac{\bar{Y} - \mu}{\sigma/\sqrt{n}} \to N(0,1)$. In electromagnetic attacks, attackers shall face a variety of noises, including the electronic noise, the thermal noise, and the algorithm noise. The algorithm noise is the most significant source of noise. It is generated by the numerous modules in the circuit that are parallel to the target leakage point. The electronic noise is mainly generated by the oscilloscope sampling and the current disturbance in the circuit. The thermal noise is mainly caused by the electromagnetic interference in the environment. The attacker sums the curves obtained by multiple samplings, and the final results meet the normal distribution, wherein the mean value of the normal distribution is the target leakage value, and the variance is the effect of noise on the target leakage. The expression (6.77) can be used to express the information composition of the sampled voltage, that is

$$e = e_{\text{signal}} + e_{A,\text{noise}} + e_{el,\text{noise}} + e_{\text{const},\text{noise}} \tag{6.77}$$

where e_{signal} is the target leak value; $e_{A,\text{ noise}}$ is the algorithm noise; $e_{\text{el, noise}}$ is the electronic noise; and $e_{\text{const, noise}}$ is the thermal noise.

The main source of the algorithm noise is that the electromagnetic wave radiated by the parallel modules inside the chip is sampled by the electromagnetic probe and converted into a noise signal superimposed on the target leakage. The schematic diagram of the sampled parallel noise is shown in Fig. 6.22.

In Fig. 6.22, the right wire is the leakage antenna, the left wire is the parallel antenna, h is the vertical distance from the center of the electromagnetic probe to the chip, and $R_{\min}$ is the shortest distance between the two antennas. When the

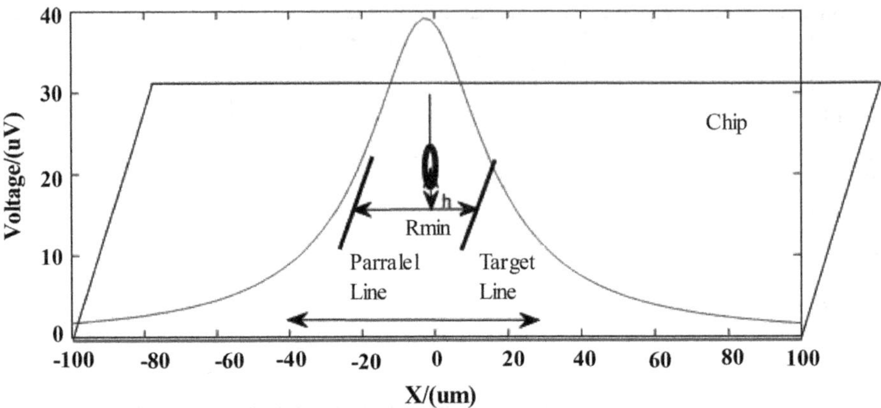

Fig. 6.22 Schematic diagram of the parallel signal sampling

vertical probe moves along the horizontal direction, the probe will sample the electromagnetic information from the two antennas. At this time, the sampled voltage of the probe is

$$V_{\text{oc}} = V_{\text{target}} + V_{\text{parallel}} \tag{6.78}$$

From the expression (6.76), we can see that the sampled information will be more accurate when the probe center is closer to the leakage antenna. However, if a parallel antenna exists around the leakage antenna, whether the probe can distinguish the leakage information of the leakage antenna from the noise information of the parallel antenna depends on the distance $R_{\min}$ between the two antennas. It is assumed that the radius of the electromagnetic probe used is 10 μm. Figure 6.23 shows the variation of the voltage curve sampled by the electromagnetic probe when the distance $R_{\min}$ between the two antennas changes.

It can be seen from Fig. 6.23 that when the distance between the parallel antennas is equal to 10 μm, the attacker cannot effectively distinguish the two parallel conductors. When the distance between the parallel antennas is equal to 40 μm, the attacker can easily obtain two spikes corresponding to the positions of the parallel wires, and 40 μm is equal to the resolution of the circular electromagnetic sampling probe. This means that the parallel signal within the resolution will be hard to distinguish with the leakage signal, and the parallel signal outside the valid range will be able to clearly distinguish with the leakage signal. Therefore, it can be seen that the closer the parallel signals are, the more difficult the attack is, and the more difficult the attack is when more parallel signals are available in the effective range of the electromagnetic probe. This is the bit-block principle in side-channel attacks [16]. In the electromagnetic attack, once the attacker determines the position of the leakage point, the electromagnetic probe is fixed at a fixed height above the leakage point to continuously sample the leakage point. The traditional electromagnetic attack countermeasure is mainly to increase the noise,

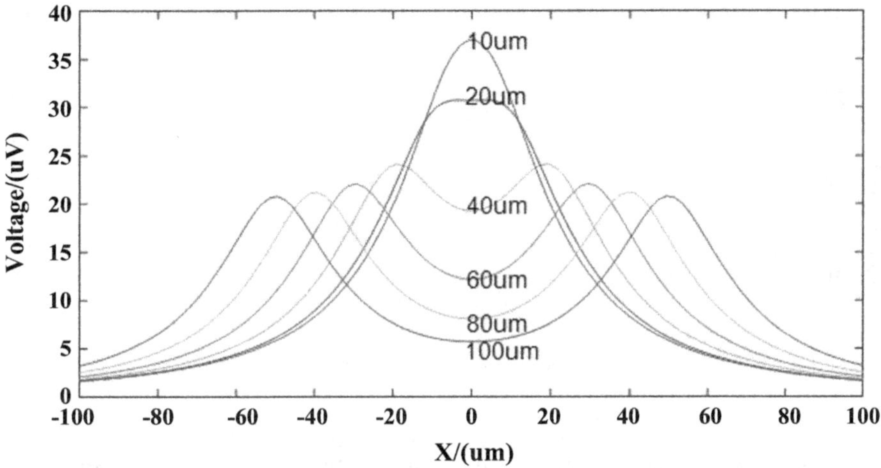

Fig. 6.23 Effect of the distance between the parallel antennas on the electromagnetic sampling

and the strategy is mainly to increase the number of parallel modules in the probe resolution to submerge the signal in the noise. Considering the increasing resolution of electromagnetic probes, designers must increase the number of parallel modules in a smaller range in order to resist electromagnetic attacks, which is obviously not a sustainable electromagnetic attack countermeasure.

The following shows the difference between the electromagnetic attack countermeasure by using the reconfigurable technology and the traditional electromagnetic attack countermeasures and analyzes the advantages of the reconfigurable technology to resist electromagnetic attacks. Figure 6.24 shows the schematic diagram of the result when the target conductor is sampled by the electromagnetic probe and the distance between the two parallel antennas is 200 μm.

In Fig. 6.24, there are three almost coincident curves, of which the upper, middle, and lower curves, respectively, represent the current flowing through the parallel antenna in the same direction of the current flowing through the leakage antenna, no current flowing through the parallel antenna, and the current flowing through the parallel antenna in the direction opposite to the current flowing through the leakage antenna. It can be seen that when the distance between the parallel conductors is larger compared to the resolution of the probe, the effect of the parallel antenna on the leakage antenna is smaller. If the parallel antenna can exchange the positions with the leakage antenna during the continuous sampling process of the electromagnetic probe, the information sampled by the electromagnetic probe mainly comes from the parallel signal, and the leakage information can hardly be sampled by the electromagnetic probe. In this circumstance, the attacker will sample a wrong signal with a great probability. The dynamic reconfiguration feature of reconfigurable technology can implement the exchange of the position of the parallel antenna and the position of the leakage antenna during the

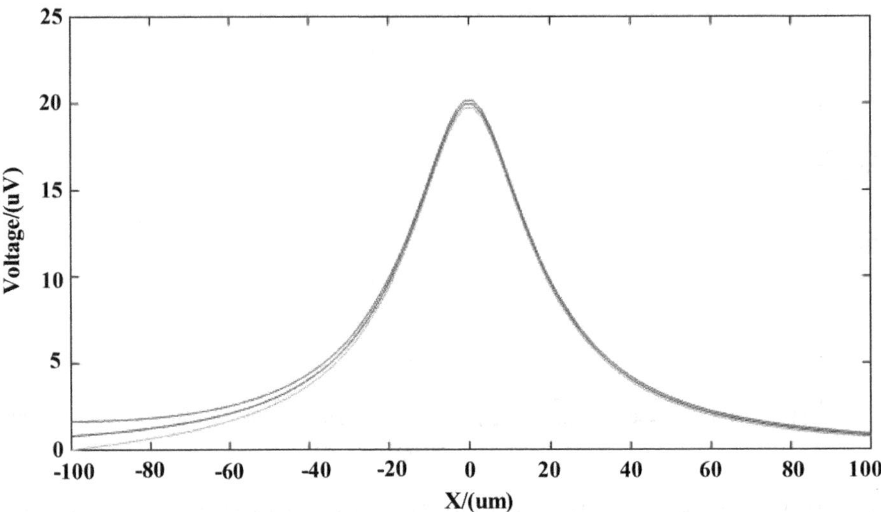

Fig. 6.24 Schematic diagram of the parallel wire sampling

operation of the cryptographic algorithm. From the above analysis, the specific guiding principle of the electromagnetic attack countermeasure based on the reconfigurable technology is that the goal of the randomization operation is to create incorrect sample data for attackers. In the process of realizing randomization through the reconfigurable technology, it is necessary to ensure that the PE replacing the leakage antenna is not idle; otherwise, it is easy to be recognized by the attacker. When the position of the leakage antenna and the position of the parallel antenna are exchanged with each other by using the dynamic reconfiguration technology, we should try to choose the parallel antenna with a longer distance. When the parallel antenna is far away from the leakage antenna, the attacker is less likely to obtain the leakage information.

Compared with the traditional electromagnetic attack countermeasure by increasing the noise, the electromagnetic attack countermeasure based on the reconfigurable technology has the following advantages:

① The reconfigurable technology does not need to add parallel modules within the resolution of the electromagnetic probe. As the resolution of the electromagnetic probe gradually increases, it becomes more and more difficult to add parallel modules within the range of the probe resolution, and the increase of parallel modules will cause additional consumption. The electromagnetic attack countermeasure based on the reconfigurable technology does not need to add parallel modules within the range of the resolution, and the distance between parallel modules is greater, and the restriction on circuit design is less.

② The electromagnetic probe will sample the incorrect data. The traditional method of increasing noise aims to submerge the leakage information in the noise. However, the attacker can extract the signal by using a mathematic method. If the reconfigurable technology is used in the electromagnetic attack countermeasure, the attacker will sample the noise information from the parallel antenna and not be able to recover the correct information by using a mathematic method.

3. Electromagnetic Attack Model

In an electromagnetic attack, the attacker's time is mainly spent on the continuous sampling process. The more the sampling curves are, the more difficult the attacker's attack is. We hereby define the number of sampling curves as the "effort" to express the difficulty level of the electromagnetic attack. The attacker obtains the correct key through a statistical method after he or she obtains the multiple sampled electromagnetic curves. The most common statistical method is the correlation coefficient. The correlation electromagnetic analysis (CEMA) is used to obtain the correlation coefficient of each possible key by performing the correlation calculation of the model output and physical output curves. The key with the largest correlation coefficient corresponds to the correct key. The correlation coefficient is

$$\rho(P,O) = \frac{\mu_{p,o} - \mu_p \mu_o}{\sigma_p \sigma_o} \tag{6.79}$$

where P represents the predicted output value of the model, and O represents the actual sampled physical electromagnetic output.

In order to obtain the correct key through the CEMA method, the attacker must sample enough electromagnetic curves. The minimum number of curves n_{min} needed to be sampled under the CEMA method can be obtained according to the statistics

$$n_{\min} = 3 + 4 \cdot \frac{z_{1-\alpha/2}^2}{\ln^2 \frac{1+\rho}{1-\rho}} \tag{6.80}$$

where $1 - \alpha$ is the confidence level; $Z_{1-\alpha/2}^2$ is the parameter factor in the case of the confidence level of $1 - \alpha$, which can be obtained by looking up the table; ρ is the correlation coefficient factor.

Let us define the effort in the continuous sampling phase

$$\mathrm{EF} = 3 + 4 \cdot \frac{z_{1-\alpha/2}^2}{\ln^2 \frac{1+\rho}{1-\rho}} \tag{6.81}$$

In order to evaluate the change in the proportion of the efforts before and after a randomization method is used, the ratio of the effort after the randomization to the effort before the randomization is defined as the attack countermeasure coefficient, i.e.,

$$R = \frac{\mathrm{EF}'}{\mathrm{EF}} = \frac{3 + 4 \cdot \frac{z_{1-\alpha/2}^2}{\ln^2 \frac{1+\rho'}{1-\rho'}}}{3 + 4 \cdot \frac{z_{1-\alpha/2}^2}{\ln^2 \frac{1+\rho}{1-\rho}}} \tag{6.82}$$

Assuming that the time randomness degree and the space randomness degree are, respectively, γ_t and γ_s when the randomization method is applied, the resolution of the probe is d, the size of each PE in the reconfigurable array is a, and the relationship between the correlation coefficient after the randomization and the correlation coefficient before the randomization is

$$\rho' = \frac{1}{\gamma_s \times \gamma_t} \cdot \frac{\pi d^2}{4a} . \rho \tag{6.83}$$

4. Randomized Electromagnetic Attack Countermeasures

The randomized electromagnetic attack countermeasures mainly include two kinds of space randomization methods, namely the absolute space randomization method and the relative space randomization method.

The absolute space randomization method and the time randomization method randomly change the space position and the time node of the leakage point so as to effectively reduce the probability that an attacker samples the leakage information. The relative space randomization method is an enhancement method, which improves the attack countermeasure by changing the position relationship between the PEs in each row. The relative space randomization method can be used together with the other randomization methods.

(1) Absolute space randomization

In order to randomly change the position of the leakage point, the array structure of the reconfigurable cryptographic processor will be reconfigured in each execution. In order to maximize the use of resources and maximize the randomized distance, the absolute space randomization implements the randomization operation of the operators at every stage in each algorithm. Assuming an encryption algorithm is composed of m-stage operators, then the degree of randomness can reach m after using the absolute space randomization operation. Figure 6.25 shows the schematic diagram of absolute space randomization.

The data flow graph (DFG) of the algorithm has three stages, and the operators of the same stage are executed at the same time, as shown in Fig. 6.25. When the algorithm is mapped onto the reconfigurable array, nine PEs in three rows are required. However, only six PEs carry out operations in each encryption process, and the other three PEs are idle (the randomized idle PEs are italicized). For an idle PE, if it does not perform any operation, an attacker can still easily distinguish it from the leakage point PE after the randomization method replaces it with the leakage point PE which does not satisfy the first requirement of the proposed

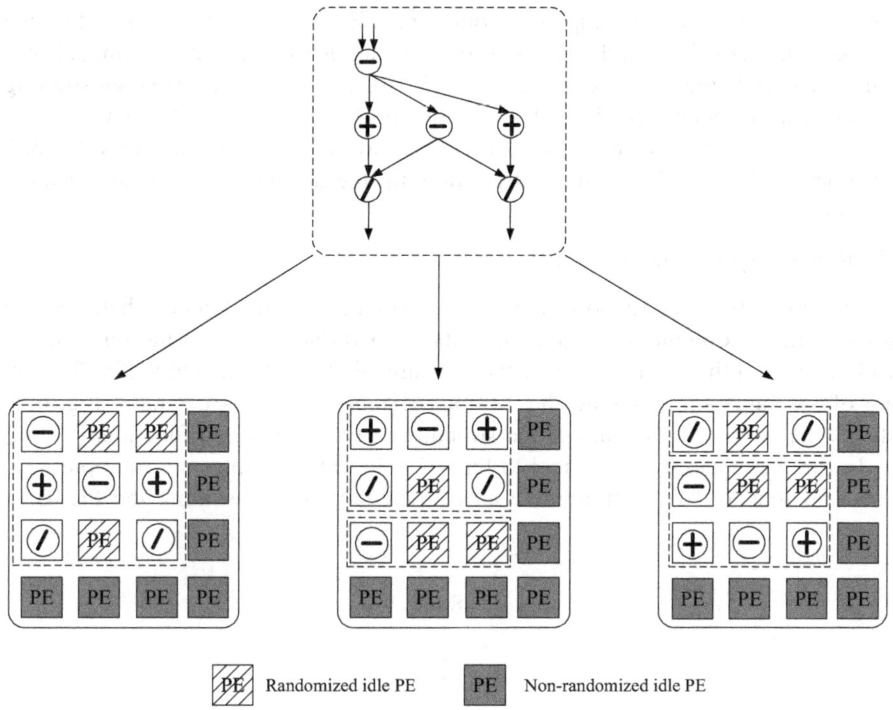

Fig. 6.25 Schematic diagram of absolute space-randomized operation

guiding ideology on reconfigurable electromagnetic attack countermeasures; that is, the goal of the randomization operation is to create the wrong data for the attacker. If an attacker can easily distinguish the wrong data from the correct data, the attack countermeasure will be meaningless. If the idle PE performs the same operation all the time, the attacker may also identify the idle PE through the leakage curves and invalidate the guiding ideology of creating erroneous data. Therefore, these idle PEs must perform some redundant operations at random. For example, the operators in Fig. 6.25 have additions, subtractions, and divisions. The idle PEs can randomly perform these three operations. The input source of the idle PEs can be the random number stored in the memory. If there are more operator types in the DFG graph of the algorithm, the idle PEs should randomly execute all the operator types in the DFG graph. If designers want to further enhance the attack countermeasure of the circuit, they can make the non-randomized idle PEs in Fig. 6.25 perform redundant operations, thus increasing the overall sampled noise and making the power consumption of the reconfigurable array constant.

In Fig. 6.25, assuming that the attacker's probe resolution is close to the PE size, the space randomness of the reconfigurable array, $\gamma s = 3$, after the designer has performed some absolute space randomization operations. The probability that the attacker samples the correct curve is only 1/3 of the original one, and the wrong curves that the attacker samples account for 2/3 of the total curves. After the attacker analyzes the correlation between the sampled curves and the model outputs, the correlation coefficient becomes 1/3 of the original one. After substituting the correlation coefficient into the expression (6.82), we can obtain the attack countermeasure coefficient R = 11.4. This shows that the attacker's "effort" becomes 11.4 times the original after adopting the absolute space randomization operation.

(2) Relative space randomization

The absolute space randomization has developed the randomness between the rows of the reconfigurable array. The following further develops the randomness within a row of the reconfigurable array. Assume that the PEs in a row of a PE array are of the same stage, taking the 128-bit AES algorithm for example; Fig. 6.26 shows the schematic diagram of the structure of the AES round function.

In Fig. 6.26, the input of the 128-bit AES algorithm is split into multiple 8-bit data and sent to the reconfigurable array for encryption (assuming the PE has a

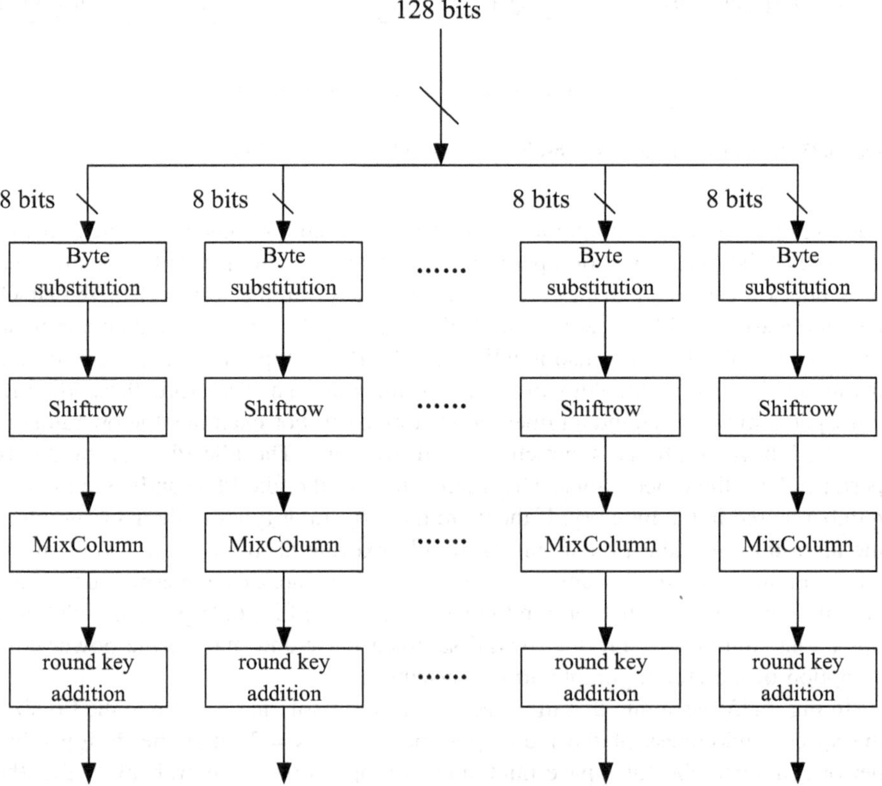

Fig. 6.26 Schematic diagram of the AES round function structure

granularity of 8). For the AES algorithm, the leakage point is in the byte substitution operation of round function. In the electromagnetic attack, the attacker's goal is to sample a piece of 8-bit electromagnetic information, and the encryption operation for the remaining bytes will be the parallel algorithm noise. The idea here is to exchange the position of the parallel algorithm noise and the position of the target leakage information so that attackers sample the wrong data. The schematic diagram of the relative space randomization operation is shown in Fig. 6.27.

In order to ensure the correctness of the algorithm function after performing the relative space randomization, the two PEs that implement the relative space randomization operation shall perform the same operation. To use the relative space randomization, it is required to add six multiplexers to each pair of PEs and use the same random number for the control, as shown in Fig. 6.27. The probability that key 1 and input 1 enter PE1 or PE2 is ½. In order to ensure the correctness of the algorithm before and after the randomization, after the computation of PE1 and PE2 is completed, the two one-out-of-two multiplexers under PE1 and PE2 are controlled by the same random number to interchange the positions of the calculated results. Assuming that the effective sampling range of the attacker's sampling probe is similar to the size of the PE, the attacker may obtain the leakage information of key 2 and input 2 during the sampling of PE1, thus obtaining erroneous information. In this case, the randomness of the space $\gamma_t = 2$, 1/2 of the data sampled by the attacker is incorrect, the correlation coefficient $\rho' = \frac{1}{2}\rho$, and the attack countermeasure coefficient R = 4; that is, after the relative space randomization operation

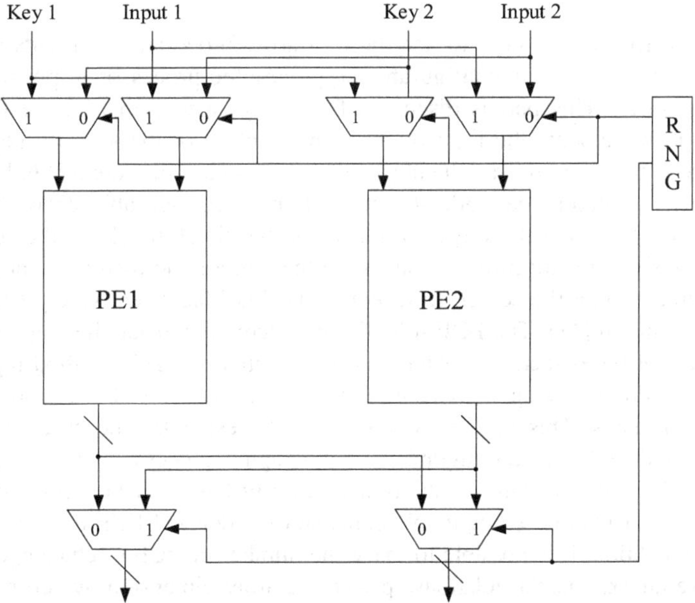

Fig. 6.27 Schematic diagram of relative space randomization

is performed, the attacker's "effort" has become 4 times of the original one. In practice, the randomness of relative space randomization can be greater than 2, depending on the number of bytes computed in parallel on the reconfigurable array.

(3) Time randomization

The time randomization operation mainly changes the time node information of the leakage point by inserting the pseudo-cycle during the running of the cryptographic algorithm. During the pseudo-cycle, the electromagnetic information of the entire reconfigurable array must be guaranteed to be similar to that of the normal execution cycle. Assuming the number of pseudo-cycles inserted by the attacker is 3, the corresponding randomness $\gamma t = 3$, then, only 1/3 of the electromagnetic curves sampled by the attacker is correct, and the attack countermeasure coefficient $R = 11.2$. It is important to point out that due to the advent of the integral method and the elastic alignment attack method, the effect of the electromagnetic attack countermeasure using the time randomization alone has been greatly diminished, so it is suggested that the time randomization operation would be used together with the space randomization operation. Since the electromagnetic attack countermeasure by inserting pseudo-periods has been discussed a lot, the interested readers can refer to the relevant literature [17].

6.2 Attack Countermeasure Technology of the Reconfigurable Processing Element Array

This section focuses on how to use the rich array processing elements and interconnect resources of the reconfigurable cryptographic processor in physical attack countermeasures. This part is discussed from two aspects, that is, the PUF technology that implements the lightweight authentication or the security key generation based on the PE units and the physical attack countermeasure technology based on the interconnection network structure of the reconfigurable array. The most powerful attack among the physical attacks is the direct attack on the secret key, i.e., the invasive or semi-intrusive attack on the nonvolatile memory that stores the keys to directly gain the secret. In recent years, PUF has been widely studied as a promising solution [18]. The PUF solidifies the secret key inside the chip circuit, and the key cannot be extracted after the chip is powered off. This method replaces the existing methods by using nonvolatile memory and has solved the problem of the attack on the keys. This section covers the PUF design for the array structure of processing elements in reconfigurable cryptographic processors. It is expected to design a PUF structure with reliability and security based on the array structure of processing elements in reconfigurable chips and the delayed PUF construction technology. In addition, it is possible to make the number of PUF's "challenge-response pairs" have an exponential relationship with the array dimension, which will greatly increase the number of elements in the challenge-response pair set if the freedom of the interconnection and processing elements in the array structure of a reconfigurable chip can be fully utilized. In addition, the PUF can perform lightweight authentication besides key-saving. What is more, there are abundant interconnection resources available on the reconfigurable cryptographic processor. If we can develop these interconnection network resources to fight against physical attacks, the resources will be reused and the cost of attack countermeasures will be reduced. This section uses the infecting fault attack countermeasure method implemented by an improved Benes network as an example to introduce the application of the interconnection network in the physical attack countermeasure.

6.2.1 Processing Element-Based PUF Technology

1. Physical Unclonable Function

PUF can characterize the uniqueness of a physical entity. It has strong similarity to human biological features such as fingerprints. PUF can use the difference between objects to produce a mutually distinguishable output, which is usually binary data. PUF has attributes such as physical, non-clonable and function. The physical attribute represents that the output of PUF is determined by physical parameters such as direction, distance, and time; the unclonable attribute means that it is very difficult and almost impossible to make two identical PUFs; the function attribute indicates that the PUF is a mapping from the input value to output value. The input and output values are generally binary data with a specific number of bits. We call the input as challenge and the output as response.

PUF can be divided into non-electronic PUF and electronic PUF [19]. The non-electronic PUFs are made using non-electronic technologies and materials. The physical one-way functions (POWFS) proposed by Pappu et al. in 2001 are the most representative non-electronic PUFs. Pappu et al. placed the transparent optical media in a three-dimensional microstructure to implement POWFS. The POWFS input is the incident laser, which produces a specific interference pattern when passing through the microstructure. The output is a fixed-length bit vector extracted from the interference pattern. The mode of the interference pattern depends on the angle and the frequency of the incident laser and the mode of distribution of the optical material. The three-dimensional microarchitecture of POWFS is not easy to integrate, and the challenge-response pairs of this structure only have tens of thousands pairs. The attacker can get all the challenge-response pairs through test and make a form to achieve the cloning of POWFS.

Some randomly changing electrical parameters such as resistance and capacitance are added to the electronic PUF in the production process. One of the most important categories of electronic PUFs is the silicon physical unclonable function (SPUF) proposed by Gassend et al. [20] in 2002. Because the current integrated circuit technology is based on silicon, SPUF can be implemented using the existing CMOS manufacturing process. SPUF can be directly connected to the standard digital circuit; thus, it can be directly used as a hardware module in the cryptographic application and conveniently integrated into the chip. SPUF has aroused widespread concern in the field of information security, and this section also focuses on SPUF. The idea of SPUF is that there are some minor differences between the circuits that have the same structure and are made using the same mask and manufacturing process due to the random variations in the integrated circuit (IC) process such as size, doping concentration, electron mobility, and oxide thickness. SPUF uses these process variations to extract the security information that distinguishes the circuits from one another. Since process variations in IC manufacturing are not artificially controlled, attackers cannot produce two identical SPUFs even if they know all the parameters of the circuit design.

As electronic devices become ubiquitous, people rely more and more on electronic devices and communications between them to accomplish some security-sensitive tasks. Many of these tasks require the electronic device to be safely certified. In order to meet such requirement, the current conventional way is to store the key in a nonvolatile memory (NVM), such as placing the key in an EEPROM, and then use cryptographic operations (such as digital signatures) for authentication protection. There are mainly two aspects of problems for this method. On the one hand, NVMs storing keys are vulnerable to direct attacks launched by attackers because such memory still holds the secret key after the power is off and an attacker could steal the key information directly via launching intrusive or semi-intrusive attack against the memory. On the other hand, the traditional entity authentication generally requires a public key algorithm, which requires a large amount of computation and results in large performance loss, area consumption, and power consumption. In contrast, SPUF, an innovative circuit structure, provides a good solution to the above problems. First, the SPUF replaces NVM and generates the key in real time using the generated response, which avoids the keys from attacking when the power of the chip is off. Meanwhile, since the intrusive attack will destroy the chip structure, which means that the inherent structure of the SPUF is damaged, the attacker cannot complete the attack against SPUF. In addition, the SPUF can perform lightweight authentication; that is, the challenge-response mechanism is used to complete the authentication operation, which has less cost than that of traditional authentication methods.

The following details the SPUF. We replace SPUF with PUF for the sake of simplicity. The following content includes the class and the instance of PUF, the PUF abstract model, and the performance evaluation metric of PUF.

First let us introduce the concepts of the classes and instances of PUF. The PUF class is a complete description of the PUF structure. From an operational point of view, the PUF class represents the design plan of the PUF and the operation procedure for completing this design, that is the PUF design file. The PUF instance refers to a PUF made from a PUF class. During the process of manufacturing a PUF instance, the manufacturing deviation has introduced entropy for the PUF instance. The finished PUF instance contains statically fixed bits that can be extracted, and the bits contained in each PUF instance are different due to the introduction of entropy. The relationship between the PUF class and the PUF instance is shown in Fig. 6.28.

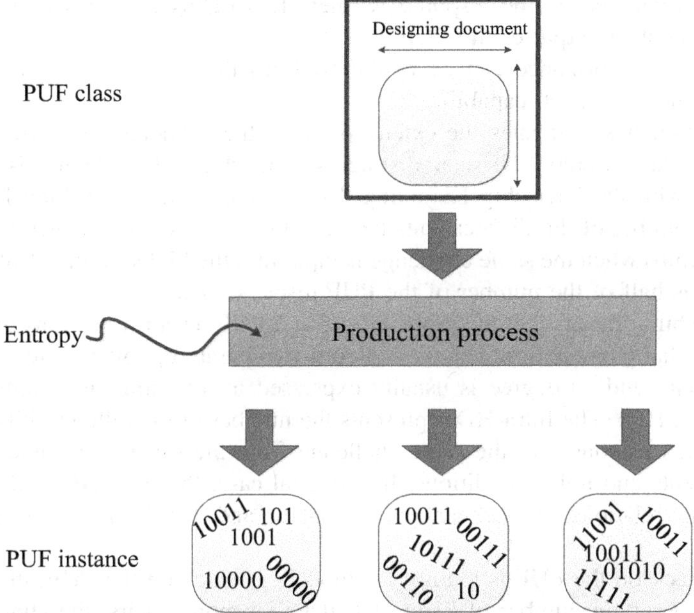

Fig. 6.28 Relationship between the PUF class and the instance

PUF can be described by using certain abstract models. An abstract model of PUF is shown in Fig. 6.29. In the current PUF application, the input and output of PUF are generally binary data, so in the PUF abstract model, the default PUF input/output is binary data. In the abstract model shown in Fig. 6.29, the PUF has n bits of input, so there are a total of 2^n possible inputs. After each input is applied to the PUF, an output with m bits can be generated by PUF's mapping. There are 2^m possible outputs. The input of the PUF is denoted by challenge *C*, and the output of

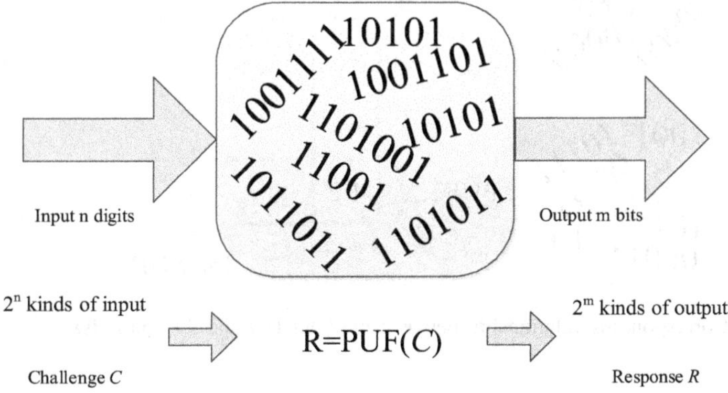

Fig. 6.29 PUF abstract model

the PUF is denoted by the response R; then, R = PUF(C) indicates the mapping from the input to output of the PUF.

The PUF performance can be evaluated from the perspective of uniqueness, stability, and anti-attack capability.

The uniqueness indicates the extent to which the challenge-response pair generated by the on-chip PUF can distinguish the chips. The degree is typically measured with the inter-chip Hamming distance (Inter-HD). The Inter-HD represents the number of the different bits between the two responses generated by two different chips when the same challenge is input into the PUFs. In the ideal case, the Inter-HD is half of the number of the PUF response bits.

The stability means that the probability that a PUF response remains unchanged under the changing environmental conditions (temperature, voltage, etc.) and noise environment, and its degree is usually expressed in the intra-chip Hamming distance (Intra-HD). The Intra-HD represents the number of the different bits between the obtained responses to the same challenge from the same PUF under different environments and noise conditions. In the ideal case, the Intra-HD is 0. The corresponding relationship between the Inter-HD and the Intra-HD is shown in Fig. 6.30.

The attack on the PUF is mainly the modeling attack on the PUF; that is, after obtaining a certain number of known challenge-response pairs, the attacker launches speculative attacks on unknown challenge-response pairs of the PUF by using various means, such as machine learning analysis and auxiliary side-channel analysis. If an attacker can infer the unknown challenge-response pairs with a sufficiently large probability, it is equivalent to successfully cloning the PUF and completing the attack on the PUF.

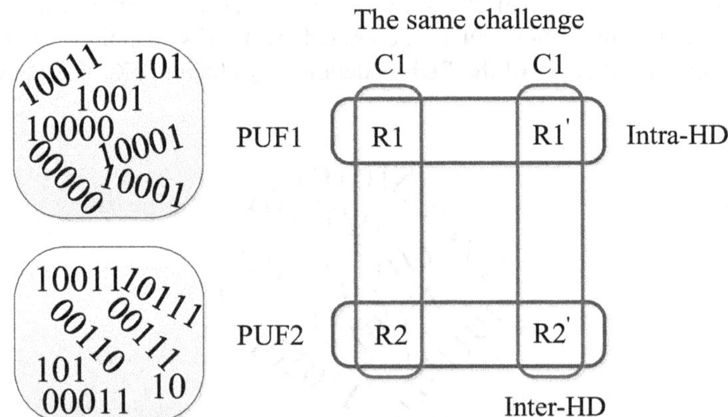

Fig. 6.30 Corresponding relationship between the Inter-HD and the Intra-HD

2. Research Status of the PUF

According to the implementation principle of the PUF, the PUF can be divided into the delay-based PUF and the memory-based PUF. The response of the delay-based PUF is obtained by comparing the delay of two same paths. Because there is a random process deviation in the CMOS manufacturing process, the delay of the same path made by the same process is not fixed but follows a certain probability distribution. Thus, the output of the PUF is also a value with a certain probability distribution. A memory-based PUF contains a bistable circuit that is composed of two cross-coupled inverters. In the bistable circuit, there is a mismatch in the parameters between the two inverters due to process variations; thus, the state of the bistable circuit tends to become 0 or 1 when the circuit is powered on.

One of the earliest delay-based PUFs was called the ring oscillator PUF (ROPUF) [21]. The output of ROPUF depends on the relative value of the frequencies of a selected pair of ring oscillators. The structure of ROPUF is shown in Fig. 6.31. In this structure, the comparator (represented as a circular) outputs 0 or 1 by comparing the frequency of the first selected ring oscillator with the frequency of the second selected oscillator. Because each bit of the output of ROPUF requires a comparison of the frequencies between two ring oscillators, obtaining a sufficiently large challenge-response space requires a large number of ring oscillators to be manufactured on the chip. Therefore, the ROPUF's challenge-response space is generally not large due to chip resource constraints. The ring oscillator consists of an inverter chain whose delay is relatively easily affected by the environmental factors, which in turn affects the output of the ROPUF and reduces the stability of the ROPUF. In order to improve the stability of ROPUF, the researchers put forward some concrete countermeasures, including the K − 1 technology proposed by Suh et al. and the configurable ring oscillator proposed by Maiti et al. [22].

Another common delay-based PUF is the arbiter PUF [22]. The structure of arbiter PUF is shown in Fig. 6.32. An arbiter PUF consists of a multiplexer and an arbiter, where the arbiter is implemented by using D flip-flops, or by using other

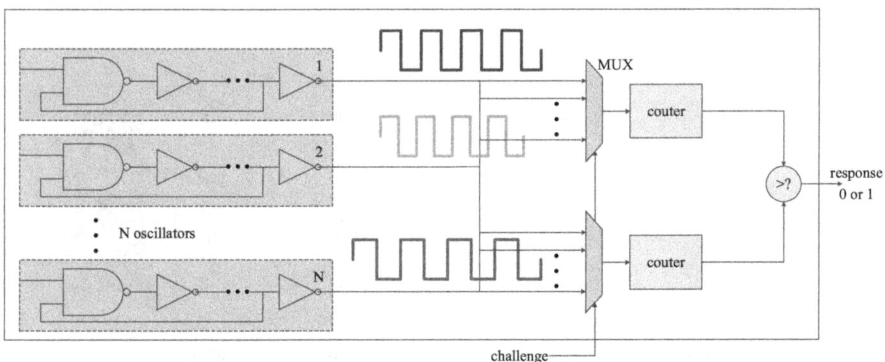

Fig. 6.31 Structure of ROPUF

circuits that perform an arbitration function, such as RS flip-flops. This circuit produces a 1-bit output from N bits of input. The output 0 or 1 is determined by the time difference of the signal propagation delay in two paths with the same length. An arbiter PUF consists of N stages of circuits, each of which consists of two multiplexers with the inputs of the first stage connected together and the two outputs of the last stage connected to the two inputs of the arbiter. The N-bit inputs of the arbiter PUF control the two multiplexers at each stage respectively: When the input is 0, the multiplexer passes the signal directly from the left to the right; when the input is 1, the signal at the top and bottom will be exchanged. In this way, each input of the circuit can be configured for a pair of delay paths. In order to get the corresponding output based on an input, a rising edge of a signal is input to the two paths in the first stage. The rising edge signal propagates through the two paths. The arbiter at the last stage will output 0 or 1 by determining the rising edge signal of which path arrives first.

The circuit structures of the two above-mentioned delay-based PUF are relatively simple, and the modeling attack can effectively attack the circuit. In a modeling attack, an attacker can build an accurate time model and calculate the delay parameters of the circuit through numerous challenge-response pairs. To counter the modeling attacks, researchers propose increasing the nonlinearity of the circuit by some means, thereby increasing the difficulty of modeling attacks. Common means of increasing the nonlinearity of the circuit include the addition of a feedforward arbiter to the arbiter PUF circuit [23] and the performing XOR operations to the output of the parallel PUF circuit [23]. However, recent studies have found that the machine learning attacks can efficiently attack these nonlinear methods and seriously threaten the security of the circuits [24].

The most representative memory-based PUF is the SRAM-based PUF proposed by Guajardo [25] in 2007. The SRAM six-transistor unit is shown in Fig. 6.33. In Fig. 6.33, PL, PR, NL, and NR represent the PMOS transistor on the left, the PMOS transistor on the right, the NMOS transistor on the left, and the NMOS transistor on the right, respectively. BLC and BL represent two-bit lines, and WL represents a word line. The four MOS transistors, PL, PR, NL, and NR, create two cross-coupled inverters. The cross-coupled inverter circuit has two stable states (0

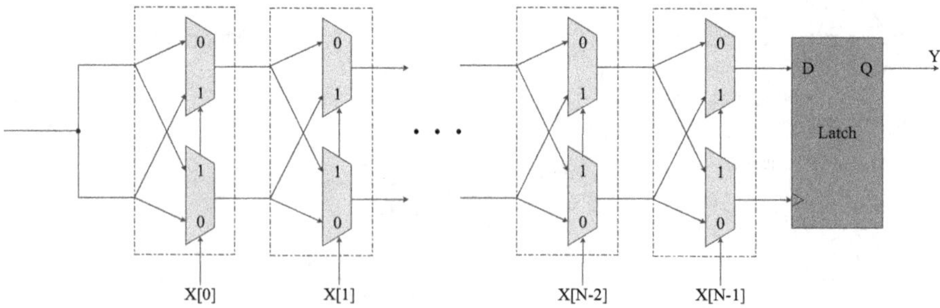

Fig. 6.32 Structure of arbiter PUF

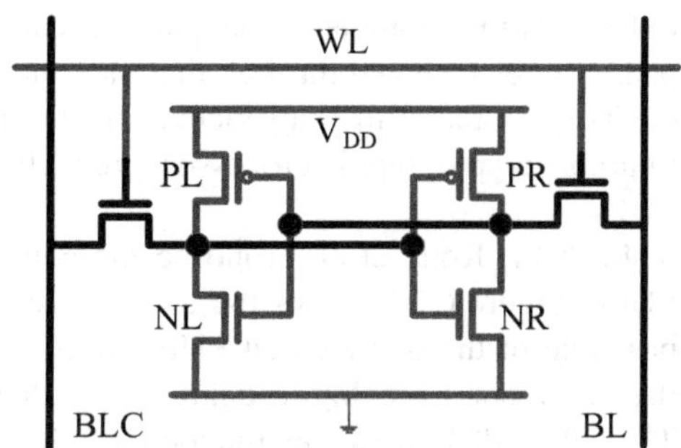

Fig. 6.33 Schematic diagram of the SRAM six-transistor unit

and 1). Due to the random process variation, there are differences in the parameters of the two inverters in the cross-coupled inverter circuit, resulting in differences in the drive capacities of the two inverters. In the process of storing data in the SRAM unit, a stable state is stored in the SRAM unit. However, when the SRAM unit is powered on, the SRAM unit will remain in a state with a high probability due to the difference between the drive capacity of the two inverters. In a SRAM array, the same SRAM unit turns into the same state with a high probability each time when it is powered on, and the states generated by different SRAM units are random and mutually independent at power-up. Therefore, the SRAM power-up operation can be utilized as a challenge for the SRAM PUF, and the response is the states of all of the SRAM units on the SRAM array after power-up. Other memory-based PUFs include latch PUF [26], DFF PUF [27], butterfly PUF [28], and buskeeper PUF [29]. Recent studies have shown that memory-based PUF is less stable [30] and less powerful in the semi-invasive attack countermeasure [31].

3. The Concrete Implementation of PUF in the Reconfigurable Cryptographic Processor

The reconfigurable cryptographic processor has a wealth of computing and interconnection resources and can easily implement the PUF structure with these resources. For example, a PE-based PUF structure may be designed in a reconfigurable cryptographic processor. It will obtain a large challenge-response space with little extra cost by reusing resources. Meanwhile, the reconfiguration and interconnection of arrays can increase the entropy of PUF and improve the security of PUF. Based on the analysis of the uniqueness and reliability of PUF, we can further propose the form of PUF based on an array structure.

From Chap. 3, we can see that the reconfigurable processing element array (PEA) is the computational part of the reconfigurable cryptographic processor. The PEA contains the reconfigurable processing elements (PEs) and the interconnections (mainly composed of multiplexors) between them. Each PE unit includes an arithmetic logic unit and registers. According to the classification of PUFs, the delay-based PUF generates 0 or 1 output by comparing the delay of two identical paths, and the memory-based PUF needs a bistable structure using SRAM or other cross-coupled inverters. The abundant computing resources on the PEA can be used to form various complex delay paths. Meanwhile, the length and the topology of the paths can be changed through the interconnection resources, thereby increasing the diversity of the paths and increasing the entropy of the PUFs. Therefore, reconfigurable cryptographic processors are well suited for implementing the delay-based PUFs.

In 2014, Kong et al. proposed the concept of processor-based PUFs and ALU PUFs. Figure 6.34 shows the structure of the ALU PUF [32]. According to the principle of the delay-based PUF, ALU PUFs use the common ALU resources in the processor to realize a delay-based PUF structure similar to the arbiter PUF. The ALU PUF uses the function of the cascade carry adder in the ALU. The specific principle is that two ALUs with the same structure will have slight differences in the time of the same carry generation when performing the same addition operation with the same data. The time difference is caused by the mismatch between the parameters of the two ALUs arising from process deviation. The differences between the time of carry generation in the bit-by-bit addition of the two ALUs drive the arbiters to generate 0 or 1 output. The implementation of the ALU PUF requires only simple synchronization logic and an arbiter made up of flip-flops between the two ALUs during the design phase of the processor, thus consuming less resources.

The PEA is rich in ALU resources. These ALUs have the same circuit structure, but their parameters are different to some extent due to the deviation of the process. These deviations cause the difference in the data computation time of the ALUs. The difference of the computation time can be used to drive the arbiter to generate 0

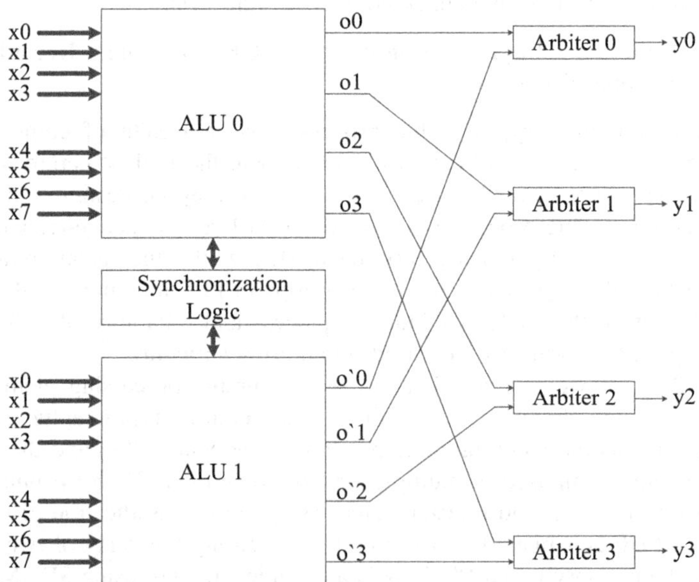

Fig. 6.34 Structure of the ALU PUF

or 1 output. In the implementation of the PE-based PUF, the ALU is used as a core component of the PUF, and a register in the PE is used as an arbiter, and a multiplexer (MUX) is used for interconnection between PEs, and different PEs can be connected to implement the structure of the PE-based PUF.

In the single-level PE-based PUF, the two delay paths are, respectively, composed of the ALUs in two PEs and the corresponding MUXs. The function configuration code and the input data of ALU are used as the PUF challenge. The time spent from input to the completion of computation on the two paths is compared. The comparison result is used to generate the response, and the comparison process is implemented through the arbiter that is composed of the registers in the PE. The structure of a single-level PE-based PUF is shown in Fig. 6.35. The single-level PE-based PUF is composed of three PE units. The ALUs in PE1 and PE2 are the main part of the delay. The time difference between the two ALUs for the same data is used to generate 0 or 1 output. During the running of the PE-based PUF, the control of the ALUs enabling can ensure that the ALUs on the two paths calculate the data simultaneously. Each time the PUF generates a challenge-response pair, and it first needs to input the challenge (including the function configuration code and the input data of ALU) into the PUF and then enable the ALU simultaneously so that the two ALUs perform the corresponding computation. The final computing results will be introduced into PE3 through the interconnection and will be arbitrated by the output register in PE3. The output 0 or 1 is determined based on the arbitration result.

When a single-level PE-based PUF is implemented in the PEA, only a few wires and MUXs are be added to the existing PEA. Therefore, the resource cost introduced is small. In the PUF structure as shown in Fig. 6.35, the two delay paths are the parts corresponding to the thick solid lines, where the black thick solid lines are the existing connection resource in the PEA and the blue thick solid lines are additional lines. These additional lines are used to direct the signals from PE1 and PE2 to the output register of PE3. The delay elements on each delay path in the single-level PE-based PUF include two MUXs and one ALU. The red MUXs and ALUs located in PE1 and PE2 are the resources already existed in the PE units, and the two yellow MUXs in PE3 are additional resources. These two additional MUXs are used to multiplex the register in PE3 and implement the arbiter through the register. The register consists of a D flip-flop. In order to implement the multiplexing of the D flip-flop, a MUX is added to the data input client of the D flip-flop in the structure of the single-level PE-base PUF to select whether the data come from the ALU in PE3 or PE2. The clock client is also added a MUX, and then, the data can determine whether to select the clock signal or the data from PE1. Figure 6.36 shows the connection of the ports of the D flip-flop after the resources are multiplexed. Thus, the D flip-flop has the PUF mode and the register mode. In the PUF mode, the D flip-flop's data client selects the data from the ALU of PE2 and the clock client selects the data from the ALU of PE1. In the register mode, the D flip-flop's data client selects the data from the ALU of PE3 and the clock client selects the clock signal from the clock network so that the D flip-flop can be normally used for storing data.

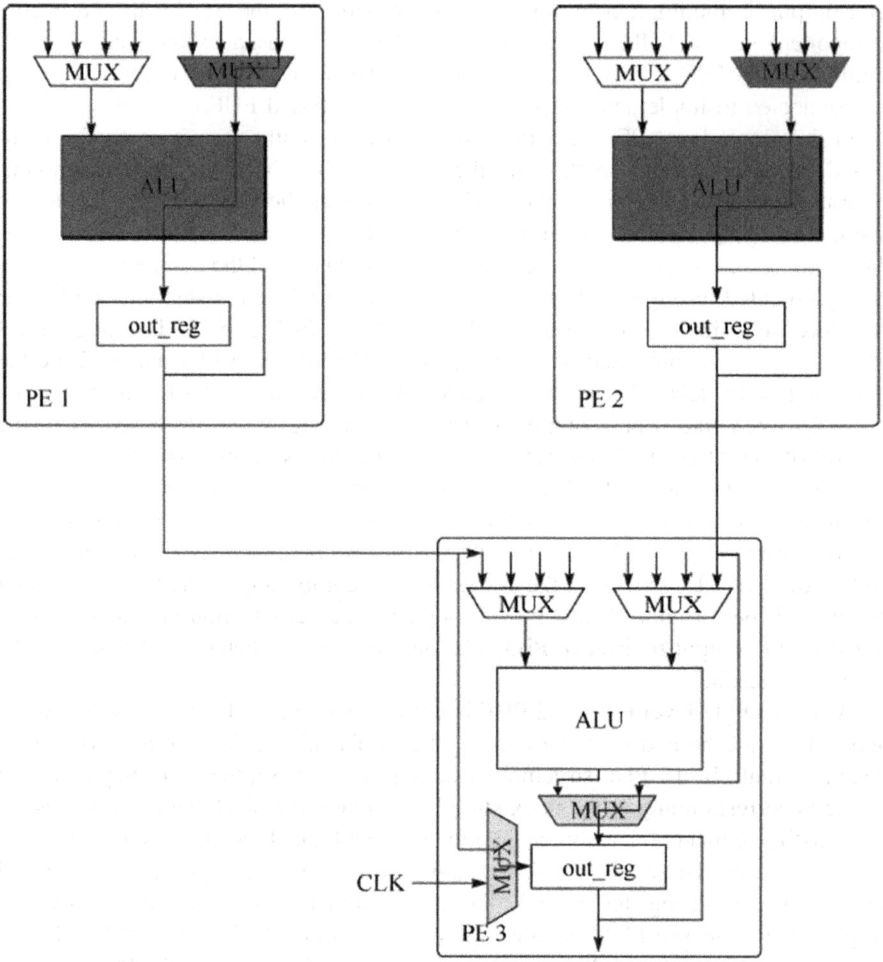

Fig. 6.35 Structure of a single-level PE-based PUF

Fig. 6.36 Port connection diagram of the D flip-flop with two modes

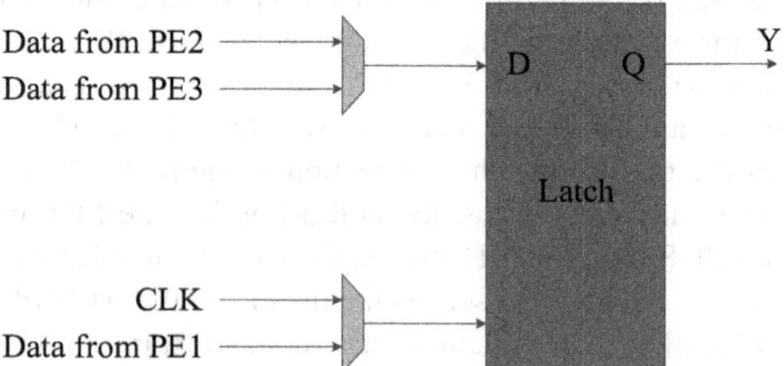

Table 6.1 ALU functions that can be used to deal with PUF challenges

Types	Function
Arithmetic	Addition and substraction
Logic	And, or, not, XOR

In the single-level PE-base PUF, the delay elements in each delay path include MUXs in addition to the ALUs. The MUX can expand the source of PUF entropy, making the entropy of the single-level PE-based PUF higher than the entropy of the ALU PUF. In the structure of the single-level PE-based PUF, the challenge of the PUF is the configuration code of the ALU and the input data of the ALU. The different configuration codes can configure the ALU into different functions. In the PE-based PUF, we can select the configuration codes of several common ALU functions as the PUF challenges. Table 6.1 illustrates several common ALU functions whose function configuration codes can be used as PUF challenges. The selection of different ALU functions also expands the source of PUF entropy, which makes the single-level PE-based PUF have a greater advantage than the ALU PUF that only use the ALU's function of cascade carry addition.

The PEA has abundant PEs and interconnection resources. To make full use of these resources, the PE units in each delay chain can be further extended through interconnections to form a multilevel PE-based PUF. The interconnection itself, as multiplexer logic, can either use its own delay as part of the total delay in the two delay paths of the PUF or greatly expand the total challenge-response space by changing the PE connection. Figure 6.37 shows the structure of a multilevel PE-base PUF (taking three-level PE-based PUF as an example). In the three-level PE-based PUF, the arbiter is implemented by the output register in the last individual PE unit, for which additional wiring and MUXs are required in the last individual PE unit, but the overall resource consumption is minimal.

In the PE-based PUF, the number of PE units contained in each delay chain is theoretically unlimited, as long as the numbers of PE units in the two delay chains are the same. In the implementation process of the PE-based PUF, the most important thing is to ensure that the two delay paths have a nominally equal delay, that is, to ensure that the two paths as equal as possible, so we can ensure that the final PUF output is mainly determined by the delay difference of the two paths arising from the process deviation. In the PE-based PUF, if there is a big deviation in the length of the interconnection, the delay of the two delay paths that finally connect to the D flip-flop will have a big time deviation, and the final output of the PUF will point to 0 or 1 with a very high probability, so that the PUF's response will have high predictability. In the implementation of a PE-based PUF, if the delay of the two delay paths has great time deviation, we can add adjustable delay circuits to the two inputs of the D flip-flop. The adjustable delay circuits can compensate the delay of the two paths and improve the unpredictability of PUF output.

The chip has some specific systematic deviations in the manufacturing process. These deviations will cause the MOS transistor parameters to change greatly with the position changes, and the effect of the random process deviation on the MOS

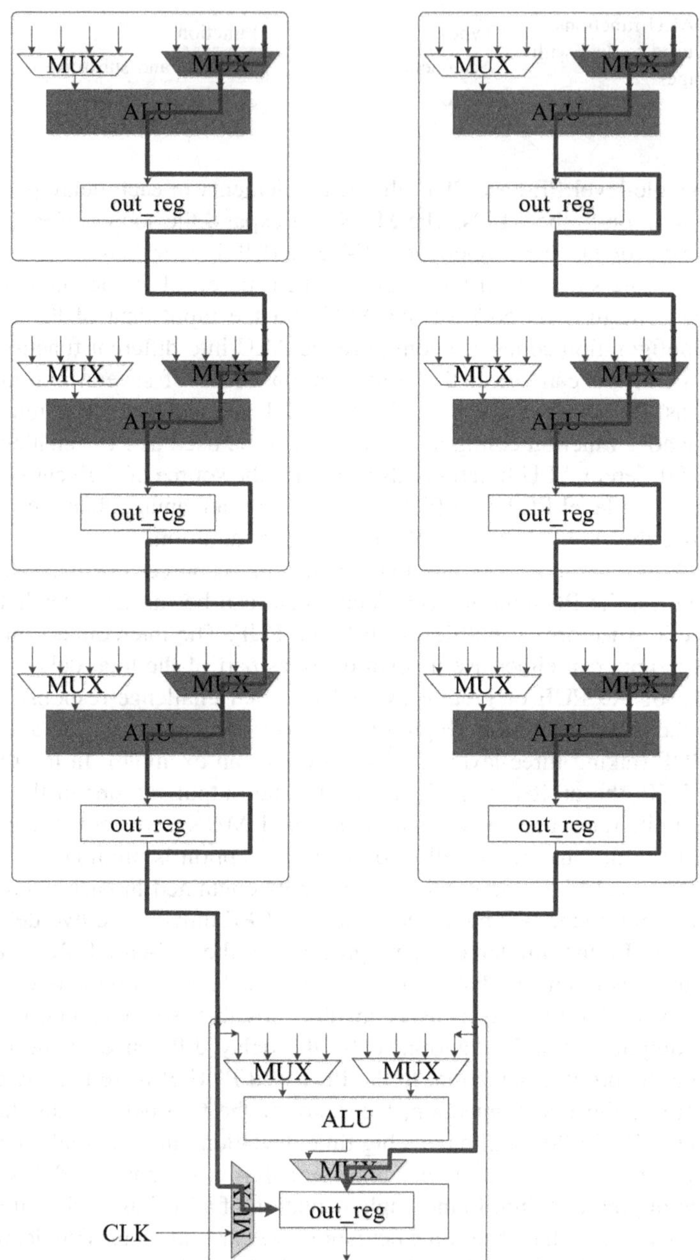

Fig. 6.37 Structure of the multilevel PE-based PUF

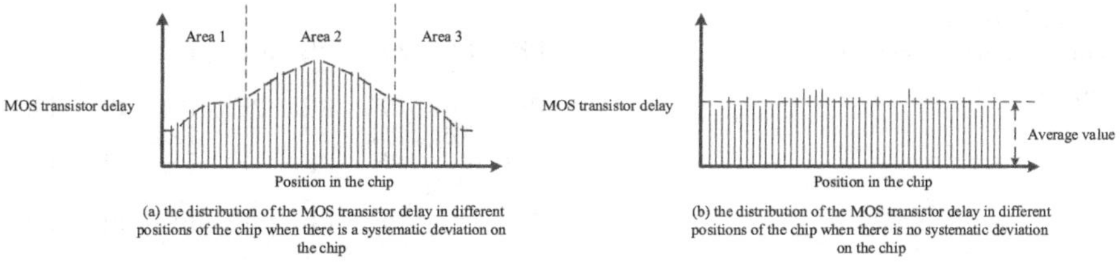

Fig. 6.38 Effect of the systematic deviation on the distribution of the MOS transistor delay in different positions of the chip

transistor parameters will be weakened. Taking the delay of the MOS transistor as an example, the effect of the systematic deviation on the distribution of the MOS transistor delay at different positions on the chip is shown in Fig. 6.38.

Due to the systematic deviation, the delay of the MOS transistor in different areas of the chip may show the scenario as shown in Fig. 6.38. In this case, if the delay elements on the two delay chains with the same length come from area 1 and area 2, respectively, the final output of the PUF will be 0 or 1 with a very high probability. In this case, the PUF output will be highly predictable, threatening the PUF's security. Meanwhile, the output of the PUFs produced on each chip has a high similarity due to the systematic deviations, which greatly reduces the uniqueness of PUFs. It can also be seen from Fig. 6.38 that there is a space correlation between the systematic deviations; that is, the closer the MOS transistors in space, the higher the similarity of their parameters, and the smaller the systematic deviation between their parameters. Based on the research on the space correlation of systematic deviations, in the design of PUF structure, in order to reduce the effect of systematic deviation on the uniqueness of PUF, the delay elements of PUF are required to be close enough [22]. In this case, the method of chip area division can be introduced to ensure that the PE units in the two delay chains of the PE-based PUF are close enough to reduce the effect of the systematic deviation on the PUF output.

The PE units on the PEA are regularly distributed throughout the entire array. In order to select the adjacent PEs to join the delay chain, the entire reconfigurable computing array can be divided into several areas. As shown in Fig. 6.39, the 8×8 reconfigurable computing array is divided into four areas with 4×4 PE units per area. After the area division is completed, when designing the PE-based PUF, the PE units added to each delay path can be selected based on the result of the area division. The specific approach is: first determine how many areas with the PE units are needed to add into the delay chain and then determine how many PE units in the selected area are needed to add into the delay chain and finally determine the topology structure of PE units in the selected area. When selecting the PE units to be added to the PE-based PUF, it shall be ensured that the numbers of the selected PE units added to the two delay chains in the same area are the same. For example, each delay chain may connect two PE units in the area 1 and the area 4, respectively, to form a four-level PE-based PUF. Alternatively, two PE units in the area 1-3 may be, respectively, selected and connected to form a six-level PE-based PUF.

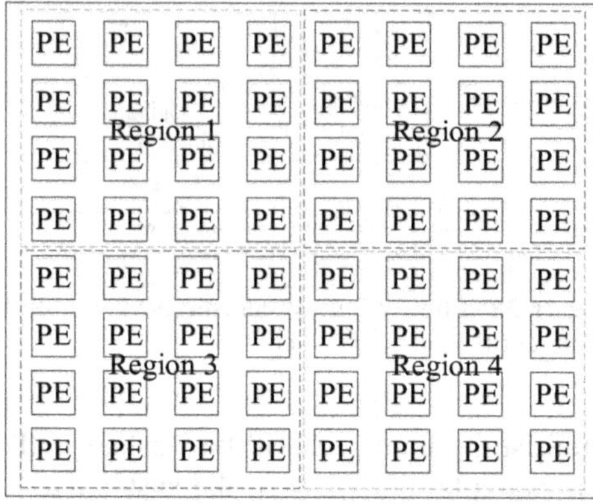

Fig. 6.39 Instance of the reconfigurable computing array division

In the chip area division-based method, how many PE units in each selected area are selected and added to the delay chain and the selection of the topology structure of the delay chain can be done freely. Figure 6.40 shows the forms of several kinds of delay chain topologies in a certain area. In the figure, red and green, respectively, represent the PE units in two delay chains, yellow represents the public PE used to compare the two delay chains, and the arrow indicates the signal propagation direction in the PE chain.

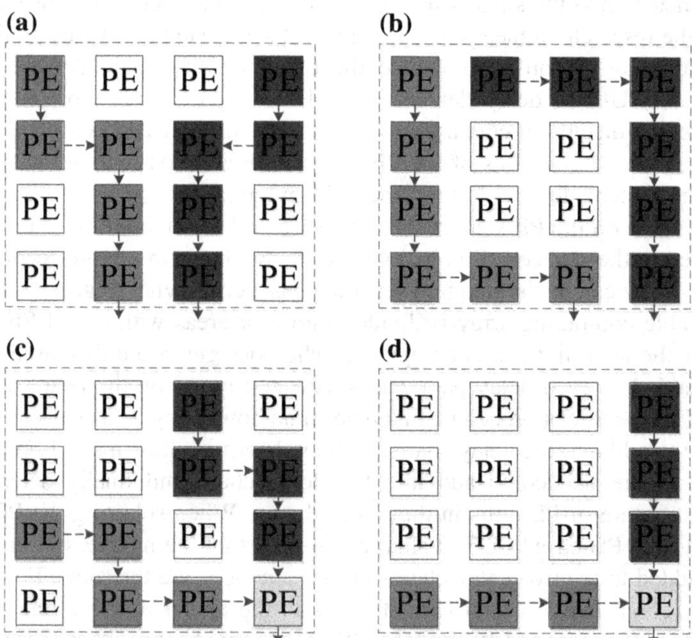

Fig. 6.40 Examples of the topology forms of PE units in an area

The principle of the area division is related to the degree of the effect of systematic deviation on the delay of the devices in different positions. Theoretically, the smaller the area after division, the smaller the number of PE units involved, the closer the distance between the PE units selected in each area, and the smaller the effect of systematic deviation on the unpredictability of PUF output, the higher the uniqueness of the PUF output.

Researches show that the random distribution of process parameters is spatially correlated; i.e., the parameters of devices that are spatially close are more likely to be similar [33]. In the design of PE-based PUF based on area division, the similarity of device parameters is high because the PE units in each delay chain are close in distance in the same area. Due to the effect of environmental change and noise, the PUF output is unstable. In the design of PE-based PUF based on area division, since the parameters of the PE units in each delay chain have a high degree of similarity, in the case of temperature or voltage variation or noise, the tendency and degree of the parameter changes of each stage of delay element in the PE-based PUF are similar, so that the change in the difference of last delay of the two delay paths is small, and the probability of the change in the response of PUF is also small, so that the stability of the PUF is improved.

The value of the entropy of an information system indicates the degree of the uncertainty contained in the information system. The greater the uncertainty of the information system, the greater the entropy, and the greater the amount of information needed to crack the information system, the more secure the system. PUF is an information system, and its security can be measured by the value of the entropy it contains. The unit of PUF entropy is bit, which is calculated as:

$$H(R) = -\sum p_i \log p_i \tag{6.84}$$

where p_i is the probability of the response R_i.

Assuming a PUF's challenge-response pair is shown in Table 6.2, the entropy of the PUF structure is 1 bit. The computation process is

$$H(R) = -(0.5 \times log0.5 + 0.5 \times log0.5) = 1\,\text{bit} \tag{6.85}$$

In the PE-based PUF, the PEA's features of abundant computing resources and flexible interconnection make the structure of the two PE chains have great flexibility. The PE chain structure is determined by the challenge of PUF. The challenges of PE-based PUF serve in four parts, namely determining the areas in which the PE units are added to the PE chain, determining which PE units in the area are

Table 6.2 PUF's challenge-response pairs

Challenge	Response
00	11
01	10
10	10
11	11

added to the PE chain, selecting which computing functions of the ALU, and determining what kind of input data is provided for ALU. Since the challenge sources of PE-based PUF are numerous, the challenge space of PE-based PUF is large. For example, the bit width of the response of PE-based PUF is determined by the ALU bit width. The ALU with a bit width of 8 can generate $2^8 = 256$ outputs, and the total response space can be very large. Therefore, the challenge-response space of PE-based PUF is large. According to the analysis of formula (6.84), the PUF with large challenge-response space will have more accumulated terms when computing the entropy, and the final calculated entropy will be large, so the PE-based PUF has high entropy. If an attacker wants to attack a PE-based PUF, a great amount of information is required. Therefore, the PE-based PUF has a strong attack countermeasure ability.

6.2.2 Network-Based Attack Countermeasure Technology

1. Benes Network

In order to implement various cryptographic algorithms efficiently, one or more multilevel interconnection networks are often added to the interconnection structure of the reconfigurable cryptographic processor to speed up multiple permutation operations in the algorithm. The multilevel interconnection network is a network that is composed of switching elements connected by a topology structure. The initial intention of its design is to the diverse application scenarios of telecommunication networks, such as the Internet, computer networks, and telephone networks [34], but later, it has found that multilevel interconnected network has some advantages over permutation operations. People begin to consider adding extra multilevel interconnection networks to cryptographic processors to improve cryptographic processor performance [35]. However, most multilevel interconnection networks are not suitable for expediting the permutation operation of cryptographic algorithm due to the following reason: first, there is insufficient number of switching elements, and second, there is a topology conflict. The following is a further explanation of the number of switching elements and topology conflicts in a multilevel interconnection network.

In a multilevel interconnection network, the number of switching elements determines the theoretical limit on the number of permutation operations that a multilevel interconnection network can perform. For a multilevel interconnection network with bi-state switching elements, supposed that the network has a total of x switching elements, and the width of the network is N, if $2^x < N!$, then this network will not be able to implement all N-bit permute without any block. Take the commonly used binary tree network [36] for example. For a binary tree network with N leaf nodes, there are $N - 1$ binary state switching elements in total. However $2^{N-1} \ll N!$, so the number of binary tree network switching elements obviously cannot implement all permutation. In addition to the number of switching elements in a multistage interconnection network, its own topology may also cause the conflict of network during the permutation operation, i.e., blockage. Taking the omega network as an example (Fig. 6.41 shows a three-level omega network) [37]. This network consists of four-state switching elements (the four states of the switch are shown in Fig. 6.42, crossover, straight-through, upper broadcast, and lower broadcast), as well as a shuffling function topology connection between various switching elements. In the omega network, the value of the corresponding output address for each input determines the configuration of the switching elements on the input–output path of the network. If there are two inputs in the permutation operation that are different in the configuration of a four-state switch element in the omega network, the permutation operation cannot be completed by one time, i.e., a blockage occurs. Although the permutation operand that is completed by the omega network is larger than $N!$, the omega network's own topology does not ensure that all permutation operations can be completed by one time, limiting the performance of the omega network for permutation operations.

Unlike most multilevel interconnection networks, Benes networks can implement $N!$ permutation (N is the width of Benes network) without any blocking, so it is widely used in the permutation operation acceleration module of the cryptographic processor [38, 39]. Figure 3.6 shows a $N = 16$ input/output Benes network, i.e., B (N), with a total of 2 ($\log_2 N$) −1 levels, each of which has $N/2$ binary switching elements (the two-input two-output switching elements are represented by a square in Fig. 3.6. These switching elements can be configured in both

Fig. 6.41 Omega network

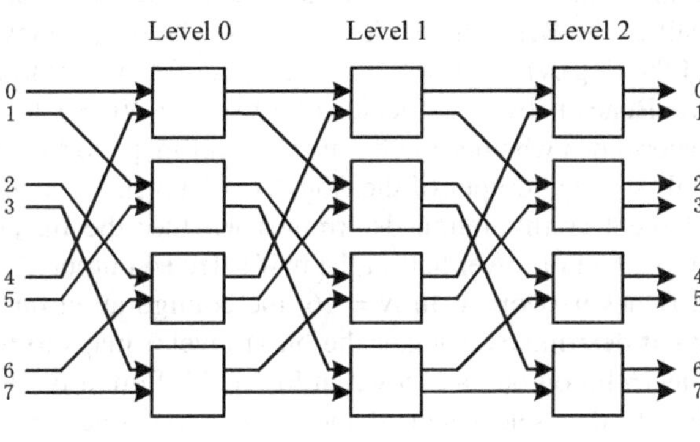

Fig. 6.42 Four states of the omega network switch

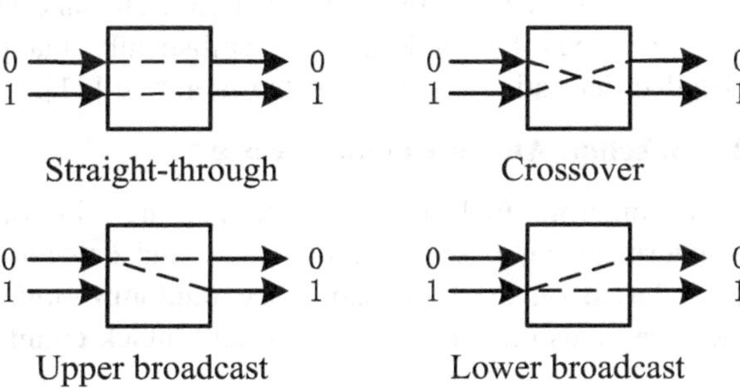

straight-through and crossover modes, as shown in Fig. 3.6b), so the total number of switching elements in the N-bit Benes network is $(N\log_2 N) - N/2$, and because $2(N\log_2 N) - N/2 > N!$, so the number of permutation operations that Benes networks theoretically can accomplish meets the requirements. The following describes the topology of the Benes network, which is symmetrically connected by these bi-state switch elements through two back-to-back butterfly networks. This connection has a recursive nature. As shown in Fig. 3.6a, the N-bit Benes network consists of two $N/2$-bit Benes networks and two-level additional stages with $N/2$ switching elements (dashed boxes highlight subnets including $B(4)$ and $B(8)$). For the permutation operation, the Benes network has provided the selection of middle position and allowed non-occlusive permutation during permutation because the Benes network is connected back-to-back with two butterfly networks, entered through a butterfly network, and passed through another butterfly network in the opposite direction [40].

In addition to the non-blocking nature of the Benes network, the Benes network also has a random characteristic; i.e., if randomness is introduced when the Benes network is configured with switching element states, the output results of permutation are characterized by randomness The following is a brief analysis of the random nature of the Benes network. Due to the symmetric nature of the Benes network, only the second half of the Benes network (corresponding to the $(\log_2 N) - 1$ to $2(\log_2 N) - 2$ levels in the network) is discussed here. For the $(\log_2 N) - 1$ level, since it is in the middle of the Benes network, the configuration of the switch element determines whether the bit at this level is mapped to the upper half or lower half of the Benes network output. For the $(\log_2 N)$ level, since it is at the next level of the $(\log_2 N) - 1$ level, the bit output of this level to the upper half or lower half of the Benes network is decided by the $(\log_2 N) - 1$ level, and the $(\log_2 N)$ level only determines whether the bit at this level maps to the upper half or lower half of the half-selected output of the $(\log_2 N) - 1$ level. By analogy, the $i + 1$ level switching element configuration determines whether the bit at this level is mapped to the upper half or lower half of the half-selected output of the i-th level. For example, for a Benes network with $N = 16$, the configuration bit for the level 3 switching element determines whether the bit at level 3 maps to the upper half or lower half of the 16-bit output, as shown in Fig. 6.43. Similarly, the level 4 switch configuration bits further select half of the 8-bit output selected through level 3. In the above description, it can qualitatively conclude that when the configuration bits of all the switching elements in the Benes network are randomly uniformly distributed, the input bits can be randomly probabilistically mapped to each output position. Detailed derivation can refer to the literature [41].

2. Infection Attack Countermeasure

In numerous fault attack countermeasures, in addition to the traditional attack countermeasures based on redundancy and detection, as well as the countermeasures based on the time and space randomization, the infection attack countermeasure is also a commonly used fault attack countermeasure.

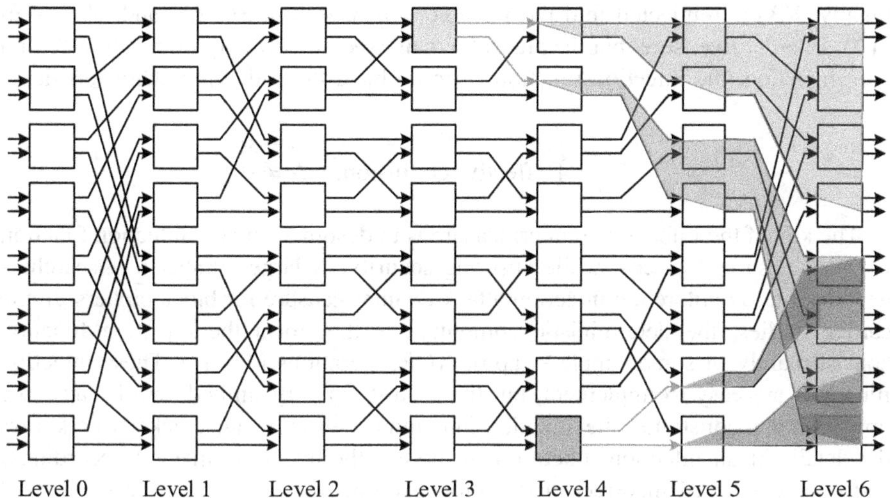

Fig. 6.43 Random characteristics of the Benes network [42]

Infection attack countermeasures are designed to disrupt the fault propagation patterns hidden by the fault attacker in the output. This kind of attack countermeasure affects the invariance of the fault propagation pattern in the output result, so that the fault output ciphertext obtained after the attacker attacks cannot be used for further deciphering. The following is a detailed description of the infection countermeasure [43]. As shown in Fig. 6.44, the infection attack countermeasure first should obtain the difference (Δ) between two intermediate variables (ω and ω′) at one or more moments during the algorithm computation, which is calculated as $\Delta = \omega \oplus \omega'$, where ω and ω′ are the values obtained during two identical executions of the algorithm (implemented either by time or by hardware redundancy). Then, the difference Δ is processed by the infection function (I) to obtain I(Δ).

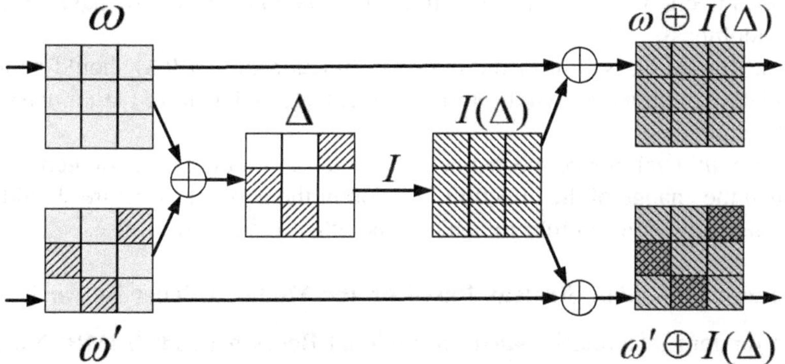

Fig. 6.44 Infection attack countermeasure [42]

Finally, I(Δ) is reinjected into the variable, where $\omega = \omega \oplus I(\Delta)$ and $\omega' = \omega' \oplus I(\Delta)$. In order to ensure that the algorithm can execute normally in the absence of a fault injection, the infection function I should have the following characteristics

$$I(\Delta) = \begin{cases} 0, & \Delta = 0 \\ \text{effective confusion}, & \Delta \neq 0 \end{cases} \tag{6.86}$$

The key of the infection countermeasure is to design a suitable infection function I, which ensures low cost while ensuring security. A large number of researchers have begun to explore the infection attack countermeasure for block ciphers. In the earlier studies, the deterministic computations that form the infection function consist mainly of some simple swap or XOR operations [44, 45]. These infection functions are easy to implement, but the security is very limited, and because the operation that constitutes the infection function is definitive, an attacker who knows the details of an infection function can modify the attacking method accordingly and obtain required incorrect propagation pattern. In other words, the security of these infection countermeasures depends on the confidentiality of the method itself. In order to solve this problem, researchers began to introduce randomness into the infection countermeasure to enhance the security of it. For example, we can add a dummy cryptographic round randomly executed in a cryptographic algorithm or add a random multiplication mask [43, 46]. However, researchers have come up with appropriate attacking method based on these infection countermeasures, which means that there are still security vulnerabilities in the enhanced infection countermeasure through randomization [47, 48]. One of the major reasons for this is that there is a tight coupling relation between the infection function and the algorithm; i.e., there is a link between the infection function and the algorithm. With such combined mathematical analysis, defects such as output bias and incomplete masking of the ciphertext present in the infection function can be easily found. Literature [42] believes that the ideal infection function should have the following characteristics:

① In order to resist the attack that makes use of the deterministic operations of the infection function, the infection function needs to have excellent randomness.

② The Hamming weight of the infection function output I(Δ) should be appropriately modified to avoid information leakage and achieve partial masking of the ciphertext.

③ If a multilevel interconnected network is used in fault attack countermeasures, then the change of the infection function in the network structure should be as less as possible in order to achieve the effect of resource reuse.

3. Random Infection Strategy Based on the Modified Benes Network

The random infection is based on modified Benes network (RIMBEN). In the RIMBEN, the infection function is implemented by an infection-oriented network structure (INS) implementing the Hamming weight (HW) balance and outputting obfuscation. The HW can be simply understood as the number of "1" s in the output, as shown in Fig. 6.45. Taking N = 16 for example, two types of modified switches (abbreviated as switching elements) have been added to the INS based on the Benes network, that is, the OR gate enhanced switch and the four-state switch. These two kinds of switches will only replace the two-state switches in the original network, but do not change the topology of the Benes network. For N-bit INSs, there are $\frac{N}{4}((\log_2 N) - 1)$ enhanced OR gate switches that are located in the first half of the INS, which are used to replace the two-state switches in the top half of Level 0 in each subnetwork of the Benes network. For example, an OR gate enhanced switch at level 2 in Fig. 6.45 is replaced with a two-state switch in the first half of level 0 of four N = 4 subnetworks. In contrast, four-state switches are located only at the $(\log_2 N) - 1$ level of the network, which is the middle level of the network, and the number of four-state switch can vary depending on the designer's security requirements. Figure 6.46 shows the circuit structure and possible states of the switches for both types of modified switches, where C1 and C2 are the two configuration bits for the switches. In the OR gate enhanced switch, the OR gate function controlled by C2 is added to the original two-state switch (Fig. 6.46a). For the four-state switch, C1 and C2, respectively, control the switch and implement the upper broadcast and lower broadcast capabilities.

For any nonzero input, the output after the HW balance is implemented is not related to the input, and the numbers of "0"s and "1"s in the output account for two halves of the total statistically. Here is how the INS implements the HW balance. First, the difference Δ between the results of two identical executions of the

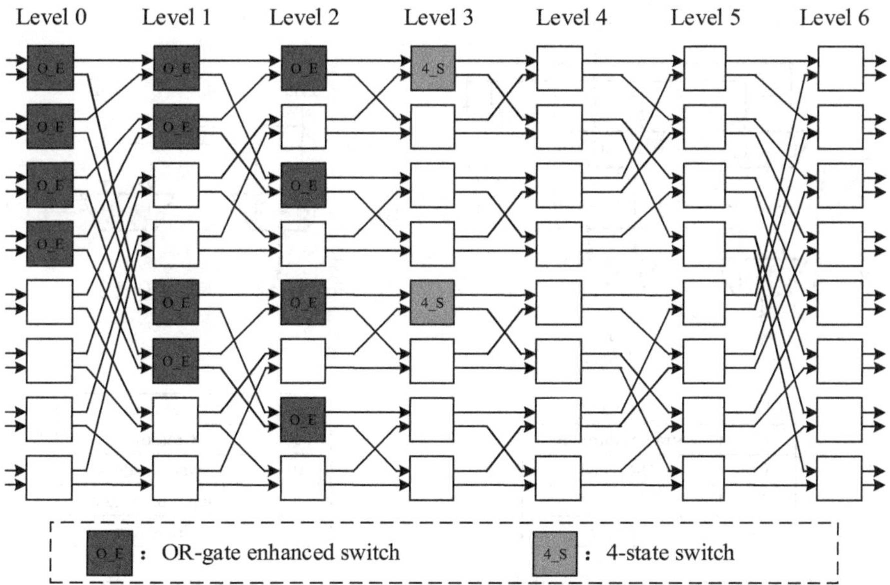

Fig. 6.45 INS structure (N = 16) [42]

algorithm is taken as the input of the upper half switch of the level 0 INS. It can be seen from Fig. 6.45 that the switch corresponding to the Δ input is an OR gate enhanced switch. The remaining switches of level 0 are the original two-state switches whose inputs are constant 0. Next, configure all OR gate enhancement switches in the INS to OR gates. In this case, as long as $\Delta \neq 0$, the output of INS at $(\log_2 N) - 2$ level is always "110011001100 …" and the numbers of "0" and "1" in the output data account for half. There is no unequal in the number of "0" and "1". Figure 6.47 shows the diffusion process of the difference Δ that is not 0 under the action of the OR gate enhanced switch. The fully balanced "0" and "1" outputs produced by the OR gate enhanced switch have strong determinacy, which are likely to be the target of an attacker's attack even though randomness is added by the output obfuscation mentioned below. In order to add randomness to the HW balanced output, the INS adds a four-state switch at its mid-stage (Fig. 6.45). It can be seen from Fig. 6.47 that the input of the INS intermediate switch is always "10" after passing OR gate enhanced switch. In this case, if C1 and C2 value are randomly selected, the four-state switch may output "00", "01", "10", or "11", and the probability of these four outputs is 25%. Therefore, INS can add the randomization in the output after passing the OR gate enhanced switch by simply randomizing the configuration bits of the four-state switch, thus increasing the randomness of the HW balanced output to a certain degree. The size of increased randomness is related to the number of four-state switches. This security of randomization has been demonstrated in literature [42]. After passing the four-state switch, the number of "0" and "1" in the HW balance output is no longer fixed, but

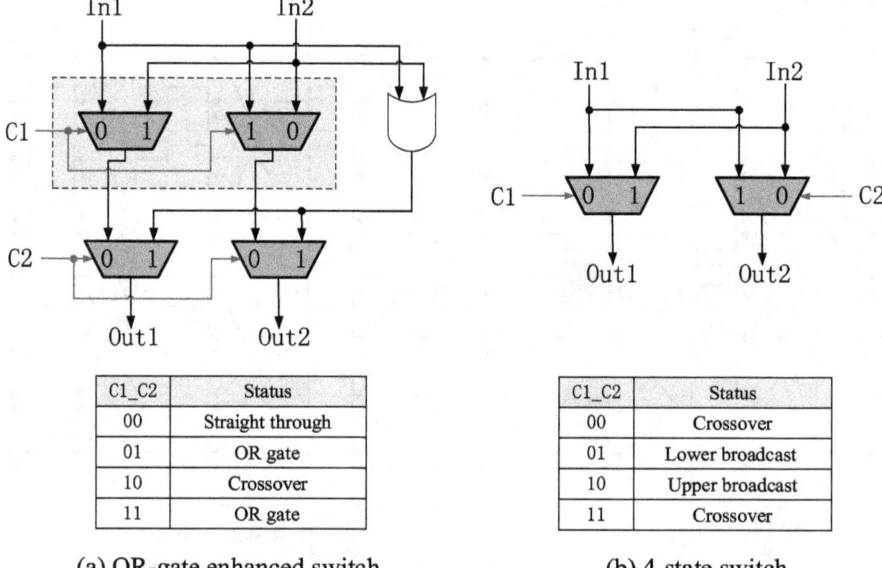

C1_C2	Status
00	Straight through
01	OR gate
10	Crossover
11	OR gate

(a) OR-gate enhanced switch

C1_C2	Status
00	Crossover
01	Lower broadcast
10	Upper broadcast
11	Crossover

(b) 4-state switch

Fig. 6.46 Two types of modified switches [42]

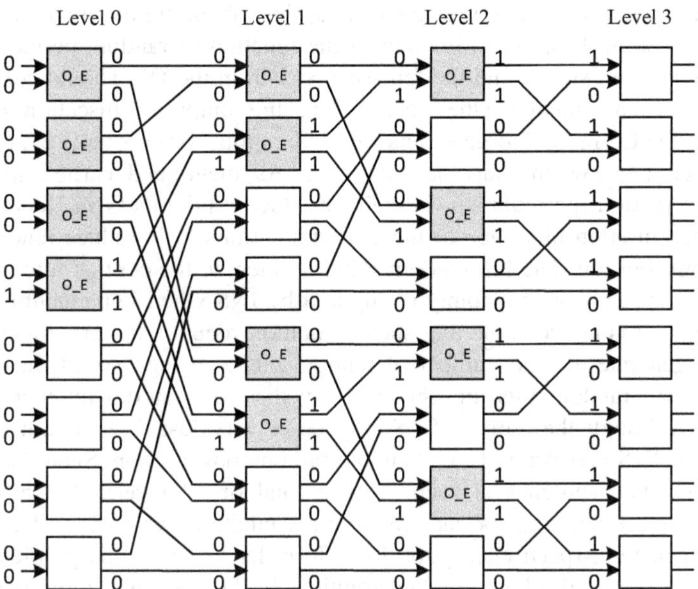

Fig. 6.47 Difference diffusion process using OR gate enhanced switches [42]

statistically speaking, the number of "0" and "1" in a large amount of statistics of HW balanced output will be counted, and the number of "0" and "1" is still equal.

The following is a brief explanation of the number of OR gate enhanced switches used by the INS. In order to guarantee that for any $\Delta \neq 0$, it is possible to output N/2 bits of "1" after passing through the OR gate enhanced switch. Consider the worst scenario where there is only 1 bit of "1" in Δ, in which case the depth of the two-input OR gate in the INS in the network should reach at least $\log_2(N/2)$. Taking into account the arbitrariness of Δ, the total number of OR gate inputs in each level of the network implementing the HW balance should not be less than the width of Δ (the width of Δ is $N/2$); i.e., the minimum required number of OR gates at each level of network is $\left(\frac{N}{2}\times\frac{1}{2}\right)$; then, the minimum number of OR gate enhanced switches required to implement HW balance is $\left(\frac{N}{2}\times\frac{1}{2}\right)\times\log_2\left(\frac{N}{2}\right)$ This number is exactly the number of OR gate enhanced switches in the INS. So the number of OR gate enhanced switches used at the INS is already the minimum number of switches required to implement the HW balance function. It should be pointed out that, in order to perform the permutation operation efficiently, the width N of the network tends to be larger or equal than the width of the difference Δ. If the width of Δ is the same as N, then an extra $N/2$ two-output OR gate should be added to reduce the width of Δ to $N/2$.

After implementing the HW balance, the INS also should complete the output obfuscation operation. The output obfuscation operation uses the random characteristics of the Benes network to perform a random permutation operation for the output after the HW balance is implemented, thus obtaining the final output of the infection function. In order to implement the function of random permutation, the OR gate enhanced switch and the four-state switch in the INS should complete the two-state switch function when performing the output obfuscation operation. Therefore, the C2 in the OR gate enhanced switch must be constant “0”, C1 and C2 in the four-state switch must be consistent. As mentioned earlier, in order to implement random permutation, it shall introduce randomness in Benes network switch configuration bits. One of the easiest solutions to introduce randomness is full random solution. The full random solution refers to the configuration bits of all switches in the network randomly configured by INS when implementing random permutation. It is conceivable that such a solution requires the cryptographic processor to generate a large number of random bits at a high speed. In the cipher processor, the random bits are generated by the true random number generator (TRNG). Although the current TRNG speed is very fast, it is assumed that the width of the INS network is 128 bits in the full random solution. Such INS network has 64 two-state switches at each level, a total of 13 levels, 832 bits random numbers that are required for each random permutation, which is still a challenge for TRNG in high-speed encryption [49]. In order to solve this problem, the literature [42] proposed a level random solution. In the case of random permutation using a level random solution, the INS randomly selects a certain level in the network at random and randomly configures the configuration bits of all the switches in the selected level. The configuration bits of other level switches of the INS are already initialized through random number when the cryptographic processor is powered on. In the level random solution, if the randomly selected random bits in certain level are ignored, the level random solution requires only N/2 random bits, and the number of random bits required is only $\frac{1}{2(\log_2 N)-1}$ of full random solution, which greatly reduces the required random number of bits when performing random permutation. In addition, each four-state switch requires two random configuration bits for the HW balance. At this point, the designer can select the appropriate number of four-state switches based on the output capacity of the cryptographic processor TRNG after meeting the basic security requirements. For the security of level random and full random solutions, readers can refer to the literature [42].

The RIMBEN’s implementation process is shown in Fig. 6.48. First, the two executing computation results (initial ciphertext) obtained by performing the encryption algorithm can be the results obtained by executing the encryption algorithm twice on the same circuit or executing it on two identical circuits, respectively. Then, XOR operations are performed against the computation results obtained by two executions, difference Δ is obtained, and I(Δ) is obtained when inputting Δ into the infection function I. Finally, XOR operation is performed against I(Δ) and one of the previous computation results, thus yielding the final output. The infection function I contains the HW balance and the output obfuscation. The upper part of the HW balance operation input is the difference Δ, and the lower part is the constant 0. The OR gate function is implemented in the OR

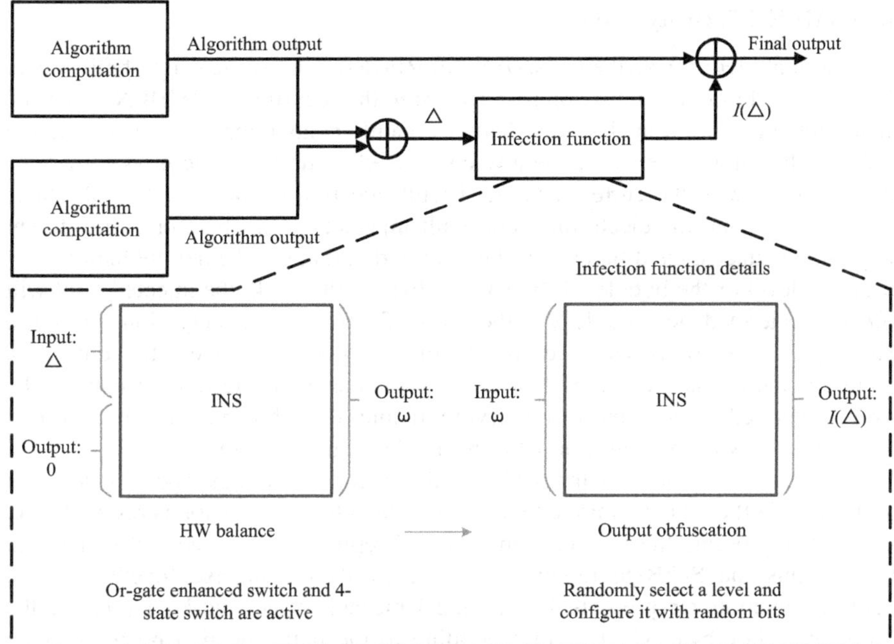

Fig. 6.48 Implementation process of RIMBEN [42]

gate enhanced switch in the INS, and configuration bits of the four-state switch are randomly generated when the HW balance operation is performed. The output of the HW balance operation is used as the input for output obfuscation. The output obfuscation operation performs a random permutation on the input. When the random permutation is performed, the random solution of the switch configuration bits is the level random solution, that is, randomly selects a certain level, and then randomly configures the configuration bit of switch on the selected level. The random permutation output with the same width of difference Δ is used as the final output of the infection function. It should be noted that the infection functions mentioned here do not depend on the particular encryption algorithm, so RIMBEN is suitable for all packet encryption algorithms and fully satisfies three features that the infection function should satisfy. First, RIMBEN's HW balance and output obfuscation operations make the output of the infection function have excellent random and satisfy the first feature. RIMBEN's inputs and outputs are then independent of each other and are able to completely mask the output ciphertext (the output of the HW balance operation is statistically half the number of "0"s and "1"s) and satisfy the second feature. At the same time, INS is exactly the same as the original Benes network topology, but only enhances the functions that individual switches can implement. Such structure not only realizes the original Benes network function, but also causes the performance loss and the area consumption of the INS within 10% compared with the original Benes network. It reuses resources as well as satisfies the third characteristic.

4. RIMBEN Security Proof

First of all, let us describe the security of RIMBEN under the single-fault attack. At the end, a brief description on how to ensure the security of RIMBEN under the multifault attack is given. The fault features mainly contain the fault values and the fault width. Fault values are generally considered random, whose physical effects are consistent with the current advanced fault injection methods such as the laser fault injection and the electromagnetic fault injection [50]. Therefore, in different single-fault attacks, the difference is the fault width, that is, whether the fault occurs at the bit level or the byte level. In a single-byte fault attack, the injected fault will affect the intermediate variable with the width of 1-byte in the encryption algorithm. This byte is randomly changed to one of 255 possible values. In contrast, a single-bit fault attack assumes that an attacker can more accurately control the wrong intermediate variable in the cryptographic algorithm, making the intermediate variable with the width of 1 bit in the algorithm go wrong.

The single-bit attack on the 128-bit AES algorithm is described in detail in Sect. 6.1. For the sake of simplicity, the single-bit attack expression is restated here. Where C represents the correct ciphertext, C' represents the incorrect ciphertext, S represents the SubByte function, S^{-1} represents the inverse function of the SubByte function, ε represents the injected 1-bit fault value, and k_{10} represents the tenth round key byte of AES corresponding to the fault injection position. After defining these variables, the single-bit fault attack expression is

$$\varepsilon = S^{-1}(C \oplus k_{10}) \oplus S^{-1}(C' \oplus k_{10}) \tag{6.87}$$

where ⊕ represents XOR operation.

For each (C, C') pair, one can find a set of k_{10} that satisfies (6.87). When an attacker repeatedly injects several single-bit faults into the cryptographic algorithm and gets several different (C, C′) pairs, the corresponding k_{10} value is obtained by taking the intersection of corresponding k_{10} set. In single-byte fault attacks, the correct key is also obtained by the equation existing between the median of the encryption algorithm and the result of fault [51–53]. A typical single-byte fault attack is to generate a fault with the width of 1-byte for the mixed column input in the second last round encryption of AES. After column mixing, the fault propagation pattern in the ciphertext becomes four bytes. According to the equation, it is possible to obtain a possible set of 4-byte keys corresponding to the second last round's round key of AES. After injecting the single-byte fault multiple times and taking the intersection of the set, the key corresponding to four bytes can be obtained [51].

The following is a formal proof of the security of RIMBEN. The proof steps contain three steps. Step 1, cite the viewpoints from the literature [54], and prove the security of infection attack countermeasure from the perspective of a cryptographic system. Step 2, by referring to the literature [55] and using the concept of mutual information, from the aspect of the information theory, prove the security of

RIMBEN under ideal conditions. Step 3, prove the security of RIMBEN under non-ideal conditions.

The basic idea of the infection countermeasure is to completely destroy the fault propagation pattern so that attackers cannot decipher the algorithm key even if they obtain some wrong ciphertext output. The literature [54] has proved that if there is a random function that can convert the result of the encryption algorithm into a random bit with the same width, then the symmetric encryption algorithm based on indistinguishability under chosen plaintext attack (IND-CPA) is also indistinguishability under chosen plaintext attack with a fault attack threat (IND$_f$-CPA). The indistinguishability, IND-CPA, and IND$_f$-CPA are described below. The definition of indistinguishability is complicated, and the reader can refer to relevant information [56]. In short, if the cryptographic algorithm is indistinguishable, then the attacker cannot distinguish the ciphertext pair from the encrypted plaintext. Therefore, the indistinguishability is an important feature of ensuring the credibility of the cryptographic algorithm. IND-CPA is a widely used definition in the cryptosystem, meaning that an attacker could not distinguish the results of different plaintexts even though they allow choosing the plaintext they are encrypting. IND$_f$-CPA is first defined in literature [54], which means that when attacker not only can choose to encrypt the plaintext, but also can introduce random faults in the encryption algorithm to observe the corresponding output, but they still cannot distinguish the encryption results of different plaintexts. The final proof of the literature [54] points out that if the fault output is completely random, the attacker's advantages obtained from the fault injection are not valid for deciphering the cryptographic algorithm key. This ensures the viability of the infection countermeasure in the demonstrable security analysis.

The following provides a formal proof of the security of RIMBEN using the concept of mutual information. The mutual information between two random variables is a measure of the interdependence between the two. The value of mutual information indicates how much information about X can be obtained from another random variable Y for the random variable X. Assuming that X and Y are two discrete random variables, the mutual information $\mathcal{I}(X;Y)$ of the two can be defined by (6.88) (in order to distinguish from the infective function, "$\mathcal{I}$" is used to replace "I"). $p(x,y)$ is the joint probability distribution function of X and Y; $p(x)$ and $p(y)$ are the marginal probability distribution functions of X and Y, respectively

$$\mathcal{I}(X;Y) = \sum_{y \in Y} \sum_{x \in X} p(x,y) \log\left(\frac{p(x,y)}{p(x)p(y)}\right) \tag{6.88}$$

Assuming that K is the used key and C_{f_out} is the final fault output after the RIMBEN infection, then $\mathcal{I}(K; C_{f\_out})$ can be used to measure how much information the attacker can have on the key from the fault attack. According to the definition of joint probability, there is $p(x,y) = p(y)p(x|y)$, where $p(x|y)$ is the conditional probability of x at a given y. Then, $\mathcal{I}(K; C_{f\_out})$ can be transformed into the equivalent of expression (6.88), that is

$$
\begin{aligned}
&\mathcal{I}(K; C_{\mathrm{f\_out}}) \\
&= \sum_{k,c_{\mathrm{f\_out}}} p(k, c_{\mathrm{f\_out}}) \log \frac{p(k, c_{\mathrm{f\_out}})}{p(k)p(c_{\mathrm{f\_out}})} \\
&= \sum_{k,c_{\mathrm{f\_out}}} p(c_{\mathrm{f\_out}}) p(k|c_{\mathrm{f\_out}}) \log p(k|c_{\mathrm{f\_out}}) \\
&\quad - \sum_{k} \log p(k) \left(\sum_{c_{\mathrm{f\_out}}} p(k, c_{\mathrm{f\_out}}) \right) \\
&= \sum_{c_{\mathrm{f\_out}}} p(c_{\mathrm{f\_out}}) \left(\sum_{k} p(k|c_{\mathrm{f\_out}}) \log p(k|c_{\mathrm{f\_out}}) \right) \\
&\quad - \sum_{k} p(k) \log p(k)
\end{aligned}
\tag{6.89}
$$

In the RIMBEN, C_{f_out} equals C ⊕ I(Δ) or C′ ⊕ I(Δ) (C ⊕ Δ⊕I(Δ)), depending on which path is injected with faults. In one case, assuming that the fault is injected into the redundant path, the output of the RIMBEN is the XOR of I (Δ) with the original computation, that is C_{f_out} = C ⊕ I(Δ). Due to the security of the encryption algorithm, C can be regarded as a random variable that is independent of K. If I(Δ) is random, then C_{f_out} is also a random variable that is independent of K. Alternatively, if an attacker injects a fault into the original path, C_{f_out} = C ⊕ Δ⊕I (Δ), Δ is a function of K. According to the secret sharing theory, if ΔI(Δ) is random, then Δ ⊕ I(Δ) is still independent of Δ (i.e., random) [57, 58], then C ⊕ Δ⊕I(Δ) is still independent at K's. Since I(Δ) of the RIMBEN output is random, $C_{\mathrm{f\_out}}$ is independent of K regardless of $C_{\mathrm{f\_out}}$, so there is p $p(k|c_{\mathrm{f\_out}}) = p(k)$. Thus the expression (6.90) can be further simplified

$$
\begin{aligned}
&\mathcal{I}(K; C_{\mathrm{f\_out}}) \\
&= \sum_{c_{\mathrm{f\_out}}} p(c_{\mathrm{f\_out}}) \left(\sum_{k} p(k) \log p(k) \right) - \sum_{k} p(k) \log p(k) \\
&= \left(\sum_{k} p(k) \log p(k) \right) \left(\left(\sum_{c_{\mathrm{f\_out}}} p(c_{\mathrm{f\_out}}) \right) - 1 \right) \\
&= 0
\end{aligned}
\tag{6.90}
$$

The result of expression (6.90) is 0, which means that an attacker cannot obtain information about the key from C_{f_out}. This proves the security of RIMBEN against single-fault attacks under ideal conditions.

It should be noted that, in actual use, in order to achieve the implementation cost and the balance of security, infection countermeasures often cannot be executed in the ideal condition, that is why these seemingly secure methods are broken in actual use. In RIMBEN, the deviation from the ideal case is reflected in the limited HW range (limited by the number of four-state switches) in the HW balance output and the level random strategy used in the output obfuscation phase. In this case, the key is to measure how much an attacker can gain valuable information about the key from the loss of randomness. Traditional infection methods inherently have random defects such as output bias and incomplete masking of the ciphertext. In contrast, the analysis of RIMBEN's loss of randomness is much more complicated because there is no obvious law of unpredictable obfuscation introduced by random permutations. Then, we introduce a quantitative security assessment of RIMBEN in practice by improving the traditional fault attack method and using statistical analysis-based attack methods for random permutation operations.

Here is an example of single-bit fault attack against 128-bit AES algorithm (a single-byte fault attack has a similar analysis method). Equation (6.87) is the corresponding attack equation. An attacker can easily get the wrong ciphertext C' from the output while performing single-bit attack on the AES algorithm without any protection. However, if the AES algorithm is protected by RIMBEN, the attacker cannot obtain the exact value of C' due to the randomness of I(Δ). After replacing C ′ in Eq. (6.87) with $C_{\mathrm{f\_out}}$, the intersection of possible sets of values for key k_{10} converges quickly to an empty set. To do this, attackers will exploit the non-ideal randomness of the infection function to obtain extra information to help decipher the key. If I(Δ) can be approximated by a deterministic variable (denoted as Iguess), and assure that I_{guess} is equal to I (Δ) with a high probability, an attacker can count the number of possible keys in different pairs $\{C$, Cf_out}, so there is a greater probability of receiving correct key. Compared to the original method of obtaining keys using the intersections, the method of counting frequency has better tolerance for errors introduced by random numbers.

To determine I_{guess}, we first need to investigate the features of I(Δ). The features of I(Δ) in the RIMBEN contain: the HW and the position distribution of each bit. Due to the obscuration process relying on random permute of the Benes network, the position distribution of each bit in I(Δ) has a good randomness, but the distribution of HW has certain regularity. Figure 6.49 shows the probability distribution of a certain byte HW in the infection function output I(Δ) obtained by a RIMBEN with a width of 128 bits without a four-state switch. Among them, the probability of occurrence of 4 in the HW is the largest, HW is $\{3, 4, 5\}$, and one of the probabilities of the occurrence of HW is 72.68%. With this set as I_{guess} to approximate I(Δ), it ensures that I_{guess} is equal to I(Δ) with a high probability.

Based on the above analysis, the modified attack method (i.e., the quantitative security assessment method) can be described in the following steps.

① Determine the set of the possible values of I_{guess}, and ensure that I_{guess} can be equal to $I(\Delta)$ with a high probability.
② Consider a pair $\{C, C_{\mathrm{f\_out}}\}$, where C is the correct ciphertext, and $C_{\mathrm{f\_out}}$ is the corresponding fault output after the infection is implemented.
③ Replace C' in the original formula with $C_{\mathrm{f\_out}} \oplus I_{guess,}$ and obtain a modified attack equation. For example, in the previously mentioned AES attack, the modified equation is

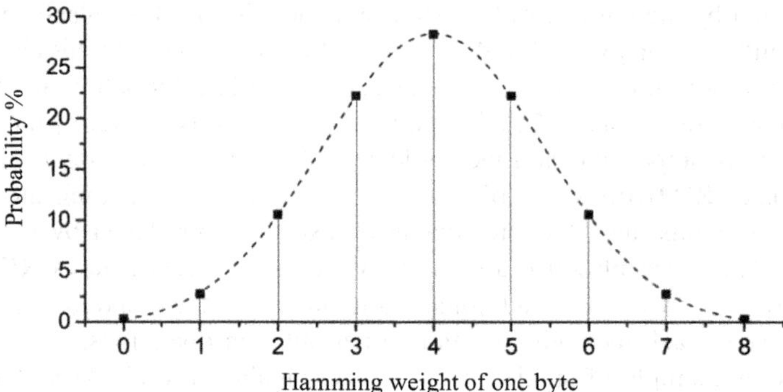

Fig. 6.49 Probability distribution of the HW of a certain byte in I(Δ) influenced by the 128-bit RIMBEN without any four-state switches

$$\varepsilon = S^{-1}(C \oplus k_{10}) \oplus S^{-1}(C_{f\_out} \oplus I_{guess} \oplus k_{10})$$

④ Select a possible value of I_{guess}, place into the equation, and get multiple k_{10} possible values, the number of occurrences of these possible values corresponding to the k_{10} will add 1.
⑤ Repeat Step ④ for each possible value in I_{guess}.
⑥ Find the value in the k_{10} value that has the maximum number of occurrence. If this value shows a stable convergence, the attack is successful; otherwise, consider a new $C_{f\_out} \oplus I_{guess}$ pair and return to Step ④.

Because I(Δ) is random, the attacker has only a certain probability of success. An increase in the number of faults (i.e., the number of unused {*C*, Cf_out}) increases the probability of success. Figure 6.50 shows two attack instances of the RIMBEN. In the case of a successful attack, the number of the times for which the correct key has appeared is significantly higher than the average compared to the 254 other possible key values. In the case of a failed attack, the key value corresponding to the maximum number of occurrences changes frequently as the number of faults increases, and in this case, the correct key cannot be found. Under the modified attack, the convergence of the correct key has been proved by the mathematical derivation in the literature [42]. By evaluating the security of the RIMBEN under non-ideal execution in this way, the result shows that it still improves the security against fault attacks by more than four orders of magnitude, indicating that even though RIMBEN is implemented under non-ideal conditions, it can still guarantee high security.

The following discusses how the RIMBEN resist multifault attacks. A multifault attack means that an attacker can precisely inject two or more faults when a single encryption algorithm is encrypted. In this type of attacks, the double-fault attack injecting two same faults into the computing path and the redundant path of the encryption algorithm has become a serious threat to the security of the encryption algorithm [43]. By using the biased fault concept, the collision probability of the

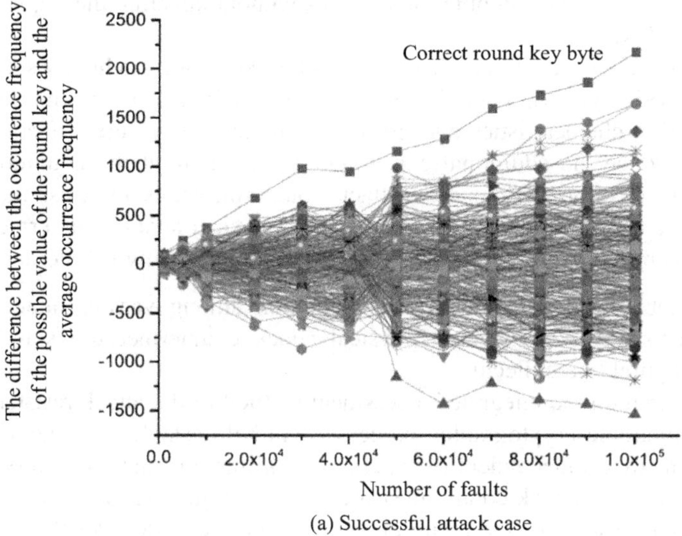

(a) Successful attack case

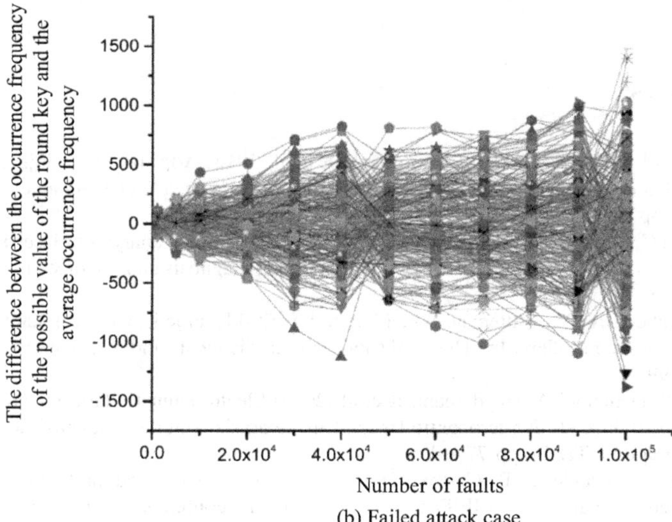

(b) Failed attack case

Fig. 6.50 Two instances of the improved attack against RIMBEN for the AES algorithm under a constraint of 100,000 faults

same fault pair is improved, thus making the double-fault attack more practical than the single-fault attack [59]. In order to resist double-fault attacks, it is necessary to increase the difficulty of attackers injecting two same faults into the RIMBEN. Therefore, it is possible to use both the fault space switching (using two different fault spaces) countermeasure and the random fake loop insertion countermeasure [60] when the original and redundant computations of the RIMBEN are conducted.

In this way, we can resist double-fault attacks without affecting the implementation of the RIMBEN.

In summary, the reconfigurable cryptographic processor has higher security than traditional ASIC and ISAP implementations. By fully using dynamic and partial reconfiguration characteristics and array computing architectures, it not only can drastically reduce the additional consumption caused by attack countermeasures, but also can develop various new attack countermeasures to resist the existing attacks that are not completely overcome. The physical attack countermeasures of the reconfigurable cryptographic processor have the following development trends.

① A collaborative research on the interaction among various attack countermeasures can help develop physical attack countermeasure solutions with reconfigurable architectures.
② A research on the integrated assessment method of the attack resistance of the reconfigurable cryptographic processor can help establish a scientific assessment method independent of specific attack implementation methods.
③ A research on attack countermeasures of reconfigurable systems can improve the security level of the entire processor from the system level.

References

1. Courbon F, Loubet Moundi P, Fournier JJ A et al (2014) Adjusting laser injections for fully controlled faults. In: International workshop on constructive side channel analysis and secure design, pp 229–242
2. Roscian C, Sarafianos A, Dutertre JM et al (2013) Fault model analysis of laser-induced faults in SRAM memory cells. In: IEEE workshop on fault diagnosis and tolerance in cryptography, pp 89–98
3. Woudenberg JGJV, Witteman MF, Menarini F (2011) Practical optical fault injection on secure microcontrollers. In: The workshop on fault diagnosis and tolerance in cryptography, pp 91–99
4. Moro N, Dehbaoui A, Heydemann K et al (2013) Electromagnetic fault injection: towards a fault model on a 32-bit microcontroller. In: IEEE workshop on fault diagnosis and tolerance in cryptography (FDTC), pp 77–88
5. Beroulle V, Candelier P, Castro SD et al (2014) Laser-induced fault effects in security dedicated circuits. In: IFIP/IEEE international conference on very large scale integration-system on a chip, pp 220–240
6. Bossuet L, Grand M, Gaspar L et al (2013) Architectures of flexible symmetric key crypto engines: a survey: from hardware coprocessor to multicryptoprocessor system on chip. ACM Comput Surv 45(4):1–32
7. Wang B, Liu L, Deng C et al (2016) Against double fault attacks: injection effort model, space and time randomization based countermeasures for reconfigurable array architecture. IEEE Trans Inf Forensics Secur 11(6):11511164
8. Kocher PC (2016) Differential power analysis resistant cryptographic processing. U.S. Patent application 15/236, 739.2016-8-15
9. Agrawal DRJRR (2003) Multi-channel attacks. In: International workshop on cryptographic hardware and embedded systems-CHES, pp 2–16

10. Sugawara T, Suzuki D, Saeki M et al (2013) On measurable side-channel leaks inside ASIC design primitives. In: International workshop on cryptographic hardware and embedded systems, pp 159–178
11. Hutter M, Mangard S, Feldhofer M (2012) Power and EM attacks on passive 13.56 MHz 13.56 MHz RFID devices. Lect Notes Comput Sci 4727:320333
12. J175212_201609 (1996) Measurement of radiated emissions from integrated circuits-surface scan method (loop probe method) 10 MHz to 3 GHz. SAE International
13. Heyszl J, Mangard S, Heinz B et al (2012) Localized electromagnetic analysis of cryptographic implementations. In: Cryptographers' track at the RSA conference, pp 231–244
14. Brier E, Clavier C, Olivier F (2004) Correlation power analysis with a leakage model. In: International workshop on cryptographic hardware and embedded systems, pp 16–29
15. Yoo HS, Herbst C, Mangard S et al (2007) Investigations of power analysis attacks and countermeasures for ARIA. Inf Secur Appl 160–172
16. Standaert FX, Malkin TG, Yung M (2006) A formal practice-oriented model for the analysis of side-channel attacks. IACR E-Print Archive 134:2
17. Shan W, Shi L, Fu X et al (2014) A side-channel analysis resistant reconfigurable cryptographic coprocessor supporting multiple block cipher algorithms. In: Design automation conference, pp 1–6
18. Herder C, Yu MD, Koushanfar F et al (2014) Physical unclonable functions and applications: a tutorial. Proc IEEE 102(8):11261141
19. Maes R (2013) Physically unclonable functions: constructions, properties and applications. Springer, Dordrecht
20. Gassend, Blaise, Clarke et al (2002) Silicon physical random functions. In: Proceedings of the 9th ACM conference on computer and communications security, pp 148–160
21. Suh GE, Devadas S (2007) Physical unclonable functions for device authentication and secret key generation. In: Design automation conference, pp 9–14
22. Maiti A, Schaumont P (2011) Improved ring oscillator PUF: an FPGA-friendly secure primitive. Springer, New York, pp 375–397
23. Lee JW, Lim D, Gassend B et al (2004) A technique to build a secret key in integrated circuits for identification and authentication applications. In: Symposium on VLSI circuits, 2004. Digest of technical papers, pp 176–179
24. Becker GT (2015) The gap between promise and reality: on the insecurity of XOR Arbiter PUFs. Springer, Berlin, pp 535–555
25. Guajardo J, Kumar SS, Schrijen GJ et al (2007) FPGA intrinsic PUFs and their use for IPl protection. In: International workshop on cryptographic hardware and embedded systems, pp 63–80
26. Su Y, Holleman J, Otis B (2007) A 1.6 pJ/bit 96% stable chip-ID generating circuit using process variations. In: IEEE international solid-state circuits conference, pp 406–611
27. Maes R, Tuyls P, Verbauwhede I (2008) Intrinsic PUFs from flip-flops on reconfigurable devices. In: The 3rd Benelux workshop on information and system security
28. Kumar SS, Guajardo J, Maes R et al (2008) The butterfly PUF protecting IP on every FPGA. In: IEEE international workshop on hardware-oriented security and trust, pp 67–70
29. Simons P, Sluis EVD, Leest VVD (2012) Buskeeper PUFs, a promising alternative to D flip-flop PUFs. In: IEEE international symposium on hardware-oriented security and trust, pp 7–12
30. Leest VVD (2012) Comparative analysis of SRAM memories used as PUF primitives. In: Conference on design, automation and test in Europe, pp 1319–1324
31. Nedospasov D, Seifert JP, Helfmeier C et al (2013) Invasive PUF analysis. In: The workshop on fault diagnosis and tolerance in cryptography, pp 30–38
32. Kong J, Koushanfar F, Pendyala PK et al (2014) PUFatt: embedded platform attestation based on novel processor-based PUFs. In: Design automation conference, pp 1–6
33. Cline B, Chopra K, Blaauw D et al (2006) Analysis and modeling of CD variation for statistical static timing. In: IEEE/ACM international conference on computer-aided design, pp 60–66

34. Kruskal CP, Snir M (1986) A unified theory of interconnection network structure. Theoret Comput Sci 48:7594
35. Bossuet L, Grand M, Gaspar L et al (2013) Architectures of flexible symmetric key crypto engines-a survey: from hardware coprocessor to multi-crypto-processor system on chip. ACM Comput Surv (CSUR) 45(4):41
36. Horowitz E, Sahni S (1978) Fundamentals of computer algorithms. Computer Science Press, New York
37. Wu C, Feng T (1980) On a class of multistage interconnection networks. IEEE Trans Comput 100(8):694702
38. Lee RB, Shi Z, Yang X (2001) Efficient permutation instructions for fast software cryptography. IEEE Micro 21(6):5669
39. Damgard I, Ishai Y, Krøigaard M (2010) Perfectly secure multiparty computation and the computational overhead of cryptography. In: Annual international conference on the theory and applications of cryptographic techniques, pp 445–465
40. Beneš VE (1964) Optimal rearrangeable multistage connecting networks. Bell Syst Tech J 43(4):16411656
41. Portz M (1991) On the use of interconnection networks in cryptography. In: Workshop on the theory and application of cryptographic techniques, pp 302–315
42. Wang B, Liu L, Deng C et al (2017) Exploration of Benes network in cryptographic processors: a random infection countermeasure for block ciphers against fault attacks. IEEE Trans Inf Forensics Secur 12(2):309322
43. Lomné V, Roche T, Thillard A (2012) On the need of randomness in fault attack countermeasures-application to AES. In: IEEE workshop on fault diagnosis and tolerance in cryptography (FDTC), pp 85–94
44. Agoyan M, Bouquet S, Fournier J et al (2011) Design and characterisation of an AES chip embedding countermeasures. Int J Intell Eng Inf 1(3–4):328–347
45. Joye M, Manet P, Rigaud J (2007) Strengthening hardware AES implementations against fault attacks. IET Inf Secur 1(3):106
46. Gierlichs B, Schmidt J, Tunstall M (2012) Infective computation and dummy rounds: fault protection for block ciphers without check-before-output. In: International conference on cryptology and information security in Latin America, pp 305–321
47. Tupsamudre H, Bisht S, Mukhopadhyay D (2014) Destroying fault invariant with randomization. Springer, New York
48. Battistello A, Giraud C (2013) Fault analysis of infective AES computations. In: IEEE workshop on fault diagnosis and tolerance in cryptography (FDTC), pp 101–107
49. Mathew SK, Srinivasan S, Anders MA et al (2012) 2.4 Gbps, 7 mW all-digital PVT-variation tolerant true random number generator for 45 nm CMOS high-performance microprocessors. IEEE J Solid-State Circuits 47(11):2807–2821
50. Leveugle R, Maistri P, Vanhauwaert P et al (2014) Laser-induced fault effects in security-dedicated circuits. In: The 22nd international conference on very large scale integration (VLSI-SoC), pp 1–6
51. Piret G, Quisquater J (2003) A differential fault attack technique against SPN structures, with application to the AES and KHAZAD. In: International workshop on cryptographic hardware and embedded systems, pp 77–88
52. Tunstall M, Mukhopadhyay D, Ali S (2011) Differential fault analysis of the advanced encryption standard using a single fault. In: IFIP international workshop on information security theory and practices, pp 224–233
53. Ali SS, Mukhopadhyay D, Tunstall M (2013) Differential fault analysis of AES: towards reaching its limits. J Cryptogr Eng 3(2):7397
54. Ghosh S, Saha D, Sengupta A et al (2015) Preventing fault attacks using fault randomization with a case study on AES. In: Australasian conference on information security and privacy, pp 343–355

55. Patranabis S, Chakraborty A, Mukhopadhyay D (2015) Fault tolerant infective countermeasure for AES. In: International conference on security, privacy, and applied cryptography engineering, pp 190–209
56. Reingold O (1998) Pseudo-random synthesizers, functions and permutations. The Weizmann Institute of Science doctor dissertation, Rehovot
57. Blömer J, Guajardo J, Krummel V (2004) Provably secure masking of AES. In: International workshop on selected areas in cryptography, pp 69–83
58. Oswald E, Mangard S, Pramstaller N et al (2005) A side-channel analysis resistant description of the AES S-Box. In: International workshop on fast software encryption, pp 413–423
59. Patranabis S, Chakraborty A, Nguyen PH et al (2015) A biased fault attack on the time redundancy countermeasure for AES. In: International workshop on constructive side-channel analysis and secure design, pp 189–203
60. Patranabis S, Chakraborty A, Mukhopadhyay D et al (2015) Using state space encoding to counter biased fault attacks on AES countermeasures. IACR Cryptology ePrint Archive 2015:806

Chapter 7
Outlook of Reconfigurable Cryptographic Processing Application Technology

With the continuous evolution of big data and globalization, the security environment has become increasingly serious. Among them, data security and cryptographic processor security issues are particularly prominent. In terms of data security, the demand for big data processing, represented by cloud computing, has brought new challenges to data security. Unfortunately, there is a significant security risk to user data stored in cloud or delegated for computing. Cloud administrators can view from "God's perspective," so the privacy of user data is difficult to protect. In order to improve the security of massive cloud data in the cloud computing mode, an expensive cost is needed for security protection. In many cases, the assurance of cloud computing security can only assume that cloud managers are not malicious. Of course, this largely limits the application of cloud computing in security-sensitive situations. In order to solve this problem, the new concept of fully homomorphic encryption has emerged in recent years. Homomorphic encryption can calculate the ciphertext without knowing the key or decrypting the ciphertext, so as to obtain the result equivalent to the result of the plaintext computation. The use of fully homomorphic encryption can enjoy cloud computing services without exposing private information to cloud servers. The emergence of quantum computing also poses a significant challenge to the cryptographic security. Shor's quantum computation algorithm proposed in 1994 can solve the factorization and discrete logarithm problems in a polynomial time [1]. From then on, the design of post-quantum cryptography against quantum computing attack becomes an important research direction of cryptography. In March 2016, quantum computers have been able to factorize 15 using Shor's algorithm in a scalable manner [2]. The rapid development of quantum computers has made the design of post-quantum cryptography more urgent. Lattice-based cryptography has been widely studied due to the characteristics of its quantum attack countermeasures. It may replace the RSA algorithm and the elliptic curve algorithm in the post-quantum cryptography era [3]. Many fully homomorphic encryption schemes are designed based on difficult problems in lattice. However, as homomorphic encryption is a developing field, its own framework and algorithms are still

L. Liu et al., *Reconfigurable Cryptographic Processor*,
https://doi.org/10.1007/978-981-10-8899-5_7

evolving. There are still many problems that need to be solved, including computational inefficiency and complicated algorithms. On the other hand, as the data processing carrier, the security of the cryptographic processor is also facing severe challenges in the new situation. In the context of global division of labor in the integrated circuit industry, cryptographic processors are not only threatened by physical attacks during their use, but also may suffer from attacks at each stage in the designing, producing, and manufacturing processes. In the entire life cycle of cryptographic processors, the intellectual property (IP) chips, outsourced design, and test services provided by untrusted third-party vendors, electronic design automation (EDA) software tools provided by untrusted vendors, and untrusted integrated circuit manufacturers, etc., are likely to pose a threat to cryptographic processor security. The kind of malicious tampering in the life cycle of the cryptographic processor collectively referred to as hardware Trojans. With the continuous improvement of processor integration, the prevention of hardware Trojans becomes more and more difficult. For example, one common form of hardware Trojan is the additional transistors to the processor, while 100 additional transistors are enough to build a hardware Trojan. It is difficult to detect the well-concealed 100 transistors in a cryptographic processor with tens of millions of transistors, which poses a serious threat to cryptographic processor security. This chapter mainly focuses on the application technology of reconfigurable cryptographic processor from the perspectives of fully homomorphic encryption and hardware Trojan and aims to provide a macroview of reconfigurable cryptography technology for readers in the future security situation.

7.1 Fully Homomorphic Encryption and Reconfigurable Computing

Since cloud computing has numerous benefits, such as reducing the requirements for client devices, saving users' software and hardware purchase cost and operation and maintenance cost, providing users with powerful computing power and massive storage capacity, and flexibly constructing basic information facilities, it is preferred by the individual users and small- and medium-sized business users. It has achieved rapid development in recent years. However, cloud security issues such as data leakage, data loss, data hijacking, sharing, and isolation have become obstacles to the further development of cloud computing. Although many people think that the cloud computing platform is more secure than the local computing platform, in fact, security incidents of cloud platforms have constantly occurred in recent years. For example, large-scale user data breaches occurred in Google Mail in 2011, large numbers of users' mails and chat records were deleted, or accounts were reset. In 2014, the cloud system vulnerabilities of Apple's iCloud led to the release of personal photographs of female celebrities on the Internet. Tens of thousands of credit card numbers stored in some Web sites such as the Amazon Web site and the

Wal-Mart Web site were exposed. In 2016, a well-known domestic forum leaked abundant account information and the information amount was amazing. These security issues are mainly caused by ineffective cloud protection and hacker attacks on the cloud, which can be reduced or avoided by improving security measures. However, the serious problem is that cloud administrators can view from "God's perspective" and can have full access to all the user data, which is the fundamental issue for cloud security. This prevents many users from migrating data and applications, especially security-sensitive data and applications to the cloud. Fully homomorphic encryption technology can make keyless parties have direct access to the ciphertexts and this feature makes it have very good application prospect in the field of information security. The emergence of fully homomorphic encryption technology can fundamentally solve the data security problems of cloud computing. In addition, fully homomorphic encryption has many benefits. For example, in a secure multi-party computing scenario, a fully homomorphic encryption scheme enables parties involved in computing to complete interactive computations without revealing data to each other. Theoretically, there is a potential application demand for fully homomorphic encryption in all untrusted computing environments.

Since Gentry proposed the first fully homomorphic encryption scheme in 2009, the theory, application, and implementation of fully homomorphic encryption have been being the hot topics in the field of cryptography. With the thorough research of theory and application of fully homomorphic encryption technology, the new homomorphic encryption scheme has been put forward and optimized constantly. However, since the complexity of the fully homomorphic encryption technology itself leads to low implementation performance of the fully homomorphic encryption scheme and huge storage space, there is still a long way to go for the application. Therefore, the implementation of fully homomorphic encryption scheme has also become the focus of academic research. Currently, there are implementation platforms and implementation methods based on general-purpose processors, graphic processing units, field-programmable gate array and application-specific integrated circuit. However, some problems still exist, such as poor performance of general-purpose processor platforms, huge power consumption of graphic processing unit, large volume of configuration data of field-programmable gate array platform, as well as the lack of flexibility for application-specific integrated circuit platforms. The reconfigurable computing technology features high performance, a range of flexibility, high energy efficiency ratio and efficient configuration, which is suitable for the current development state and characteristics of fully homomorphic encryption technology, and makes it most likely to become the homomorphic encryption technology platform. The following describes the concept and application, history and status, characteristics of fully homomorphic encryption and implementation of reconfigurable computing technology.

7.1.1 Concept and Application of Fully Homomorphic Encryption

Let us introduce the concept of somewhat homomorphic encryption before introducing fully homomorphic encryption. Homomorphic encryption means that two ciphertexts En (a) and En (b) are, respectively, obtained after two plaintexts a and b are encrypted, and then perform a certain operation $\odot$ on the ciphertexts and obtain the result En (a) $\odot$ En (b). If an operation $\oplus$ is performed on the plaintexts directly and encryption is performed on the result a $\oplus$ b, then the result becomes En (a $\oplus$ b), which is equal to En (a) $\odot$ En (b), that is, En (a) $\odot$ En (b) = En (a $\oplus$ b), where $\oplus$ is a kind of operation in the plaintext field, $\odot$ means the corresponding operation in the ciphertext field. If the plaintext operation $\oplus$ means addition, the homomorphic encryption is called homomorphic addition; if the plaintext operation $\oplus$ means multiplication, the homomorphic encryption is called homomorphic multiplication. The RSA algorithm introduced in Sect. 2.4.1 is a typical encryption algorithm of homomorphic multiplication, because multiplying two ciphertexts of an RSA yields the ciphertext obtained by multiplying the two plaintexts followed by encrypting the result. Homomorphic addition, homomorphic multiplication, finite homomorphic addition and multiplication are collectively referred to as somewhat homomorphic encryption.

Fully homomorphic encryption is relative to somewhat homomorphic encryption. It refers to the encryption function satisfying both the homomorphic addition property and the homomorphic multiplication property which can perform any number of addition operations and multiplication operations, that is

$$f(\mathrm{En}(m1), \mathrm{En}(m2), \ldots, \mathrm{En}(mk)) = \mathrm{En}(f(m1, m2, \ldots, mk)) \tag{7.1}$$

f can be the combination of any number of additions and multiplications. In logical operations, we can use XOR to represent addition, and AND to represent multiplication, XOR and AND to form a minimal complete set, form arbitrary logic functions with any number of XORs and ANDs. In other words, any function can be expressed by a certain number of XORs and ANDs operations as long as it can be expressed by a logic circuit or a computer program. In this way, the homomorphic encryption that satisfies both addition homomorphic and multiplication homomorphic is equivalent to the implementation of any logic function.

The homomorphic encryption technology can be used to protect the privacy of user data; that is, the plaintext is encrypted, calculation is done based on the ciphertext, and then the ciphertext result is decrypted; that is, the computation of the ciphertext can be implemented by the keyless parties,

$$\mathrm{Dec}(f(\mathrm{En}(m1), \mathrm{En}(m2), \ldots, \mathrm{En}(mk))) = f(m1, m2, \ldots, mk) \tag{7.2}$$

It can be seen that fully homomorphic encryption can perform any operation on a ciphertext when the third party does not know the key and the data are not decrypted, and the operation result is equivalent to that of the same operation performed on the plaintext. With the fully homomorphic encryption, the operation result of the ciphertext is the same as that of the plaintext while the plaintext is not exposed. This feature enables fully homomorphic encryption to be widely applied in the field of secure cloud computing, secure multi-party computing, and ciphertext retrieval.

The following takes the cloud computing scenario as an example to introduce the fully homomorphic encryption process and fully homomorphic scheme components, as shown in Fig. 7.1. Users hope to process data using a fully homomorphic encryption method through cloud computing. The whole data processing process is roughly divided into five steps.

① The user generates a group of fully homomorphic encryption parameters.
② The user generates a public key and a private key. Some somewhat homomorphic encryption schemes may also need to generate a set of computation keys and participate in the ciphertext computation.
③ The user encrypts his or her plaintext data with the public key, sends the encrypted ciphertext data, the algorithms to be executed, the parameters and the computation keys to the cloud. Arranging for the parameters and algorithms to be executed in advance is also possible.
④ The cloud processes the ciphertext according to the algorithm requirements. The computation process may need the participation of the computation key, and the processed ciphertext computation results should be sent to the user.

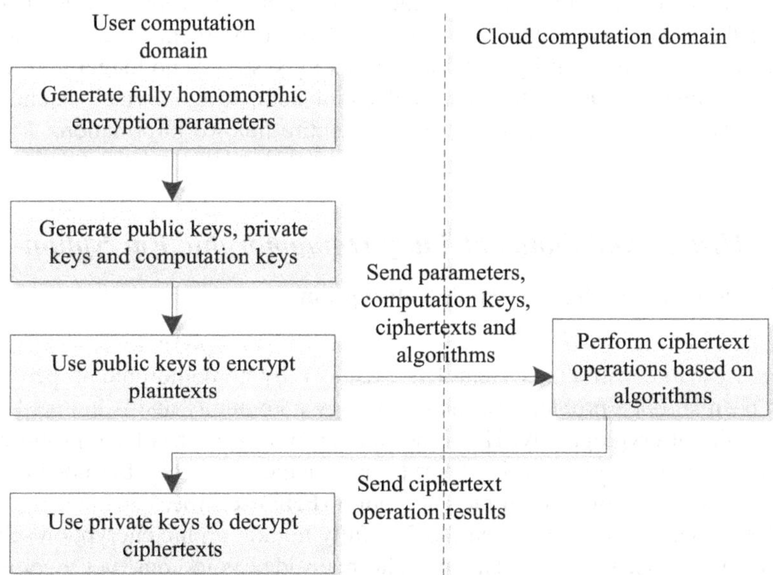

Fig. 7.1 Fully homomorphic encryption process

⑤ The user applies the private key to decrypt the ciphertext computation result to obtain the expected result, which should be consistent with the result of processing the plaintext data with the algorithm.

According to the above data processing flow, a fully homomorphic encryption scheme can be intuitively obtained, which generally includes several functional modules such as key generation, encryption, homomorphic computation, and decryption. The key generation module should be operated by the user to generate the public key used by the encryption module, the private key used by the decryption module, and the computation key required for the ciphertext computation. Here, usually an asymmetric cipher system, the public key used to encrypt plaintexts and the private key used to decrypt ciphertexts are adopted. The encryption module should also be operated by the user, and the public key is used to encrypt the user's plaintext data (*data*) and obtain the ciphertext. The homomorphic computing module is operated by the cloud. The data processing algorithm *f* given by the user can be used to operate on the ciphertext to obtain the operation result of the ciphertext. The operation result is equivalent to the result of encrypting *f*(*data*) with the public key. The decryption function module is operated by the user and is used to decrypt the ciphertext processed by the cloud, and the decryption result is *f* (*data*).

Because the cloud computation domain does not need to decrypt the ciphertext and does not know user's data information, there is no need to worry about the leakage of user data, so that the cloud can be regarded as a non-trusted computing environment. This feature provides security for user information in secure cloud computing and trusted computing, which is of great significance for protecting user information security. Fully homomorphic encryption can enable keyless party to compute the ciphertext, and fundamentally solve the problem of data security in cloud computing. The security of cloud computing system is no longer a necessary condition, which can make big data and cloud computing services extend from non-security-sensitive application field to the entire field of information.

7.1.2 History and Status of Fully Homomorphic Encryption

1. History of Fully Homomorphic Encryption

The idea of fully homomorphic encryption was first proposed by Rivest et al. [4] in 1978. How to construct a system that satisfies fully homomorphic properties has always been an open problem for the cryptology community and is honored as the "Holy Grail" of cryptography. Homomorphic cryptography has been proposed for 30 years and many homomorphic encryption schemes appeared, but none of these schemes was a fully homomorphic encryption (FHE) scheme.

In 2009, Gentry [5] constructed the first fully homomorphic encryption scheme. The scheme is based on ideal lattices. The main idea is to construct a somewhat homomorphic encryption scheme, which can support a certain depth of homomorphic computation, then compute the ciphertext homomorphically by using the decryption function. If the computational depth of the decryption function is less than that supported by the homomorphic encryption of this part, then the scheme can be transformed into a fully homomorphic encryption scheme. The key of the construction process is to find a somewhat encryption scheme, and its supported depth of the homomorphic computation is higher than that of its decryption function or the computational depth of the decryption function can be reduced by using compression techniques to meet this requirement. Once f in the homomorphic formula meets the requirements of the decryption function, the scheme is called bootstrapping and the ciphertext can be updated without knowing the key, thus reducing the noise and realizing the fully homomorphic encryption. The bootstrapping procedure is time-consuming, and the decryption will occupy a certain depth of homomorphic computation, so that the depth of homomorphic computation that can be used in practice is greatly reduced. The GH scheme [5] is a major breakthrough in the cryptology and computer science communities, solving the open problem that has plagued people for more than 30 years and has opened the curtain of the research on homomorphic encryption. As for the construction method with the GH scheme in the past few years, some researchers proposed several fully homomorphic encryption schemes [6–8]. Similar to the GH scheme, these schemes all adopted the bootstrapping procedure in the construction, belonging to the first generation of fully homomorphic encryption schemes.

The first-generation fully homomorphic encryption scheme uses a bootstrapping procedure, which is a boundless bootstrapping design scheme and theoretically can perform infinite-depth homomorphic operations. However, the boundless bootstrapping fully homomorphic encryption scheme has complex bootstrapping steps, huge computing costs, large key sizes, and ciphertext sizes. The existence of the bootstrapping procedure causes the performance of the fully homomorphic encryption to be low and does not have practical usability. Therefore, eliminating bootstrapping procedures becomes the focus of the research on fully homomorphic encryption and enhances the development of second generation of homomorphic encryption schemes. The second-generation homomorphic cryptosystem is a compromise between seeking the homomorphic operating depth and the resulting noise management, i.e., the leveled fully homomorphic encryption. Such schemes should first give the required depth of homomorphic computation L and perform polynomial-level homomorphic operation with the depth of L. This scheme is designed to put the FHE into practical application as soon as possible. The second generation of homomorphic encryption scheme breaks the framework of Gentry's first-generation fully homomorphic encryption scheme. It does not follow an infinite number of homomorphic computations. Instead, it attempts to satisfy the limited number of homomorphic computation that satisfies the application requirements by controlling the noise growth rate. The second-generation homomorphic scheme firstly constructs a somewhat homomorphic encryption scheme. After the ciphertext computation, it does not perform the bootstrapping procedure. Instead, it uses the key exchange technology to control the expansion of ciphertext noise. It constructs a hierarchical fully homomorphic encryption scheme that performs polynomial-level finite depth circuits to satisfy most applications. The second-generation fully homomorphic scheme started the BGV scheme [9], which is the first homomorphic scheme to eliminate the bootstrapping procedure. The scheme constructs a fully homomorphic circuit model and performs the homomorphic computation of the AES algorithm [10]. The second generation of homomorphic encryption schemes also includes the GSW scheme based on the learning with errors (LWE) [11], and the FV [12], YASHE [13] and LTV [14] schemes based on the ring learning with errors (RLWE). As the homomorphic schemes with higher efficiency such as the FV scheme and the LTV scheme were proposed, the computational time unit of the complex fully homomorphic circuit was changed from hour to second. At present, the fully homomorphic application research mainly focuses on the hierarchical FHE. With the continuous development and optimization of the fully homomorphic encryption schemes, its applied research and its theoretical research have become hot topics in the cryptographic research field, and its practical prospect is becoming increasingly clear.

2. Introduction to Fully Homomorphic Scheme LTV

The following introduces the second generation of hierarchical homomorphic encryption scheme LTV, taking it for example. The LTV scheme is a variant of the NTRU scheme [15]. The LTV scheme uses relinearization techniques and modular switching techniques to control the noise. The LTV scheme supports the multi-user environment and each user has its own key. To simplify the statuses, we pay close attention to the single-user situation. The basic operational element of LTV is a polynomial on the ring $R_q = Z_q\,[x]/<x^N + 1>$, where N is the order of the polynomial and q is a prime number and the coefficients of the polynomial are all the results of modulo q. χ is a truncated discrete Gaussian distribution, which is used to sample random polynomials. The LTV, a fully homomorphic encryption scheme, has the following six algorithms at the algorithm level.

① Key generation (KEYGEN): Select polynomials $u^{(i)}$, $g^{(i)} \leftarrow \chi$ for $i = 1, 2, \ldots, d$ and take $f^{(i)} = 2u^{(i)} + 1$, satisfy $f^{(i)} = 1 \bmod 2$; if $f^{(i)}$ is irreversible, then we reselect $f^{(i)}$ and compute $h^{(i)} = 2g^{(i)}\left(f^{(i)}\right)^{-1} \in \mathrm{R}_{q_i}$, generate the public key $pk = (h^{(i)}, q)$ and the private key $sk = f^{(i)} \in Rq_i$; for i = 1,2, …, d, τ = 0, …, $\lfloor \log q_i \rfloor$, we select $s_\tau^{(i)}, e_\tau^{(i)} \leftarrow \chi$ randomly, compute $\zeta_\tau^{(i)} = h^{(i)} s_\tau^{(i)} + 2e_\tau^{(i)} + 2^\tau \left(f^{(i-1)}\right)^2 \in \mathrm{R}_{q_{i-1}}$, and generate the auxiliary public key $evk = \{\zeta_\tau^{(i)}\}\mathrm{GS}_\tau^i$.

② Encryption algorithm (Enc): In order to encrypt a 1-bit m, we randomly select the polynomial s, e $\leftarrow \chi$ and compute the ciphertext $c^{(0)} = h^{(0)}s + 2e + m \in \mathrm{R}_{q_0}$ using the public key$(h^{(0)}, q_0)$.

③ Decryption algorithm (Dec): Decrypt the ciphertext c, use the private key $sk = f^{(i)}$, and compute $m = cf^{(i)} \bmod 2$.

④ Homomorphic addition (EVALUATE.ADD): Suppose $c_1^{(0)} = Enc(m_1)$ and $c_2^{(0)} = Enc(m_2)$, compute $Enc(m_1 + m_2) = c_1^{(0)} + c_2^{(0)}$.

⑤ Homomorphic multiplication (EVALUATE.Mult): Similar to the above homomorphic addition, compute $\tilde{c}^{(i-1)} = c_1^{(i-1)} \cdot c_2^{(i-1)} (\mathrm{mod}\emptyset(x))$, and then convert $\tilde{c}^{(0)}$ module to the ciphertext $\tilde{c}^{(1)}$ by relinearization and modular transform.

⑥ Relinearization: To convert $\tilde{c}^{(i-1)} = c_1^{(i-1)} \cdot c_2^{(i-1)} (\mathrm{mod}\emptyset(x))$ into $\tilde{c}^{(i)}(x)$, first express $\tilde{c}^{(i-1)}(x)$ in a binary form $\tilde{c}^{(i-1)}(x) = \sum_\tau 2^\tau \tilde{c}_\tau^{(i-1)}(x)$, where the coefficients of $\tilde{c}_\tau^{(i-1)}(x)$ are 0和1, we note it as $\tilde{c}^{(i)}(x) = \sum_\tau \zeta_\tau^{(i)} \tilde{c}_\tau^{(i-1)}(x) \in R_{q_i}$, then we can get

$$
\begin{aligned}
\tilde{c}^{(i)}(x) &= h^{(i)}(x)\left[\sum_{\tau=0}^{\lfloor \log q_i \rfloor} s_\tau^{(i)}(x)\tilde{c}_\tau^{(i-1)}(x)\right] + 2\left[\sum_{\tau=0}^{\lfloor \log q_i \rfloor} e_\tau^{(i)}(x)\tilde{c}_\tau^{(i)}(x)\right] \\
&\quad + \left[f^{(i-1)}\right]^2 \sum_{\tau=0}^{\lfloor \log q_i \rfloor} 2^\tau \tilde{c}_\tau^{(i-1)}(x) \\
&= h^{(i)}(x)s(x) + 2E(x) + \left[f^{(i-1)}\right]^2 \tilde{c}^{(i-1)}(x) \\
&= h^{(i)}(x)s(x) + 2E(x) + \left[f^{(i-1)} c_1^{(i-1)}(x)\right]\left[f^{(i-1)} c_2^{(i-1)}(x)\right] \\
&= h^{(i)}(x)s(x) + 2E(x) + m_1 m_2
\end{aligned}
$$

3. Research Status of Fully Homomorphic Encryption Implementation

At present, the computation amount of the fully homomorphic encryption scheme is very large and the performance of the fully homomorphic encryption is poor. There is a huge gap between the performance of the fully homomorphic encryption and the actual application requirements. To improve the implementation performance of the fully homomorphic encryption is the necessary condition for giving full play to the huge application value of the fully homomorphic encryption. At present, the research in this aspect has become an important field of research in the academic circles. One of the research directions is to develop a new fully homomorphic encryption scheme to reduce the computational complexity. Another research direction is to improve the implementation algorithm to reduce the computation complexity of the fully homomorphic encryption. The third research direction is to improve the processing power of the fully homomorphic encryption hardware and enhance the processing performance of the fully homomorphic hardware using parallel processing, pipelines, storage optimization, and other technologies. At present, the research on the fully homomorphic encryption implementation mainly focuses on the two latter aspects, that is, the implementation algorithm optimization and the hardware platform and architecture optimization, so as to enhance the performance of the fully homomorphic encryption. At present, the fully homomorphic encryption schemes are implemented respectively based on different hardware platforms such as general-purpose processors, graphics processors, field-programmable gate arrays, and application-specific integrated circuits. Table 7.1 shows some hardware platforms and performance of implementing fully homomorphic encryption. There is a big gap between the performance and the application requirements of implementing these fully homomorphic encryptions. For example, the fastest fully homomorphic AES circuit has a processing speed of only 10 B/s [16], far short of the application demand. The hardware processing

Table 7.1 Performance of fully homomorphic encryption implementation

Implementation object	Fully homomorphic scheme and parameter	Hardware platform	Performance
General-purpose processor			
Single-bit addition [17]	FV scheme, (n = 4096, p = 127)	Xeon CPU E5-2666v3 @ 2. 9 GHz, 15 GB memory	10 μs
Single-bit multiplication [17]			17200 μs
SIMON encryption circuit [18]	FV scheme, (n = 10 500, p = 570)	IntelI7-2600 @ 3.4 GHz	526 s
	FV scheme, (n = 27 000, p = 1024)		3062 s
Graphics processor			
Single-bit addition [19]	BGV scheme	NVIDIA GTX 980 @ 1126 MHz, 4 GB of memory	200 μs
Single-bit multiplication [19]			3447 μs
AES encryption circuit [20]	LTV scheme, (n = 32 768, p = 1271)	NVIDIA GTX 690 @ 915 MHz, 4 GB of memory	73 s
PRINCE encryption circuit [20]	LTV scheme, (n = 16 384, p = 575)		12.8 s
FPGA			
AES encryption circuit [16]	LTV scheme, (n = 16 384, p = 575)	Virtex-7 XC7VX690T @ 250 MHz	4.4 s
PRINCE encryption circuit [16]			0. 5 s
Encryption [21]	FV scheme, (n = 4096, p = 141)	Virtex-6 XC6VLX240T® 125 MHz	0. 275 ms
Decryption [21]			0. 153 ms
Re-encryption [21]			0. 428 ms
ASIC			
Encryption [22]	GH scheme, n = 2048)	IBM 90 nm	18. 1 ms
Decryption [22]			16. 1 ms
Re-encryption [22]			3. 1 s
Multiplier [23]	768 K bit	IBM 90 nm	0. 206 ms

capability will limit the fully homomorphic encryption technology, which will hinder the practice application.

The GPP-based hardware platform uses the general-purpose processor as the computing resource and it belongs to the serial instruction execution architecture. This platform can implement the instruction-level parallelism (ILP) in the multi-stage pipeline condition and it can also implement the thread-level parallelism in multi-core situations, thus improving the FHE processing performance. Because the implementation and verification of the GPP platform are easy and fast, most of the homomorphic schemes have general processor-based implementation versions, such as the GH scheme [10], the YASHE scheme, the FV scheme [18], and the LTV scheme [24]. In addition, these platforms can apply some open-source libraries or design and optimize some new libraries for some types of fully homomorphic encryption. For example, the generic number theory library (NTL) [25] and the fast library for number theory (FLINT) [26] are, respectively, C++-based and C-based libraries. They can support arbitrarily large integers and polynomial computing based on integers or finite fields, which are usually used for the cipher implementation based on ideal lattices. The HELIB library is a library written in C++, which is built based on the NTL library. The library is optimized for the fully homomorphic encryption schemes based on the cyclotomic polynomial ring $Z_q[X]/F[X]$. It is used to optimize the implementation of the BGV scheme, the GH scheme, etc. [27, 28]. The NFLIB library, written in C, is optimized for the polynomial ring and it can be used to replace the NTL library content used in the HELIB library [17]. In addition, some open-source libraries are designed and optimized for specific homomorphic schemes. For example, the first version of the open-source library SEAL designed by Microsoft is optimized for the YASHE scheme and the second version of the open-source library SEAL is optimized for the FV scheme [29]. These libraries are the results of the researches on homomorphic encryption implementation. They further facilitate and promote the research on the homomorphic encryption implementation algorithms.

The homomorphic encryption computation performance of the general-purpose processor is insufficient and it needs to be accelerated by other computing platforms. As the GPU-based cloud computing platforms are common, the application of graphics processors to accelerate homomorphic encryption scheme is a better choice. With the continuous improvement of the GPU architecture in recent years, the GPU has evolved into a system with a lot of parallelisms, multi-threading, and multi-core computing ability. The GPU-based hardware platform can utilize the programmable stream processor in the GPU to perform in the single instruction stream and multiple data stream, so as to obtain the data-level parallelism (DLP) and improve the processing efficiency of the FHE. The compute unified device architecture (CUDA), which includes the CUDA instruction set architecture and the parallel computation engine inside the GPU, enables researchers to write programs in C that can run in the GPU. By using it, the GPU can solve complex computational problems, further facilitates and speeds up the implementation of the homomorphic encryption application, etc., other than image processing, in the GPU. Many homomorphic encryption schemes are accelerated based on the GPU platform, such as the GH scheme [30], the BGV scheme [19], and the LTV scheme [31]. Some researchers also designed an open-source library for fully homomorphic encryption cuHE [31] based on GPU, which further facilitates the research on the fully homomorphic encryption implementation.

Another option for accelerating homomorphic encryption is using the FPGA-based hardware platform. The FPGA-based implementation can spatially unfold algorithms by using parallelable hardware resources so as to implement the operation-level parallelism (OLP) and improve the processing efficiency of the FHE. Since the FPGA resources are limited, it is difficult to fully implement the fully homomorphic encryption scheme. In fact, it is not necessary. It generally acts as a heterogeneous architecture consisting of a coprocessor and a general-purpose processor so as to jointly complete the homomorphic encryption scheme. The FPGA can generally implement large integer multiplication or polynomial multiplication, or even only implement the NTT and coefficient multiplications. For example, a 768 Kbits multiplier is implemented based on the StratixV FPGA [32]. The homomorphic multiplication, homomorphic addition, and key switching of the RLWE-based fully homomorphic encryption schemes such as the YASHE are implemented based on the Stratix V FPGAs [33]. The homomorphic ciphertext data operation of the YASHE scheme is implemented on the Virtex7 XC7V1140T and the SIMON64/128 block cipher is also implemented [34]. The AES and Prince algorithm of the LTV scheme is implemented based on the Virtex7 XC7VX690T [16]. The encryption, decryption, and re-encryption operation of the FV scheme is implemented based on the Virtex 6 FPGA [21].

The best method to accelerate the homomorphic encryption operation is to use an ASIC hardware platform and design special hardware for accelerating a certain fully homomorphic encryption method and specific parameters. The ASIC's positioning is similar to that of the FPGA, which acts as a coprocessor to accelerate the computation of homomorphic cryptography. Compared to the FPGA, we can further improve the processing performance of fully homomorphic encryption and reduce power consumption. The goal of ASIC implementation also concentrated on large integer multiplication and polynomial multiplication. For example, the literature [23] implements a 768-Kbit integer multiplication using IBM 90 nm process; the literature [35] implements a million-bit integer multiplication using the TSMC 90 nm process; the literature [22] implements the encryption, decryption, and re-encryption of the GH fully homomorphic scheme using the TSMC 90 nm process, which is the only complete operation of implementing fully homomorphic encryption by using the ASIC method.

The above-mentioned several implementation platforms have optimized the fully homomorphic encryption implementation method based on their own hardware architectures. However, due to the computational complexity of the fully homomorphic encryption and the defects of the hardware platform, the performance still cannot meet the application requirement or there are other problems. Based on the general-purpose processor, software can be used to implement the fully homomorphic encryption, which has high flexibility and can implement any parameters of any fully homomorphic scheme. However, the performance of the homomorphic scheme is very poor and the distance does not satisfy the performance requirements of the fully homomorphic encryption. The GPU-based hardware platform can greatly improve the performance of the fully homomorphic encryption while the flexibility is kept high. However, the typical power consumption of the high-end GPU is as high as 200–400 W [23], and it occupies 900 million equivalent logic gates of the hardware resources [36]. The FPGA-based hardware platform also greatly enhances the performance of the fully homomorphic encryption. Because its hardware reconfigurability also has some flexibility, the power consumption has been reduced compared to that of the GPU. However, the FPGA is a common device and has not optimized for fully homomorphic applications. FPGAs also have some drawbacks. For example, their configuration data amount is large and they cannot be dynamically reconfigured due to their fine-grained reconfiguration characteristics. The FPGA development lacks the high-level language support and is relatively hard for CPUs and GPUs. The ASIC-based hardware platform can implement the optimization of homomorphic encryption in many aspects such as performance, area, power consumption, and energy efficiency. However, the ASIC design is the most difficult and the development cost is the highest. The ASIC hardware platforms have poor flexibility and they cannot be programmed or reconfigured once they are completed, making it difficult to adapt to the constantly evolving scheme and potential diverse applications of fully homomorphic encryption.

7.1.3 Fully Homomorphic Encryption Based on Reconfigurable Computing

Fully homomorphic encryption features large parameters, intensive computation, high flexibility and rich parallelism. All of these features make the fully homomorphic encryption be suitable to be implemented with the reconfigurable computing technology. This section firstly analyzes the features of fully homomorphic encryption and then designs a reconfigurable key module.

(1) Fully homomorphic encryption features large parameters and intensive computation

The fully homomorphic encryption scheme features large parameters, which determines that the fully homomorphic encryption is computationally intensive application and the typical application scenario of reconfigurable computing. Fully homomorphic encryption requires a compromise among the practicability, the computational complexity and the security in terms of parameter selection. The practicality of fully homomorphic encryption contains two meanings: one is the depth of the fully homomorphic circuit, the larger the depth is, the wider the range of application is. Theoretically, the homomorphic operation can be carried out for infinite times according to the maximum depth, but the consumption, key size, and ciphertext size of the homomorphic computation are very large. Therefore, the main research at present concentrates on the fully homomorphic circuits with a depth of 40–80 levels. With such a depth, the fully homomorphic encryption can complete the homomorphic computation of AES with 10 rounds of iterative. The higher depths require that homomorphic schemes use larger parameters. When selecting RLWE-based fully homomorphic encryption scheme parameters, it is required to specify three parameters (dimension n, module q, Gaussian parameter r). In the case of fixed number of dimensions n, q/r is inversely proportional to the difficulty of the RLWE problem, i.e., the larger the q/r, the more insecure it is. The r in the fully homomorphic encryption scheme based on RLWE is generally fixed in advance. The larger the value of q is, the lower the security of the homomorphic scheme is. However, the circuit depth supported by the homomorphic scheme is also proportional to q, hoping that the value of q should be as large as possible to obtain the computation of deeper circuits. Therefore, in general, in order to ensure the security and computational depth of the scheme, it increases n while increasing q. Therefore, it is necessary to balance the values of q and n in the homomorphic encryption scheme to ensure the security of the homomorphic scheme and the capability of homomorphic computation. As shown in Table 7.1, in order to support the homomorphic computation of the AES algorithm, the maximum of the parameter n is 32 768 and the maximum of q is 1271; that is, the fully homomorphic encryption scheme needs to perform polynomial operation with the order of 32,768 and coefficient with 1271 bits module. In most cases, the order of the polynomial is at 2^{15} or 2^{16}, and the coefficients are about 1200 bits or about 2500 bits [34]. Therefore, the polynomial multiplication of high-order large coefficients is very computationally intensive. The key generation, encryption, decryption, homomorphic multiplication, and relinearization and other operations in the homomorphic scheme will use polynomial multiplication, and an application generally has multiple levels, corresponding to multiple homomorphic multiplication and relinearization operation. This determines that there are many computationally intensive polynomial multiplication operations in fully homomorphic encryption, which are computationally intensive algorithms.

(2) Fully homomorphic encryption features high flexibility requirements

Fully homomorphic encryption has high requirements for flexibility of the implementation platform, which makes reconfigurable computing a very attractive technology of implementing fully homomorphic encryption. First of all, the fully homomorphic encryption schemes are diverse and are still evolving. From the perspective of time dimension, since the first fully homomorphic encryption scheme was proposed by Gentry in 2009, a new scheme or an optimized version of the previous scheme has been proposed each year and the fully homomorphic encryption scheme is still a hot topic in the field of cryptographic theory and application. From the perspective of mathematical difficulties, there are currently lattice-based homomorphic encryption schemes, integer-based homomorphic encryption schemes, and the LWE (or RLWE)-based homomorphic encryption schemes [37]. From the perspective of the basic units of the bottom-level processing, there are integer-based fully homomorphic encryption schemes and polynomial-based fully homomorphic encryption schemes. These fully homomorphic encryption schemes are still under development and have advantages and disadvantages in terms of computational complexity, key size, ciphertext size, and noise growth, and none of them has absolute advantages or is generally approved by the filed of academy or industry. In the scenarios with different users and applications, there is a need to select different fully homomorphic encryption schemes so that the implementation platform will meet the requirements of adapting to various fully homomorphic encryption schemes. Second, even in the case of the specific fully homomorphic encryption schemes, different levels of security and applications have the flexibility requirement for the implementation platform. This is mainly because the fully homomorphic encryption scheme should select suitable parameters according to the security and the specific application. In terms of security, just as the traditional algorithm parameters and security requirements, the higher the security requirement, the greater the choice of parameter. For example, the different key lengths 128, 192, 256 of the AES determine different security levels, and the different key lengths 1024, 2048 of the RSA also correspond to different security levels. In the RLWE-based fully homomorphic encryption scheme, the most important parameters are the polynomial module and the coefficient module, and some of the parameters are shown in Table 7.2. The corresponding parameter pairs (n, log q) are (1024,35), (2048,60), (4096,116), (8192,226), and (16384,435), respectively. The choice of these parameters is not only related to the level of security, but also related to the specific application. Specifically, it is related to the increase of noise in the calculation of ciphertexts. The ciphertext generated by the fully homomorphic encryption scheme is a ciphertext with noises. The ciphertext that has just been encrypted will have noises even if it has not been calculated. Moreover, the ciphertext computation leads noise in the ciphertext to increase, and when the noise exceeds a certain limit, it cannot be correctly decrypted. The upper limit of the noise is determined by the parameters of the fully homomorphic encryption scheme. Therefore, when selecting parameters, it must ensure that the upper noise limit is greater than the noise that grows with the ciphertext computation. Parallel computation has no data dependency and does not result in an additional increase in the noise of the final result, so only computations involving serialization are concerned. It shall also be noted that the contributions of homomorphic addition and homomorphic multiplication to noise growth are different, and the homomorphic multiplication brings much more noise than that

Table 7.2 Examples of some parameters

Polynomial module	Coefficient module
$x^{1024}+1$	$2^{35}-2^{14}+2^{11}+1$ (35 bit)
$x^{2048}+1$	$2^{60}-2^{14}+1$ (60 bit)
$x^{4096}+1$	$2^{116}-2^{18}+1$ (60 bit)
$x^{8192}+1$	$2^{226}-2^{26}+1$ (60 bit)
$x^{16384}+1$	$2^{435}-2^{33}+1$(60 bit)

brought by homomorphic addition, so the depth of multiplication basically determines the noise growth in homomorphic computation. In general, it is generally necessary to estimate the noise reached by the homomorphic computation based on the multiplication depth in the application, and then select the appropriate parameters to ensure that the upper noise limit is greater than the noise level that can be reached by the homomorphic computation. Therefore, the security requirement and the application determine various parameter choices. The different parameters directly determine the number of integer digits involved in the operation, or the number of the polynomial orders and the number of the polynomial coefficient digits. This requires that the platform be able to adapt to different security requirements and meet the requirements for a variety of parameters. Under normal circumstances, after the security level and the application are determined, the parameters are determined accordingly. However, some fully homomorphic encryption schemes still require the flexibility of parameters even under the determined security levels and applications. For example, in the LTV scheme, q is different at different levels. q is the largest in the first level, and then it is reduced in each level and q is the smallest in the last level. This requires that the modification of the value of q is simultaneous as the level of homomorphic computation advances. The above only lists two important parameters, polynomial module and coefficient module. In fact, there may be many other parameters in the fully homomorphic encryption scheme, such as the plaintext space coefficient module, and standard deviation and limit of truncated discrete Gaussian distribution, basic of decomposing ciphertext in relinearization. These parameters may change with different homomorphic programs, different security levels, different applications, and different ways of implementation. In a word, the different homomorphic schemes, homomorphic applications, different security levels, and different parameters require the greater flexibility for the hardware platform.

(3) Fully homomorphic encryption features rich parallelism

The computation of fully homomorphic encryption scheme features rich parallelism and is suitable for the processing element array architecture of reconfigurable computing. Polynomial multiplication or large integer multiplication is the most important commonality logic of major fully homomorphic encryption schemes. It is also the most time-consuming computing element in the fully homomorphic encryption scheme, which is the focus of academic research at present. Because there are many similarities between large integer multiplications and polynomial multiplications, many of the algorithms that implement polynomial multiplication also have corresponding integer multiplication versions. For the sake of convenience, we use polynomial multiplication as an example for parallel analysis. Before analyzing the parallel of polynomial multiplication, we first analyze the polynomial features and multiplication methods in the homomorphic encryption. As mentioned earlier, the practical fully homomorphic encryption scheme uses larger parameters, so the polynomial used by it is a high-order large coefficient polynomial. If it is multiplied directly, the computational complexity is $O(n^2)$; that is, the modular multiplication of the polynomial coefficient needs to be carried out for n^2 times. Although the modular multiplication of the coefficient can be carried for n^2 times in parallel, there are two practical problems. One problem is that the multiplication is carried out for too many times. For a parameter of $n = 2^{15}$, the multiplication of coefficients needs to be carried out for times of the order of magnitude of one billion. Since the computation amount is too large, an algorithm with lower computational complexity is required. Another problem is that the coefficients are also large, reaching one or two thousand bits, which makes the storage and operation of multiplying a single coefficient hard. As a result, it requires a large-scale hardware or much longer time, as well as limits the multiplication of coefficient as a basic parallel unit. To overcome these two problems, researchers use the Chinese remainder theorem (CRT) transform and the number theoretical transform (NTT) to optimize the polynomial multiplication [20].

First, let us introduce the CRT algorithm. Due to the large polynomial coefficients, if a bit-parallel multiplier is used for coefficient multiplication when performing polynomial multiplication, a multiplier of up to 2,500 × 2,500 bits not only consumes a large area but also lowers the operating frequency. On the other hand, the speed will be too slow if a serial multiplier is used. In this case, we can consider using the CRT to process each coefficient A_j of Polynomial A and transform a larger number A_j into a set of small numbers, that is,

$$\mathrm{CRT}: A_j \rightarrow \{A_j \bmod p_0, A_j \bmod p_1, \ldots, A_j \bmod p_{l-1}\} \tag{7.3}$$

where p_i is a prime number smaller than q, ($i < l$); l is the number of these prime numbers and satisfies $q = \prod_{i=0}^{i=l-1} p_i$ A total of l polynomial is obtained through the CRT transform, i.e., $\{A^{(0)}(x), A^{(1)}(x), \ldots A^{(l-1)}(x)\}$, $A^{(i)}(x) \in \mathrm{R}_{p_i} = Z_{p_i}[x]/x^N + 1$; that is, the ith polynomial $A^{(i)}(x)$ belongs to the polynomial ring R_{p_i}. The order of R_{p_i} is N and the coefficients of the polynomial are all modulo p_i; i.e., the coefficients of the ith polynomial are less than p_i. These small coefficient polynomials make polynomial computations faster and more efficient. The operation between two polynomials will be performed between two polynomials whose coefficients are small and have the same superscript. For example, the computation of $C(x) = A(x) \cdot B(x)$ is transformed into $\{A^{(0)}(x) \cdot B^{(0)}(x), A^{(1)}(x) \cdot B^{(1)}(x), \ldots A^{(l-1)}(x) \cdot B^{(l-1)}(x)\}$. After the computation is completed, the coefficients of the resulting polynomial $C(x)$ are derived from the inverse Chinese remainder theorem (ICRT)

$$\mathrm{ICRT}(C_j) = \sum_{i=0}^{l-1} \left(\frac{q}{p_i}\right) \cdot \left(\left(\frac{q}{p_i}\right)^{-1} \cdot C_j^{(i)} \bmod p_i\right) \bmod q \tag{7.4}$$

Then, let us introduce the NTT algorithm. In polynomial multiplication, if $a(x) = \sum_{i=0}^{N-1} a_i x^i, b(x) = \sum_{i=0}^{N-1} b_i x^i$, and a, b $\in R_q$, $R_q = Z_q[X]/F[X]$, R_q is a polynomial ring, and the polynomials in the set Rq are all modulo F [X] whose coefficients are modulo q. When calculating the polynomial multiplication c (x) = a

(x) b (x), the coefficient $c_i = \sum_{j=0}^{i} a_j b_{i-j}$, which can be seen as a circular convolution of two integer sequences. The classical textbook multiplication is a square progressive complexity, i.e., O(N^2); that is, multiplication of two polynomials requires that the number in Z(p) is multiplied by N^2 and equivalent number of additions or subtractions is performed. In addition to the classic textbook multiplication, we can also use the number theoretic transform to calculate the product of two polynomial elements in Rq. When the number theoretic transform is applied against a(x) and b(x), it will obtain A(x) and B(x), respectively. The polynomial multiplication between a(x) and b(x) is calculated by multiplying the corresponding coefficients of A(x) and B(x), and then inverse number theoretic transforms (INTTs) are performed against the result. The formula can be expressed as c(x) = INNT (NTT(a(x)) ⊙ NTT(b(x))), ⊙ represents the multiplication of two corresponding polynomial coefficients. The NTT is similar to the Cooley Tukey algorithm [38] of Fast Fourier Transform (FFT), as shown in Algorithm 7.1, except that FFT is a complex field computation and the NTT is a computation in the finite field GF(P). The inverse number theoretic transform (INTT) is the same as the number theoretic transform except that ω is replaced by $\omega - 1 mod q$ [38]. The polynomial multiplication based on the number theoretic transformation can implement quasi-linear asymptotic complexity, i.e., O(N logN log logN). This NTT can greatly reduce the number of coefficient multiplication. Compared with the classical algorithm, it has greatly reduced the amount of computation, particularly effective for large N.

Algorithm 7.1 NTT Transform

Input: Polynomial a (x) ∈Zq [X], number of a (x) is N-1, $N = 2^n$; ω

Output: A (x) ∈Zq [X] = NTT (a)

1. A ← BitReverse (a)
2. *for* m=2 *to* N *by* m=2m *do*
3. $\omega_m = \omega_N^{N/m}$
4. ω ← 1

   for j=0 *to* $\frac{m}{2}$ *do*

   for k=0 *to* N−1 *by* m *do*

5. $t \leftarrow \omega \cdot A[k+j+\frac{m}{2}]$
6. u ← A [k + j]
7. A [k + j] ← u + t
8. $A[k+j+\frac{m}{2}] \leftarrow u - t$
9. $\omega \leftarrow \omega \cdot \omega_m$

The NTT-based polynomial multiplication greatly reduces the number of coefficient multiplication and also results in the consumption of the NTT. The NTT is applied to the polynomial a(x) with the upper order of $2^n - 1$ of the polynomial ring

Zq[X]/F[X]. Since the NTT-based order of the resulting polynomial after the multiplication is 2^{n+1}, a(x) needs to be zero padded so that the order becomes 2^{n+1}, and the coefficients of padding are all zero. The result is $A(x) = \sum_{i=0}^{2(n+1)-1} A_i \cdot x^i$ after the NTT is applied to the polynomial a(x), where the coefficient $A_i \in Zq$, and the value is $A_i = \sum_{j=0}^{2(n+1)-1} a_j \cdot \omega^{ij} mod\, p$. The twiddle factor $\omega \in$ Zq and satisfies $\omega^{2^{(n+1)}} \equiv 1 \bmod p$, whereas $\omega^i \neq 1$ mod p for any case $i < 2^{(n+1)}$. In order to quickly implement the NTT, the Cooley Tukey algorithm is used, which can divide the NTT into the odd part and the even part. After the NTT is applied to the polynomials of the two parts, the transform results will be re-integrated into the complete NTT, that is

$$A_i = \sum_{j=0}^{2^n-1} a_{2j} \cdot \omega^{i(2j)} mod\, p + \sum_{j=0}^{2^n-1} a_{2j+1} \cdot \omega^{i(2j+1)} mod\, p \tag{7.5}$$

It can be simply described as $A_i = E_i + \omega^i O_i$, where E_i and O_i represent the *i*th coefficients of 2^n point NTT corresponding to odd and even coefficients of a(x), respectively. Due to the periodicity of NTT operations, we know $E_{i+2^n} = E_i$, $O_{i+2^n} = O_i$. Therefore, $A_i = E_i + \omega^i O_i$ when $0 \leq i < 2^n$ and $A_i = E_{i-2^n} + \omega^i O_{i-2^n}$ when $2^n \leq i < 2^{(n+1)}$. For the twiddle factor ω, $\omega^{i+2^n} = \omega^i \cdot \omega^{2^n} = -\omega^i$ is satisfied. Combining these conditions, we can split a larger $2^{(n+1)}$ point NTT into two smaller 2^n point NTTs, namely

$$\begin{aligned} A_i &= E_i + \omega^i O_i \\ A_{i+2^n} &= E_i - \omega^i O_i \end{aligned} \tag{7.6}$$

With this method, we can continue to split E_i and O_i into $2^{(n-1)}$ point NTT, respectively, and continue splitting the 2-point NTT continuously with this method. This 2^{n+1} point NTT can get 2-point NTT after making n splitting and transform. The 2-point NTT can be done with only one addition and one subtraction with only two coefficients.

Finally, the polynomial multiplication combing CRT and NTT is shown in Algorithm 7.2. The two polynomials a and b are multiplied, and the CRTs are first transformed into two sets of small coefficient polynomials $\{a_i\}$ and $\{b_i\}$, and then the NTT is applied to these two sets of polynomials, which yields two sets of polynomials $\{A_i\}$ and $\{B_i\}$. A set of polynomials {Ci} is obtained by multiplying each two corresponding coefficients of {Ai} and {Bi}. The inverse NTT is applied to each $\{C_i\}$, and then the inverse CRT transform is performed to obtain the product of polynomials a and b.

Algorithm 7.2 Polynomial multiplication optimization algorithm

Input: Polynomials a, b, parameter is (n, *log*q)

Output: Polynomial c, parameter is (2n, *log*nq^2)

1. $\{a_i\}=CRT(a)$, $\{b_i\}=CRT(b)$
2. $\{A_i\}=NTT(\{a_i\})$, $\{B_i\}=NTT(\{b_i\})$
3. $\{C_i\}=\{A_i\}\cdot\{B_i\}$
4. $\{c_i\}=INTT(\{C_i\})$
5. $c=ICRT(\{c_i\})$

After introducing the concrete implementation method of polynomial multiplication, we analyze its parallelism. Even if classical textbook polynomial multiplication has stronger parallelism and the two polynomial coefficients can be stronger parallelism. After the CRT transform is adopted, the polynomials with large coefficients are transformed into polynomials with small coefficients while the orders of the polynomials remain unchanged. This polynomial multiplication can transform the large coefficient multiplication into a number of small coefficient multiplications. This not only increases the frequency of the multiplier and reduces the area, but also creates a great deal of parallelism so that all small coefficient polynomials can be operated in parallel. Compared with the polynomial multiplication without CRT, the degree of parallelism improves by l times (l is the number of small coefficients into which the parameter q is separated). In addition, for parameters q with different values, they can be separated into small primes with similar values. Select a set of small primes p_i with similar sizes, and the corresponding numbers of small primes l will be different. In this way, the CRT method can make the sizes of coefficient multipliers used in the polynomial multiplication corresponding to the fully homomorphic encryption scheme with different parameters q consistent. This feature has positive significance for almost all implementation platforms. For general-purpose processor platforms and graphics processor platforms, pi generally takes 32-bit or 64-bit primes, which is consistent with the maximum multiplication instructions supported by the platform. For the FPGA platform, it is consistent with the multiplication operation supported by the DSP; for the ASIC platform, it is more flexible, and we can flexibly select the size of p_i. For reconfigurable computing platforms, massive parallel computations that are created by such rules as CRT are well suitable for computing with reconfigurable array architectures. p_i is generally consistent with the granularity of the processing element. In fact, because the processing element and the array should be designed with specific condition, the size of the array and the granularity of the processing element should be designed with reference to the size of q and prime p_i.

The polynomial multiplication parallelism based on the NTT manifests itself in two aspects: one is the parallelism of the NTT and the INTT, and the other is the multiplication parallelism in the NTT domain after the NTT. The multiplication after the transform is the pair-wise multiplication of corresponding coefficients, and the parallelism of which is obvious. The coefficient multiplication can be performed in parallel for at most $2^{(n+1)}$ times. There are n + 1 layers of butterfly operations in the NTT. These layers are data related and can only be executed serially. Within the same layer, there are 2^n butterfly operations, each of which contains one addition, one subtraction, and one multiplication. There is data dependency between the multiplication operation and the addition/subtraction operation. We must first complete the multiplication, and then perform the addition/subtraction operation. All these 2^n additions and 2^n additions in the same layer can be carried out in parallel. All the 2^n multiplications in the same layer can be carried out in parallel.

(4) Reconfigurable design of fully homomorphic key modules

It has been previously analyzed that polynomial multiplication (or large integer multiplication) is the most important basic computation for fully homomorphic encryption, so it is also the most key module to which various acceleration platforms pay close attention. We take polynomial multiplication based on NTT as the object and discuss the implementation method based on reconfigurable computing technology.

First, we discuss the design of reconfigurable array size and topology. For polynomials with the order of 2^n, reconfigurable arrays have three design options. The first kind of structure of the array requires n + 1 columns of processing elements, and each column has 2^{n+1} processing elements in total. There is no interconnection between the processing elements within each column, and the processing elements in each column are connected with the configurable interconnection resources. The advantage of this approach is that it is very convenient and efficient to map the complete NTT algorithm, and the NTT operation can be completed after making just one configuration. The disadvantage is that it occupies a very large amount of resources and requires (n + 1) 2^{n+1} processing elements in total. When the value of n is relatively large, the resource consumption is unacceptable. For example, when n is 15, the reconfigurable array requires 1,048,576 processing elements. The second kind of structure of the array requires only one column of processing elements, which has 2^{n+1} processing elements in total. After each configuration, this array can compute a complete layer of NTT operations, and the entire NTT requires n configurations and operations. When n is 15, the reconfigurable array requires 65,536 processing elements, which is 1/16 of those of the first scheme. Due to the large number of configurations, this method is less efficient compared with the first method. We can reduce the amount of configurations and improve efficiency by using partial configuration methods. The third kind of structure of the array is to implement a smaller-scale array. For example, let the array have six columns and each column has 64 processing elements. Then, the array has 384 processing elements in total, which is 1/2370 of those of the first scheme. This scheme uses less hardware resources and can process multiple sets of data once the configuration is finished. For example, when configuring a NTT with n = 15, it can process 2^{16-6} = 1024 sets of data at a time, and these operations can run smoothly. The disadvantage of this structure lies in the fact that there are many configurations and runs; on the other hand, the mapping of operations after the sixth layer of NTT is more difficult than the first two schemes. The access of the coefficients is no longer sequential and should be adjusted based on number of layers, which results in a larger total amount of data configured. Figure 7.2 shows the schematic diagram of three kinds of reconfigurable computing unit arrays. PE is a reconfigurable logic unit, and each column of PE is connected through a reconfigurable interconnection, solid arrows indicate data flow, while dotted arrows

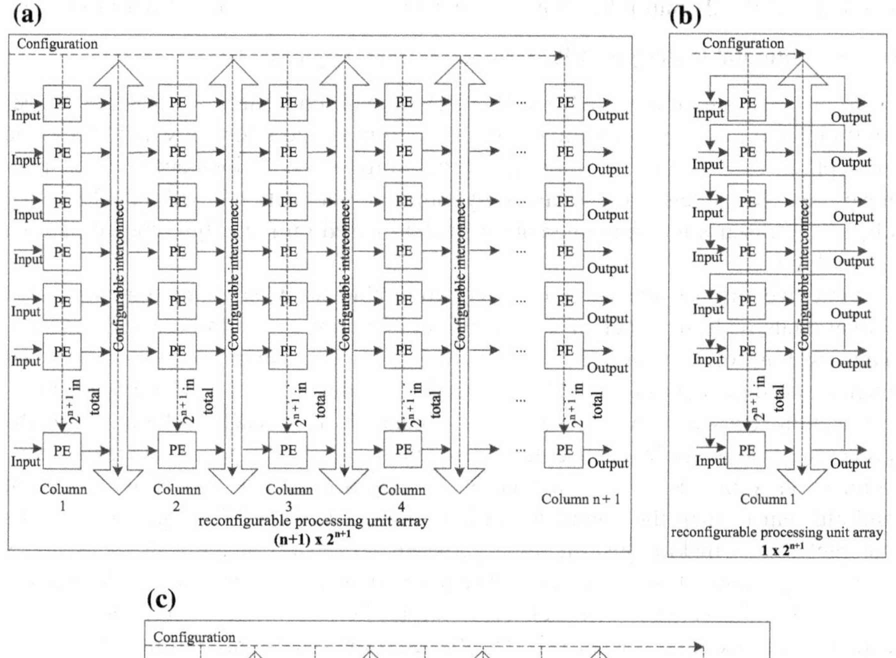

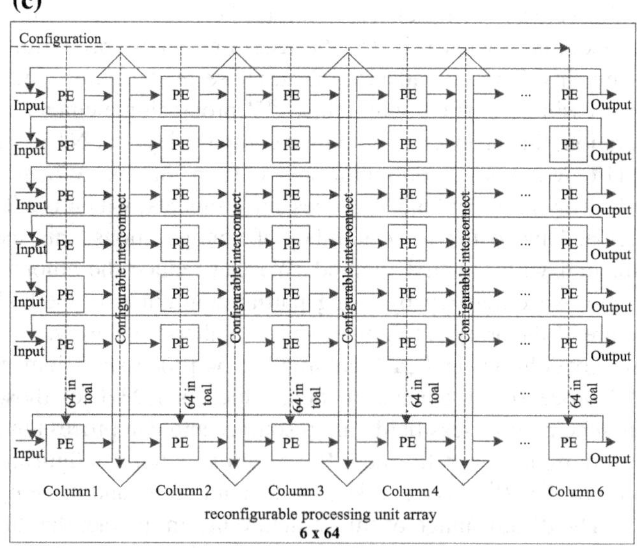

Fig. 7.2 Reconfigurable processing element array structure

indicate the configuration flow. Figure 7.2a shows the first array structure, and the results are directly output as it is able to complete the NTT. Figure 7.2b shows the second array structure. NTT requires multiple configuration and iteration and also requires feedback to the input to continue the operation of the next iteration in addition to the output. Figure 7.2c shows the third array structure with the same structure as Fig. 7.2a except that it is small in size and NTT cannot be configured and completed at once. The output should also be fed back to the inputs. In practical applications, different array structures and sizes can be selected according to available hardware resources.

Second, let us discuss the functional design of the processing element. From the functional point of view, the processing element should cover all the basic operations of polynomial multiplication. It mainly includes the modular multiplication of the coefficient and the twiddle factor, the modular addition in the butterfly operation, and the modular subtraction in the butterfly operation in the NTT; the modular multiplication of the corresponding coefficient after the NTT; the modular multiplication of the coefficient and the inverse twiddle factor, the modular addition in the butterfly operation, the modular subtraction in the butterfly operation and the division of resulting coefficients in the inverse NTT. Since the parameter q does not change frequently, the twiddle factor, the power of the twiddle factor, the inverse twiddle factor, and power of inverse twiddle factor used in the NTT and inverse NTT can be obtained through pre-computation after the parameter q is determined. Due to the NTT, the corresponding coefficient multiplication and inverse NTT are serial and will not be executed in parallel, and all the processing elements need only be configured with one modular multiplication unit. For modular addition and subtraction in butterfly operations, only one of them is actually used when computing a coefficient, and since the logics of modular addition and subtraction are very close, a modular addition/subtraction unit can be designed. The division of the result coefficients requires dividing the points by the number of NTT points; here, it is 2^{n+1}. The division is equivalent to multiplying by the inverse of the divisor since the modular computation is required. Since parameter n is also infrequently changed, the inverse of 2^{n+1} module q can be pre-computed once it is determined. Then, the modular multiplication is performed by multiplying with the coefficient. Also, since this part of the operation is serial with the butterfly operation of previous NTT, coefficient multiplication and multiplication used in the butterfly operation of the inverse NTT, the same modular multiplier in the processing element can be shared. One of the two modulo-multiplied operands is a coefficient and the other may be a coefficient, a power of the twiddle factor, an inverse power of the twiddle factor, and an inverse of the 2^{n+1} module q. The result of modular multiplication is returned to the storage unit where the coefficients are stored. One of the two operands for modular addition and subtraction is the coefficient and the other is the output of the mathematical modular multiplication unit. Figure 7.3 shows the structure of a reconfigurable processing element, where the solid line is the data flow and the dotted line is the configuration flow. PE can perform the coefficient modular multiplication, modular addition, modular subtraction,

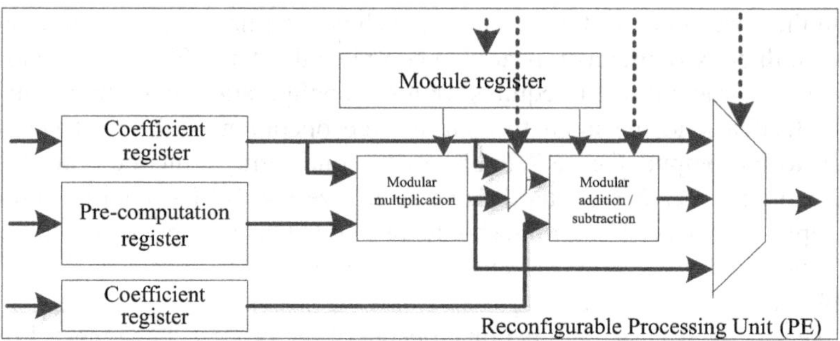

Fig. 7.3 Reconfigurable processing element structure

coefficient modular multiplication + modular addition, coefficient modular multiplication + modular subtraction and direct output. The coefficient register holds polynomial coefficients or intermediate results of the coefficients during the NTT. The pre-computation register stores the power of the twiddle factor, the power of the inverse twiddle factor, and the inverse of 2^{n+1} module q, etc. The module register stores the value of the module q.

Again, we will analyze the granularity of the reconfigurable processing element. The granularity of reconfigurable processing element mainly refers to the granularity of modular multiplication and modular addition/subtraction. Because they are modular computing, in fact, they depend on the size of the module. Although the parameter module q of the fully homomorphic encryption differs according to different schemes, applications, and security requirements of the fully homomorphic encryption, the Chinese remainder theorem is mainly used to decompose the polynomial coefficients into a series of smaller prime number module. These small prime numbers can select prime numbers with the same number of bits, so that the granularity of participating in the operation is consistent. The hardware module is very regular and suitable for reconfigurable arrays. Generally, it will select 32-bit or 64-bit primes, preferably it has the same data width as that of system, so that the entire system can be consistent and improve the data processing and transmission efficiency.

Then, we will analyze the interconnect structure of the reconfigurable array. The relationship between the processing elements in the same column in the reconfigurable logic array is parallel and they do not interconnect with each other. The interconnection between the adjacent column processing elements of the reconfigurable logic array is regular, so there is no need to design fully connected interconnect structure between the adjacent columns. Let us observe the m-th level and i-th coefficient of 2^{n+1} point NTT, where $1 \leq m \leq n + 1$, $1 \leq i \leq 2^{n+1}$. If $i\%2^m \leq 2^{m-1}$, its value comes from the i-th coefficient of the m − 1th level and the $i + 2^{m-1}$ th coefficient, otherwise its value comes from the i-th coefficient and the $i - 2^{m-1}$ th coefficient of the $m - 1$th level. It can be seen that each processing element only needs to be connected with n + 1 processing elements in the previous column except the processing element in the same position; that is, each processing element needs to be connected with n + 1 processing elements in the previous column. For example, the first processing element needs to be connected to the first, and the second, third, fifth, ninth, …, 2^{n+1}th processing elements in the previous column. The i-th processing element needs to be interconnected with the i-th processing element in the previous column; when $i\%2^m \leq 2^{m-1}$, it is interconnected with $i + 2^{m-1}$ processing element; when $i\%2^m > 2^{m-1}$, it is interconnected with the $i - 2^{m-1}$ processing elements, where $1 \leq m \leq n + 1$. In addition, in order to implement the larger NTT efficiently for smaller reconfigurable arrays, the processing elements in the last column need to be interconnected with the processing elements in the first column in addition to being directly output. Figure 7.4 shows a reconfigurable interconnecting structure with eight processing elements per column, and two processing elements in two adjacent columns are shown as gray lines. Each PE has two inputs, one of which is directly from the PE output of the same position in the previous column and the other input is obtained from the output of three different PEs of the previous column via a multiplexer, which is controlled and selected by the configuration information.

Finally, let us discuss the configuration of there configurable array. To reduce the amount of configuration data and support dynamic partial reconfiguration better, a hierarchical configuration information structure needs to be designed. The information to be configured by the reconfigurable processing element array includes the functions of the reconfigurable processing element, the pre-computed values, moduli, and interconnection structure, etc. Reconfigurable processing element with modular multiplication, modular addition, modular subtraction, modular multiplication + modular addition, modular multiplication + modular subtraction and other functions needs to be configured. The pre-computed values contain the twiddle factors and their powers, the inverse of the twiddle factors and their powers,

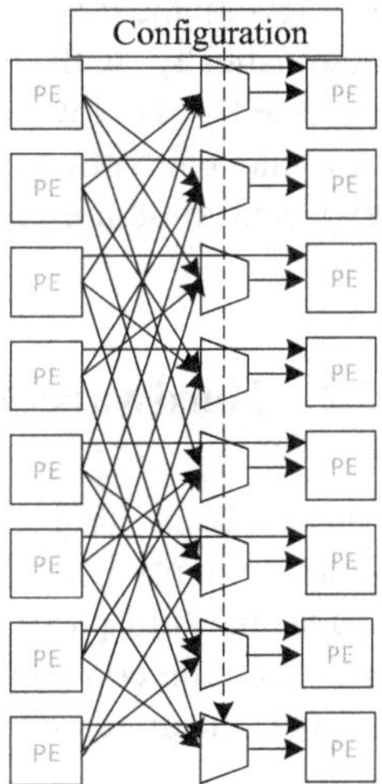

Fig. 7.4 Schematic diagram of reconfigurable interconnecting structure

the inverse of the 2^{n+1} module q, etc. These three types of numbers are all involved in the modular multiplication operation. The data that is involved in modular multiplication can be loaded into the pre-computed register when configuring. In addition, modular multiplication, modular addition, and modular subtraction also need to be pre-calculated (as determined by Chinese remainder theorem) and configured into the processing element. The interconnection structure is also an important part of the configuration, and where the input from the processing element comes from and where the output is sent depending on the configuration information. In order to reduce the configuration volume and speed up the configuration, the configuration information needs not be configured completely each time. Instead, it is hierarchically configured, and the configuration information is divided into different levels according to the change frequency. For example, the module of modular operation is the level of low-frequency variation, and the computing functions and interconnections are the configuration levels of relatively high-frequency variations. By setting NT transformation, NTT domain multiplication, inverse NT transformation of high-level function selection parameters, it can extract the configuration levels with relatively low-frequency change. Similarly, the new configuration level can be further divided and extracted and structure of configuration data can be optimized by setting the level configuration information of the NTT (or the inverse NTT).

In short, the reconfigurable computing technology can achieve a better balance among the aspects of performance, power consumption, area, flexibility, and development difficulty, etc., and obtain a higher energy efficiency, which is expected to implement a better fully homomorphic encryption platform. The fully homomorphic encryption features intensive computing, rich parallelism, and flexible demands, which makes it suitable for being implemented with reconfigurable computing technology. The reconfigurable computing technology can bring higher performance and energy efficiency than GPP, GPU, FPGA, as well as higher flexibility than ASIC. With the development of automatic compilation technology, the development work based on reconfigurable computing technology can be done automatically with the high-level language, and its development difficulty is expected to be lower than that of FPGA and ASIC. Therefore, the fully homomorphic encryption scheme based on reconfigurable computing technology is expected to obtain comprehensive advantages and better application prospects.

7.2 Hardware Trojans and Reconfigurable Computing

The hardware Trojan (HT) has become one of the major security issues for integrated circuits [39]. Since the business model of existing integrated circuits has made the third-party design and manufacture the circuits, which makes the integrated circuit companies lose most of their control over the design and manufacture of their integrated circuit products, so the integrated circuits are very vulnerable to

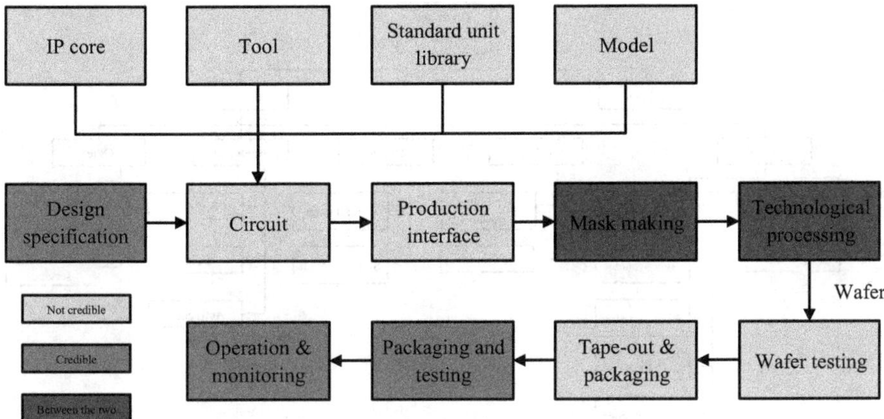

Fig. 7.5 Vulnerable phases of the integrated circuit life cycle [40]

the damage of the maliciously planted circuits. Figure 7.5 shows the level of trust at various stages in the typical integrated circuit life cycle [40]. It can be seen from Fig. 7.5 that each of the parties involved in the design and manufacture of integrated circuits may be an adversary who potentially and maliciously inserts modification circuits. A maliciously modified circuit inserted by an adversary is called a hardware Trojan [40]. In the following, we first introduce the classification of hardware Trojans and several typical examples of hardware Trojans, then introduce the defense technology for hardware Trojans, and finally discuss the feasibility of dealing with threat of hardware Trojans through reconfigurable cryptographic processors.

7.2.1 Classification and Examples of Hardware Trojans

Hardware Trojans are generally composed of trigger logic and payload logic [40]. When the circuit status meets the conditions of triggering a hardware Trojan, the trigger logic is activated and drives the payload logic of the hardware Trojan. Figure 7.6 shows the classification according to the trigger and payload logic of hardware Trojans.

According to the different trigger conditions, hardware Trojans can be divided into two types: the analog trigger and the digital trigger. The analog trigger hardware Trojan activates the hardware Trojan through analog circuit statuses such as temperature, delay, or device aging effects. The digital trigger hardware Trojan activates the hardware Trojan via Boolean logic functions. The digital trigger hardware Trojans can be divided into the combinational trigger hardware Trojans and the sequential trigger hardware Trojans.

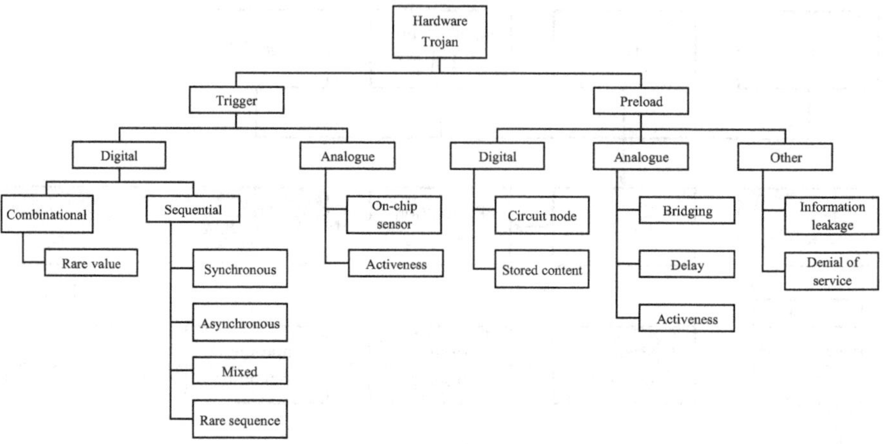

Fig. 7.6 Classification of hardware Trojans based on the trigger and payload [40]

Figure 7.7a shows an example of the combinational trigger hardware Trojan. The trigger conditions are A = 0 and B = 0. When the trigger conditions are satisfied, the payload logic will change the original output C of the circuit into $C_{modified}$. Usually, the attacker chooses an extremely rare condition as the trigger condition of the hardware Trojan. In the traditional chip test stage, these hardware Trojans are very difficult to trigger, and the testers also cannot find the hardware Trojan hidden in the chip. The sequential trigger hardware Trojan, also known as the time bomb Trojan, is activated by a sequence of or some continuous operations. The simplest sequential trigger hardware Trojan is a synchronous independent counter. The hardware Trojan is triggered when the counter reaches a specific count. Figure 7.7b is a synchronous k-bit counter. When the number of counter reaches $2^k - 1$, the hardware Trojan is triggered and it changes C into C_{modified}. The hardware Trojan can also be triggered by a simulated condition. In Fig. 7.7c, the hardware Trojan is triggered by the circuit activity and the temperature sensor [41]. The trigger circuit consists of two parts: the heat-generating part indicated by a circle and the temperature sensor indicated by a dashed box. The heat-generating part consists of a set of inverter-based ring oscillators. These ring oscillators share an enable signal. When an enable signal comes, all ring oscillators operate at a high frequency, generating a significant amount of heat. The temperature sensor uses two ring oscillators, one is located near the heat-generating part, which is called the test ring oscillator and the other is located far from the heat-generating part, which is called the reference ring oscillator. The outputs of the two oscillators are sent to two counters as the clock signals. Subtracting the results of the two counters gives the difference between the ring oscillators. The difference is then sampled using a low-frequency clock and the result is stored in Register 1. After delaying one cycle, the result in Register 1 is stored in Register 2, and the difference between Register 1 and Register 2 is stored in Register 3. If the value in Register 3 is greater than a

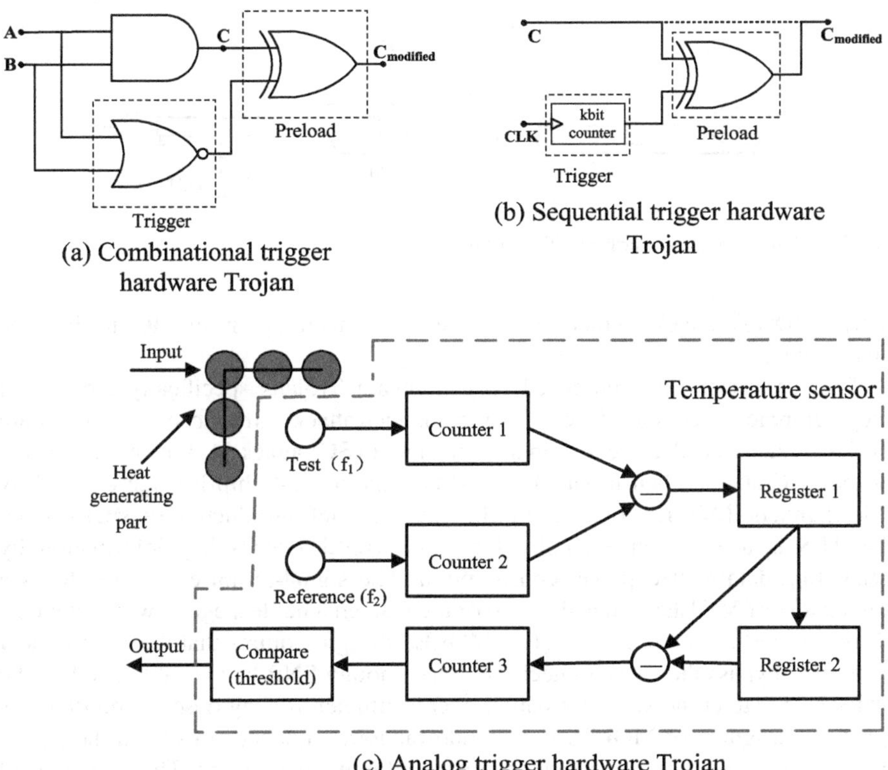

Fig. 7.7 Hardware Trojans with different trigger mechanisms [40]

pre-set threshold, then the trigger signal is set to "1," otherwise the trigger signal remains "0."

Hardware Trojans can also be divided into digital payload hardware Trojans and analog payload hardware Trojans according to their different payload effects. In addition, there are other hardware Trojans with other payload effects. Digital payload hardware Trojans mainly affect the logical values of the circuit, such as logic values of circuit nodes or memories. Analog payload hardware Trojans affect the circuit parameters, such as performance, power consumption, and noise margins. Figure 7.8a shows an example of introducing a bridging fault by inserting resistors. Figure 7.8b shows an example of how the path delay can be affected by increasing the capacitive load. Another form of analog payload effect is to increase the activitiness of the circuit (similar to Fig. 7.7c) to accelerate the IC's aging process without affecting its function. In addition, the hardware Trojan payload effect can also be presented in other two forms. One form is information leakage, and secret information in the IC is leaked by a hardware Trojan through the transmitted radio signal or the serial data port interface. Another form is denied of

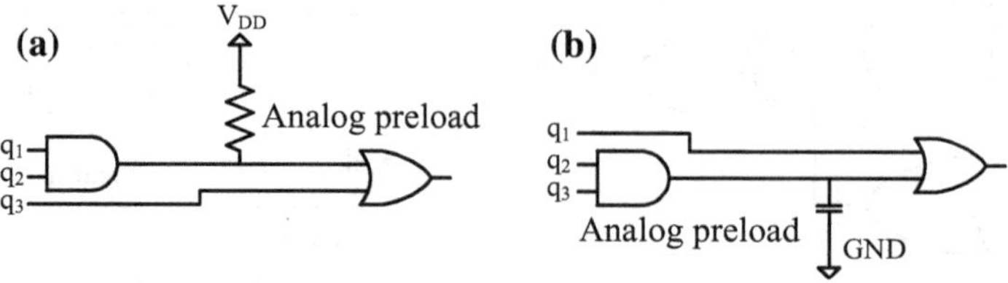

Fig. 7.8 Analog payload hardware Trojan [40]

service (DOS) attack, which can cause some features in the IC to become unavailable.

The researchers also proposed two hardware Trojans specifically targeted at cryptographic processors. One is a side-channel attack introduced by a hardware Trojan, which is called the Trojan side-channel (TSC) attack [42]. Its earliest name of the TSC attack when it was designed is malicious off-chip leakage enabled by side channels (MOLES) [43]. The following is a brief introduction to MOLES and the AES is used as an example (Fig. 7.9). MOLES leaks key information by generating data-related power consumption. The signal-to-noise ratio (SNR) is a key feature of MOLES and is the ratio of the side-channel leakage power to the host IC power consumption. An effective MOLES design requires that the SNR be low enough to bypass the tester's checks and that enough SNR be available to allow the attacker to extract the key information over a sufficiently long observation time. As a result, designers of MOLES use pseudo-random sequences to spread the power consumption of side-channel leakage over multiple clock cycles. The side-channel power consumption is low enough per clock cycle, and attackers can extract key information by comparing a large number of clock cycles. As shown in Fig. 7.9, the key bus is cryptographically connected to the XOR gate (only k_0 is shown in Fig. 7.9). A pseudo-random number generator (PRNG) generates a pseudo-random number r_{k0} (t) that is XOR'ed with key k0. The output of the XOR gate is

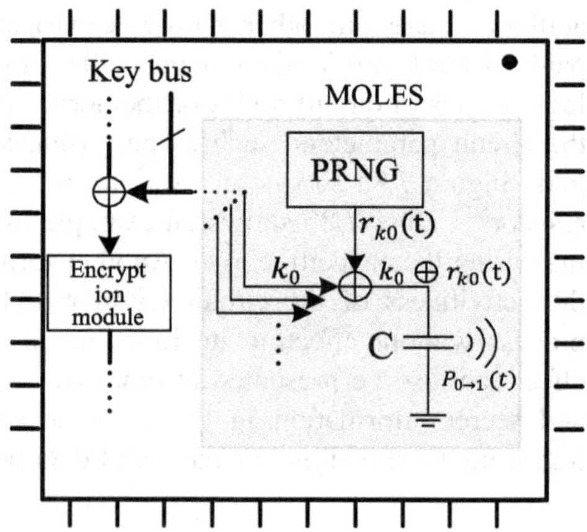

Fig. 7.9 MOLES circuit embedded in a cryptographic processor [43]

connected to a load capacitor. When the logic of the XOR gate changes from 0 to 1, the load capacitance leaks a small amount of power. The size of the load capacitor is an adjustable design parameter of MOLES, which determines the size of the side-channel power leakage, thus identifying the signal-to-noise ratio. For different key bits, MOLES uses different pseudo-random sequences to carry out XOR operations. If the detector does not know the pseudo-random sequence, the power curve cannot be analyzed to find the side-channel leakage power consumption.

The other is to launch fault attacks on a cryptographic processor through a hardware Trojan [44]. Figure 7.10 shows a hardware Trojan injected in a 10-round

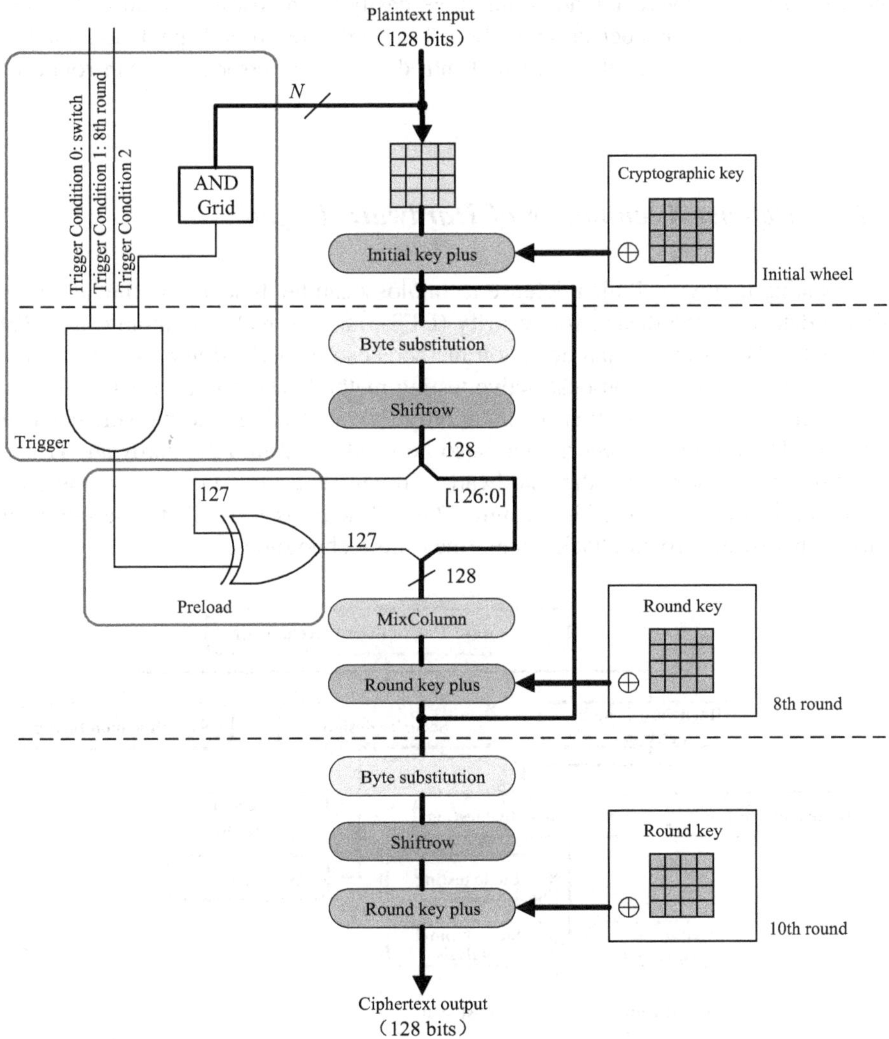

Fig. 7.10 Hardware Trojan that initiates the AES fault attacks [44]

AES. After the inserted hardware Trojan is triggered, a fault is injected into the 8th round of the AES and an attacker can use it to launch differential fault attacks [45]. The hardware Trojan trigger condition consists of three parts. First of all, it has an external switch, the on/off of the switch can determine whether to enable the hardware Trojan; and then, the hardware Trojan will detect whether the cryptographic round counter is indicating the eighth round, so that the fault injected through the hardware Trojan can recover all values of the last round of round key; finally, the hardware Trojan needs to detect whether there is an attacker's specified mode in the input plaintext. The mode length N is a variable. The longer the length N is, the lower the probability of triggering the Trojan $\left(\frac{1}{2N}\right)$ is. However, the larger the size of the hardware Trojan is, the more easily the hardware Trojan is detected via the hardware Trojan detection method. The hardware Trojan's payload logic has only one XOR gate, and its result will introduce a 1-bit error in the 8th round of AES.

7.2.2 Defense Technology of Hardware Trojan

The existing hardware Trojan defense technology can be divided into the hardware Trojan detection, the design for security (DFS), and the real-time monitoring [39] (Fig. 7.11). Hardware Trojan detection methods can be divided into the destructive testing method and the non-destructive testing method. The non-destructive testing can be divided into the IP trust verification, the logic testing, and the side-channel analysis. The design for security can also be subdivided into the hardware Trojan insertion prevention and the hardware Trojan detection facilitation measure according to the purpose of its design. The following is a detailed description of various hardware Trojan attack countermeasure technologies.

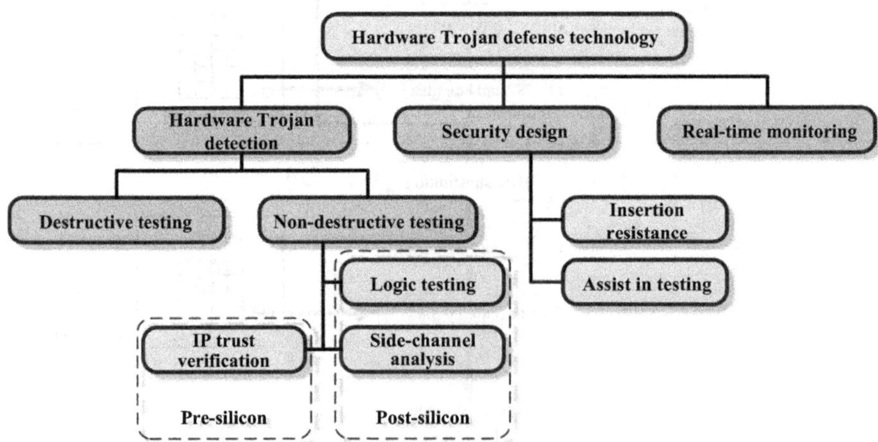

Fig. 7.11 Classification of hardware Trojan defense technologies [39]

Destructive testing methods are used to perform destructive tests for integrated circuits. When there are a large number of integrated circuits that require testing, the tester can only destructively test some of the integrated circuits. An attacker may only inject hardware Trojans into some of the integrated circuits; testers are most unlikely to detect the hardware Trojans. Therefore, destructive testing methods are not suitable for testing a large number of integrated circuits. Destructive testing methods are often used to verify whether an integrated circuit is a trusted circuit to obtain the circuit feature of an ideal integrated circuit and to test other integrated circuits using that circuit's feature as a standard.

IP trust verification can be divided into three categories. One is to perform IP trust verification through directed testing and verification. For example, testers may use a method called suspect-signal-guided sequential equivalence checking (SEC) to verify a certain degree of trust in IP in the gate-level netlist [46]. The SEC can be divided into the following three steps.

Step 1, generate the test vector and get suspicious signals by employing automatic test pattern generation (ATPG).
Step 2, let the system reach the status that some functions cannot reach, to further filter suspicious signals.
Step 3, trigger suspicious signals, and compare a suspicious circuit and a standard circuit to find whether they are identical.

The second type of IP verification method is to ensure the reliability of IP through some standardized behaviors. For example, researchers design a credible IP transport protocol based on a proof-carrying code (PCC) [47]. Figure 7.12 is a diagram of credible IP transport protocols between IP vendors and consumers. After the IP vendor has designed the IP, it sends the hardware description language

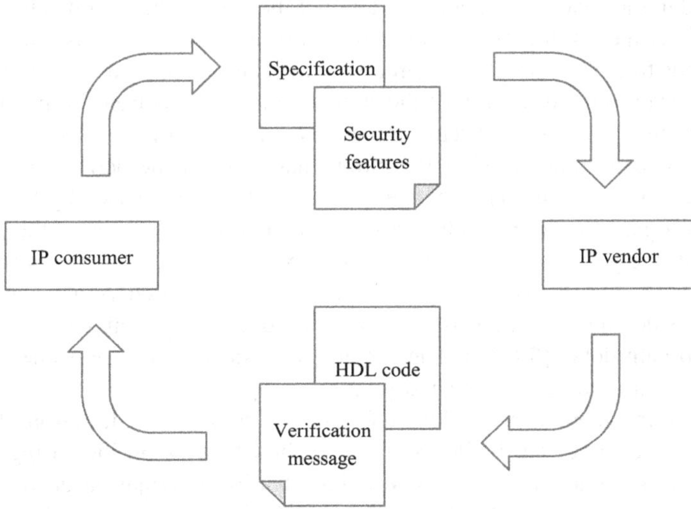

Fig. 7.12 IP capture and transfer protocol in PCC [48]

(HDL) code that carries the verification information to the IP consumer. After the IP consumer gets the code, it extracts the security-related features in the IP and sends it to the IP vendor for verification. These security-related features are given by the IP provider to ensure that any changes made in the IP may cause the IP to violate these security-related features, thus ensuring that the IP has not tampered. The final method is to use the knowledge base (KB) to detect the IP security [48]. Such methods first integrate existing trusted designs into one or more KBs and then validate the IPs using the rules of the KB.

The logic test method is another kind of method in the hardware Trojan detection methods. The logic test method is to trigger the hardware Trojan in the integrated circuit by outputting the test vector and the payload effect of the hardware Trojan needs to be observed. Due to the optional hardware Trojan trigger condition and various payload effects after the hardware Trojan is triggered, the logic test method cannot generate the corresponding test vectors to detect all the hardware Trojans. One method to solve this problem is to detect hardware Trojans from the statistical point of view. For example, a tester can try to excite hardware Trojans through rare behaviors for multiple times, which is known as multiple excitation of rare occurrence (MERO) [49]. The basic idea of this method is to detect low-probability events of internal nodes and to derive an optimal set of test vectors, which not only minimizes the test time and cost but also maximizes the coverage of Trojan detection. In addition, some researchers use Boolean functions to test circuits and determine whether the basically unused modules in those circuits are creditable [49].

In addition to logic test methods, side-channel analysis techniques have also been extensively studied. Side-channel analysis techniques can be used to detect hardware Trojans because malicious modification of the integrated circuit during the design and manufacturing phases potentially affects the power consumed by the circuit, and in most cases, it affects the delay of the path in the circuit. The impact of a hardware Trojan on the power consumption of a circuit is because the hardware Trojan needs to add or modify the circuit elements to implement specific functions, which will lead to the deviation of the static and transient current in the circuit and thus affect the power of the circuit. Because hardware Trojans constantly monitor the circuit to determine whether the circuit has reached the activation condition, they also consume power. The changes of circuit delay are caused by the addition/deletion of logic gates in the path performed by hardware Trojans, or the change in the capacitance of some path. According to the features of the side-channel information change introduced by hardware Trojan, researchers have proposed a variety of side-channel analysis methods based on quiescent current, transient circuit, and path delay [50, 51]. These methods basically have the same ideas, and we will not introduce them repeatedly here.

The detection method of hardware Trojans focuses on the detection after hardware Trojans are injected, while the DFS method focuses on increasing the difficulty of injecting hardware Trojans. Blocking hardware Trojan insertion is one of the typical DFS practices. Blocking hardware Trojan insertion mainly consists of the fuzzification-based method and the layout-filling method. The fuzzification-based method prevents the hardware Trojan from being inserted by obfuscating the features of functions and structures in the design. The fuzzification-based method will use the key-based obfuscation technique to modify the state transition function of a given circuit [52]. Such method provides two different modes of the circuit: the normal mode and the fuzzy mode. The normal mode produces the desired output, while the fuzzy mode produces some erroneous functional behavior for some input modes. Figure 7.13 is an example of a modification of the state transition diagram through the fuzzification method. The fuzzification method enables some hardware Trojans to trigger only in an obfuscated mode, thereby making the hardware Trojans become benign. The layout-filling method is designed to fill gaps in the circuit, preventing attackers from inserting extra unit in the design. If an attacker identifies designer's additional parts for filling, the attacker can replace the filling parts with hardware Trojan circuit, so the designer needs to hide the filling part. The built-in self-verification (BISA) is a technique proposed by researchers to fill up the unused space in integrated circuits by using standard units, which cannot only fill up the unused space in the integrated circuit but also prevent the attackers from easily identifying the filling part [53].

Another DFS method is to design a method that can reduce the difficulty of detecting hardware Trojans. For example, designers can reduce hard-to-detect connections by inserting dummy scan flipflops (dSFFs) at the connection point with low flipping over probability [54]. Figure 7.14 shows the structure of the dSFF and

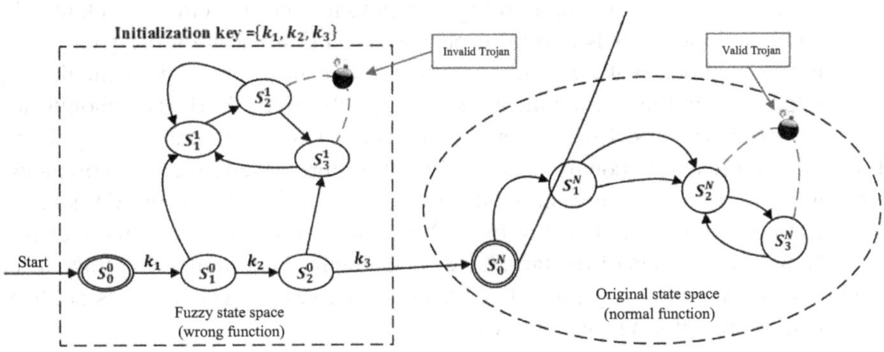

Fig. 7.13 Fuzzification of the state transition diagram [52]

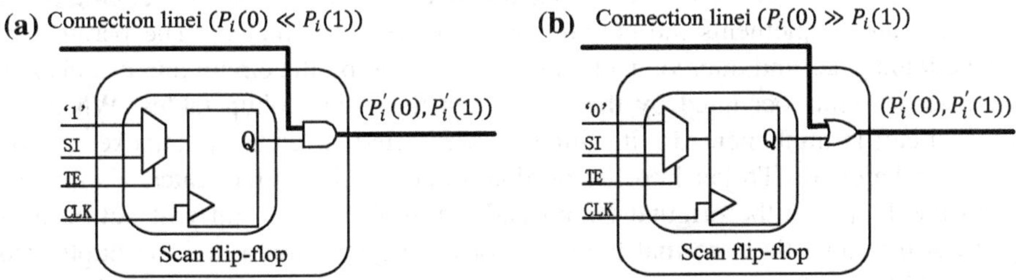

Fig. 7.14 dSFF structure [54]

the logic gate (AND gate or OR gate) added to ensure the functionality of the circuit. The TE signal in Fig. 7.14 represents the test enabling signal. During the test phase, the TE signal is activated. The multiplexer (trapezoid) selects the scan signal (SI signal) as the output of the dSFF, which in turn tests the connection i. The proportion of logic "0" and "1" in the SI signal typically accounts for 50%, respectively. The following explains the rules for selecting additional logic gates added to the structure. If the probability Pi(0) for logical "0" in Fig. 7.14 is much smaller than the probability Pi(1) for logical "1," then as shown in Fig. 7.14a, the AND gate is added behind the dSFF to increase the probability of logical "0" on connection i during the test phase and the TE signal is low during normal use. In Fig. 7.14a, the multiplexer selects logic "1" as the output of dSFF, and the logical value of connection i in normal use phase is unaffected. Similarly, as shown in Fig. 7.14b, when Pi(0) is much larger than Pi(1) in connection i, an OR gate needs to be added behind dSFF to increase the probability of logical "1" on connection i in the test phase, and the logical value of connection i in normal use phase is unaffected.

Although the researchers put forward a variety of methods for detecting hardware Trojans and increasing the difficulty of hardware Trojans injection, the existing methods cannot guarantee that the circuit is absolutely secure. Therefore, it is necessary to monitor the status of the circuit in real time during the operation so as to prevent the trigger of the hardware Trojan or to minimize the impact after the hardware Trojan is triggered. The real-time monitoring can be divided into two categories, one is to design a monitoring system to monitor the circuit module units in real time and the other is a kind of parallel execution based on the parameters. For the first method, Waksman and Sethumadhavan have proposed a monitoring system to monitor the communication between the units [55]. The monitoring system is proposed based on the prediction/reaction monitor triangle (Fig. 7.15). The prediction/reaction monitor triangle consists of three different parts: a predictor, a reactor, and a target (the monitored units shown in Fig. 7.15). In addition, the system also contains a monitoring unit. During the monitoring, the predictor predicts the possible events of the target and sends them to the monitoring unit. If the reactor does not receive any predicted events or receives an event that has not been predicted, the monitoring unit will alarm.

For real-time monitoring method based on parallel execution of parameters, the most commonly used is the triple modular redundancy (TMR). The parameter-based TMR is a circuit that produces two copies of the circuit to be protected and implements the copying with different parameters. The parameters can be logic gates and interconnection structures used by the circuit and can also be the memory units occupied by the circuit and the used lookup tables. When the same circuit is implemented with different parameters, even if the attacker inserts the same hardware Trojan into the original circuit and the duplicated circuit, the hardware Trojan in the original circuit and the duplicated circuit will not be triggered at the same time, so that the behavior of original circuit and the duplicated circuit will be different, thus detecting the hardware Trojan. However, the TMR needs to produce two copies of the circuit to be protected, which results in a huge

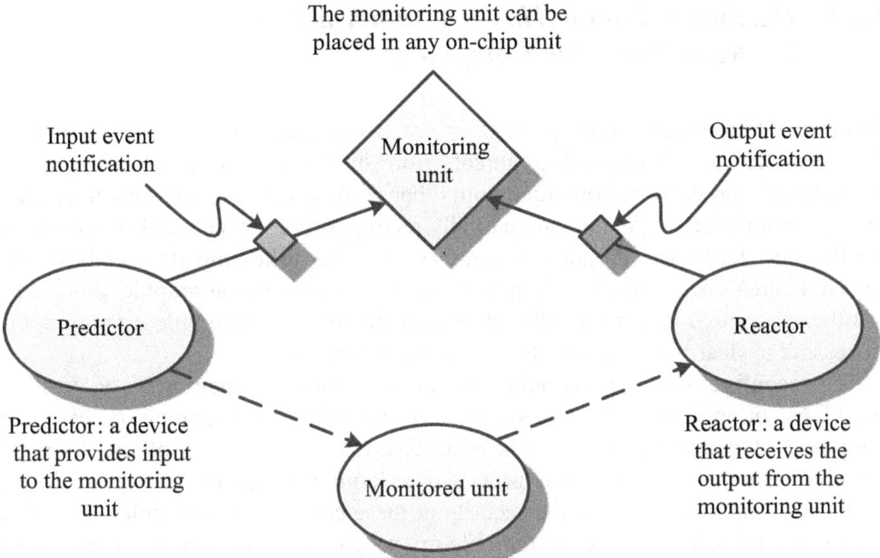

Fig. 7.15 Prediction/reaction monitoring triangle [55]

resource consumption. Therefore, researchers propose an adaptive triple modular redundancy (ATMR) technology to reduce the huge resource consumption brought by the traditional TMR technology [56]. Figure 7.16 shows an example of the TMR and ATMR operation, where Node c is a comparator. It can be seen from Fig. 7.16 that in the absence of errors, the ATMR only needs to copy the circuit once, the third circuit operates only when an error occurs, and the TMR runs in the three circuits at all times.

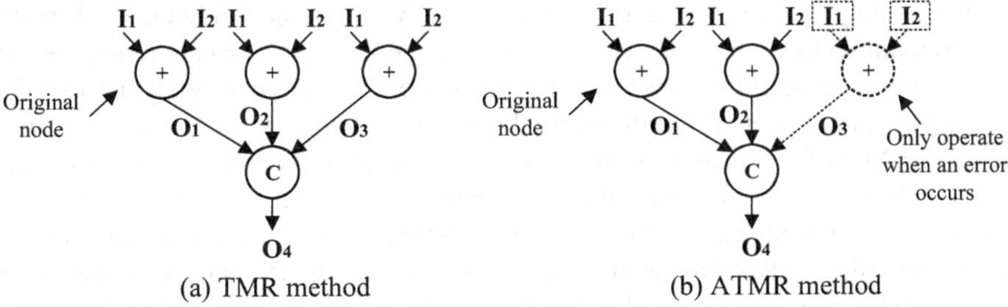

Fig. 7.16 TMR and ATMR operation examples

7.2.3 Hardware Trojan Threat Countermeasures for Reconfigurable Computing

Reconfigurable cryptographic processors are being used more and more widely. However, they are facing serious threats from hardware Trojans. Reconfigurable cryptographic processors, compared with other cryptographic processors, have their own characteristics—dynamic and partially reconfigurable, which makes it possible to offer new hardware Trojan countermeasures. The following first analyzes the potential threats of hardware Trojans to reconfigurable cryptographic processors and then discusses how to use the characteristics of reconfigurable cryptographic processors to deal with the threats of hardware Trojans.

The reconfigurable cryptographic processor consists of memories, controllers, and PEAs. In addition to the PEAs, the reconfigurable cryptographic processor is composed of conventional modules such as memories and controllers. As mentioned earlier, researchers have proposed a large number of testing, real-time monitoring methods to ensure the security of these conventional modules. The PEA is a unique module in the reconfigurable cryptographic processor that needs to be protected in conjunction with its characteristics. PEA, on the other hand, occupies a large amount of processor area of reconfigurable cryptographic processor and runs the majority of operations on the reconfigurable cryptographic processor, which gives attackers more free space to insert hardware Trojans, and provides more rare circumstances to choose as trigger conditions for hardware Trojan. For two above-mentioned reasons, attackers are more likely to choose PEA as attack position of hardware Trojan when attacking the reconfigurable cryptographic processor.

As for PEA-based hardware Trojans, the academic community has not yet published the corresponding research results, but we can briefly analyze the hardware Trojans injected into PEAs from the perspective of attacker's purpose intention of injecting hardware Trojans and his or her implementation method. First of all, it is easy to know that the attacker's intention of attacking the reconfigurable cryptographic processor is to obtain the key in it. As mentioned in Sect. 7.2.1, there are two main ways in which an attacker can use the hardware Trojan to acquire the key of the cryptographic processor at this stage. One is to directly access the key-related components through the hardware Trojan, such as memory for storing the key and bus for transmitting the key. After obtaining the key-related information, the hardware Trojan uses the side-channel technology to transmit the key information to the attacker. This method is similar to the traditional side-channel attack method, but the side-channel attack introduced by the hardware Trojan is more threatening compared to the traditional side-channel attacks. This is because the traditional side-channel attack method often needs to obtain side channels (power consumption, delay, and electromagnetic, etc.) curves to get the key when multiple cryptographic processors are operating. As for the side-channel attacks introduced by hardware Trojans, the attacker clearly knows the condition for triggering the hardware Trojan and can easily obtain the key information of the cryptographic processor by just triggering the hardware Trojan. Another method is to introduce the fault in the cryptographic processor through the hardware Trojan. Afterwards, the attacker obtains the key by means of fault analysis. This method is also more threatening because an attacker can trigger the hardware Trojan by himself or herself. For the first method that leaks key information using side channel, it has not only protected by the existing method [57], and the hardware Trojan related to it will not be injected into the PEA of the reconfigurable cryptographic processor because the method needs to access key-related units, so that the hardware Trojan injected into the PEA is more likely to introduce a fault while the reconfigurable cryptographic processor is operating, and the attacker will analyze and use the fault to obtain the key.

The following will briefly analyze the hardware Trojans injected into the PEA from the perspective of the implementation method of hardware Trojans. There are three main ways to inject hardware Trojans, that is, modify the original design of the integrated circuit, modify the third-party IP or standard unit library, and change the manufacturing process parameters of the integrated circuit. The hardware Trojan introduced by changing the manufacturing process parameters of the integrated circuit cannot generate an attacker-controlled fault and therefore cannot attack according to the specific steps in the PEA. Therefore, the attacker is most likely to choose the first two ways to inject Trojans.

In terms of the possible implementation method of hardware Trojans, this section proposes two possible types of hardware Trojans, that is, the PE-independent Trojan of the reconfigurable computing unit and the PE-dependent Trojan of the reconfigurable computing unit. The PE-independent Trojan of the reconfigurable computing unit refers to the trigger signal of the hardware Trojan that is not entirely from the PE. Trigger signals can come from several different parts, such as lookup tables, configuration information memories, and various internal controllers in the special function module of the reconfigurable cryptographic processor. The PE-independent Trojan of the reconfigurable computing unit is often injected by modifying the original circuit design in the design and manufacturing phase. Figure 7.17a shows an example of reconfigurable computing unit PE-independent Trojan. Figure 7.17 also considers the interconnection structure in PEA as a special PE, denoted by PE_I, and the PE in the usual sense is denoted by PE_P. It should be pointed out that when all PEs in PEA are isomorphic, an attacker will need to add many extra connections when inserting reconfigurable computing unit PE-independent Trojan in PEA, resulting in a large deviation between PEA and the original circuit design and making the inserted hardware Trojans easily detected by conventional detection methods. However, in order to increase the processor speed and reduce the area consumption of the processor in the existing reconfigurable cryptographic processor, PEs in the PEA are often heterogeneous [58], which also provides the attacker with a chance of injecting a PE-independent Trojan of the reconfigurable computing unit. The PE-dependent Trojan of the reconfigurable computing unit indicates that the trigger signals of the hardware Trojan are entirely from the PE. Such Trojans can be injected through untrusted third-party IP cores or

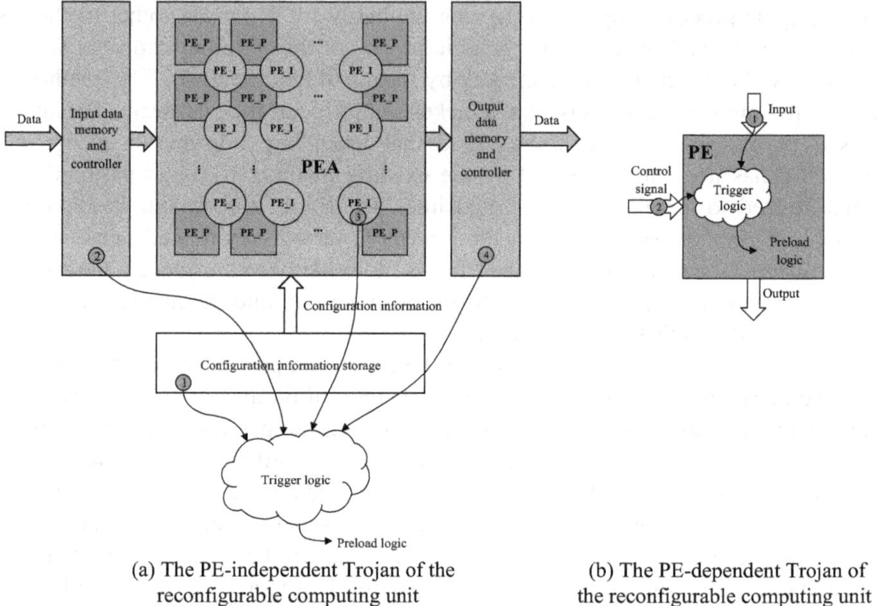

(a) The PE-independent Trojan of the reconfigurable computing unit

(b) The PE-dependent Trojan of the reconfigurable computing unit

Fig. 7.17 Hardware Trojan on PEA

standard unit libraries. Figure 7.17b shows an example of the PE-dependent Trojan of the reconfigurable computing unit.

For the PEA-based hardware Trojans designed for reconfigurable cryptographic processors, the testing methods proposed by the academic circles cannot guarantee 100% detection rate of hardware Trojans. The DFS design can only provide a degree of protection for PEA. Therefore, it is very important to add real-time monitoring technology to the reconfigurable cryptographic processor, which can provide a final barrier to the safe operation of the reconfigurable cryptographic processor. Section 7.2.2 mentions that the existing real-time monitoring methods are mainly divided into real-time monitoring based on monitoring system and parallel execution based on parameters. The monitoring system proposed by researchers at this stage cannot detect the hardware Trojans that only modify the processor data [55]. Section 7.2.3 mentions that hardware Trojans injected into PEA are most likely to achieve the purpose of obtaining the key by introducing a fault in the execution of the cryptographic algorithm so that the real-time monitoring method based on the monitoring system cannot remove this Trojans. In contrast, parameter-based parallel execution seems to be an effective manner to eliminate the threat of hardware Trojans on PEA. It should be noted that the existing parameter-based parallel execution method (ATMR) with minimum resource consumption also needs to consume at least two times of resources, and the resource consumption thereof is derived from the resource consumption of the duplicated circuit, and the results of the original circuit and duplicated circuit are compared in the consumption. If ATMR is used to protect all the operations that are mapped to PEA, it will seriously affect the efficiency of PEA execution. To do this, we need to consider how to protect PEA in the case of resource constraints. Depending on the dynamic and partially reconfigurable nature of the PEA, there are two possible schemes to protect PEA with resource constraints.

The first scheme is to dynamically modify the granularity of PE execution. In this way, regardless of whether the reconfigurable computing unit PE-independent hardware Trojan or whether the reconfigurable computing unit PE-dependent hardware Trojan, the data related to the PE in the hardware Trojan trigger information is randomly changed when the attacker triggers the two types of hardware Trojans injected in advance. From an attacker's point of view, the triggering of hardware Trojan is changed from deterministic event to stochastic event, and the greater the amount of PE data on which the triggering of hardware Trojan relies, the lower the probability of triggering the hardware Trojan. However, this method essentially only increases the difficulty of triggering hardware Trojans and does not guarantee that hardware Trojans never trigger. For the reconfigurable cryptographic processor, the attacker can obtain the key when the hardware Trojan triggers only once. Therefore, this method cannot guarantee absolute security.

The second option is to implement the parameter-based dynamic parallel execution plan for the PEA. First of all, a parameter-based parallel execution protection for PEA is implemented under the acceptable resource consumption constraint. At this time, we shall ensure that which hardware Trojans are most likely to inject, and protect the positions that are most vulnerable to attack, which not only ensures the security of the PEA, but also improves the PEA's operating efficiency. When the parameter-based parallel protection detects the triggered hardware Trojan, it indicates that there is a hardware Trojan in the PEA and the first triggered hardware Trojan has been triggered. In this case, the PEA needs to be more highly protected so as to be able to detect subsequent hardware Trojans which will be triggered during the execution. Therefore, it is necessary to increase the resources that can be consumed based on the parallel execution of parameters, thus protecting more operations performed in PEA. If the triggered hardware Trojan is still detected in the PEA, the resources available for parameter-based parallel execution are constantly increasing. When consumed resources exceed certain threshold, the PEA stops running to ensure the security of the key.

For the second scheme, how to protect the position into which hardware Trojans are most likely to be injected and which would be first attacked under the given resource consumption constraints is the focus and the difficult point of the scheme implementation. At the present stage, there is no research on this aspect in the academic circles. This section presents a possible implementation plan called dynamic resource management based on security value (DRMaSV). The basic idea is to propose security value metric and measure the security value of the operations performed in the PEA, thus determining the probability of the attack from attacker. The dynamic resource management strategy is used to select the operation to be protected on the PEA based on the security value under limited resource constraints.

The DRMaSV consists of security value (SV) metric and the dynamic resource management (DRM) policies. The SV is composed of four basic definitions. Because the DRMaSV is protected by the PEA here, the operation performed on the PEA is compiled and mapped by the DFG. Therefore, when introducing the SV definition, we will take a node of the DFG for example. The detailed definition of SV is as follows.

Definition 1: Impact value The impact value of Node i refers to the number of output ports that can be connected by Node i, denoted as n_i. The impact value is a basic parameter of SV. The following introduces the reason why the impact value is regarded as a SV parameter.

For the PE-independent hardware Trojan of the reconfigurable computing unit, each PE in the PEA is probably attacked by a hardware Trojan, and the probabilities of triggering the hardware Trojans for each PE are the same. Suppose there are Node A and Node B. The impact value of Node A is 2, and the impact value of Node B is 4. Suppose Node A and Node B are injected into the same hardware Trojan, and the payload effect of the hardware Trojan is to modify the output of the corresponding node. If an attacker triggers a certain hardware Trojan at some time, then the probability of the triggered hardware Trojan being in Node A and the probability of the triggered hardware Trojan being in Node B are, respectively, 50%. In this case, if Node A is protected, then the expected number of output ports affected by the hardware Trojan will be $2\left(4 \times \frac{1}{2}\right)$; when protecting Node B, the expected number of output ports affected by the hardware Trojan will be reduced to one. Therefore, the attacker hopes that the triggered hardware Trojan is located in Node B, so as to increase the key information that can be acquired once by the hardware Trojan, thus first triggering the hardware Trojan in Node B. Therefore, the node with high impact value needs to be preemptively protected.

For the PE-dependent hardware Trojan of the reconfigurable computing unit, such hardware Trojans are located in a third-party untrusted IP core or a standard unit library. The numbers of inputs/outputs of a third-party untrusted IP core or standard unit library have been determined, so the number of outputs affected by the hardware Trojan that is located in it does not change with the operations performed. Furthermore, the number of outputs affected by the node where the hardware Trojan is triggered is also determined. In this case, if the node has a large impact value, the attacker is more likely to trigger the hardware Trojan in that node. Therefore, when preventing the PE-dependent hardware Trojan of the reconfigurable computing unit, the node with a large impact value should be protected first.

In general, in order to hide the Trojan function, an attacker needs to inject nodes that have little impact on other nodes. For hardware Trojans injected into a PEA, the attacker cannot predict the operations that are going to be performed on the PEA. To do this, the attacker needs to inject a large amount of hardware Trojans into the PEA to increase the probability of triggering the hardware Trojans, and all the PEs may be injected with the hardware Trojans. However, attackers tend to trigger hardware Trojans in nodes with larger impact value. In this way, the triggered hardware Trojans can produce more destructive power, so nodes with high impact value need to be protected first.

Definition 2: Untrustworthiness The untrustworthiness of a node i refers to the probability that a hardware Trojan is contained in a PE mapped by a node i, which is marked as p_i.

For the PE-independent hardware Trojans of the reconfigurable computing unit, they are likely to be distributed throughout a PEA, and triggering of hardware Trojans does not have significant differences. Therefore, it is reasonable to assume that all PEs in a PEA have the same degree of untrustworthiness. For the PE-dependent hardware Trojans of the reconfigurable computing unit, the untrustworthiness of each PE in a PEA can be calculated according to the third-party IP core and the standard unit library used by the configured PE. For example, third-party IP cores and standard unit libraries can be graded. Well-known companies' commonly used IP cores and standard unit libraries are defined as level 1; unknown companies' commonly used IP cores and standard unit libraries are defined as level 2; well-known companies' uncommonly used IP cores and standard unit libraries are defined as level 3; unknown companies' rarely used IP cores and standard unit libraries are defined as level 4. The determined level is the PE's untrustworthiness. p_i can also be calculated according to the hardware Trojan's logic test method.

Definition 3: Node security value (NSV) The node security value of Node i refers to the expectation of the number of untrusted outputs connected by Node i, marked as NSVi, that is,

$$\mathrm{NSV}_i = n_i p_i \tag{7.7}$$

Definition 4: Security Value The security value is the sum of the NSV values of nodes protected by the real-time monitoring strategy. For a given set of DFG nodes, it is marked as V, and the set of protected nodes is marked as V_p (Obviously $V_p \subseteq V$); the security value is

$$\mathrm{SV} = \sum \mathrm{NSV}_i = \sum_{i \in Vp} n_i p_i \tag{7.8}$$

The above definition gives a detailed description of the security value, and it can be seen that the time complexity of the security value depends on the time complexity of n_i and p_i. For a DFG with N nodes, the time complexity of computing the impact value of a node is O(N). Then, the time complexity of calculating the impact value of the entire DFG node becomes $O(N^2)$. When calculating the untrustworthiness of a node, we take the grading method for example, ignore the time for grading each node. Then, the time complexity of calculating the untrustworthiness of the entire DFG node is O(N). From the expression (7.8), we know that the time complexity of SV is determined by more complicated sub-items (n_i), so the time complexity of SV is $O(N^2)$.

The security value represents the degree of importance of the protected node. The higher the security value is, the higher the probability that the hardware Trojan in the protected node is triggered preferentially. Therefore, the problem of selecting the nodes to be protected (operation) according to the security value under the condition of resource constraint is changed to how to get the maximum security value under certain resource constraints. For the convenience of description, the definitions of the relevant variables are given as follows: the number of DFG nodes (N), the maximum resource (R_{total}) that a protection strategy can consume, the resource consumption (R_i) for protecting a single node—Node i—and the node security value (NSV_i). The security value optimization problem is similar to the 0/1 knapsack problem, which can be described in the expression (7.9). The following introduces how to solve the security value optimization problem through a dynamic resource management strategy, that is

$$\begin{gathered} V_p = \arg\max V_p[\mathrm{SV}(V_p)] = \arg\max V_p[\sum_{i\in V_p} \mathrm{NSV}_i] \\ \text{s.t.} \sum\nolimits_{i\in V_p} R_i \le R_{\text{total}} \end{gathered} \tag{7.9}$$

Before introducing the dynamic resource management strategy, we need to define a two-dimensional matrix SV[i][R]. SV[i][R] represents the maximum SV that can be achieved by selecting a node from Node 1 to Node i for protection under the condition that the protection consumption is R. The dynamic resource management strategy can be described by the expression (7.10). The expression (7.10) describes the dynamic resource management strategy recursively. For a given resource consumption R, we can see that SV[0][R] = 0. Then for any node i, SV[i − 1][R] may choose to protect Node i or not protect Node i compared to SV[i][R] and choose the larger of the two schemes as its result. The final result of the dynamic resource management strategy is stored in SV[N][R], i.e.

$$\begin{cases} \mathrm{SV}[i][R] = \max\{\mathrm{SV}[i-1][R], \quad \mathrm{SV}[i-1][R-R_i] + \mathrm{NSV}_i\} \\ \mathrm{SV}[0][R] = 0 \end{cases}, \quad i = 1, 2, \ldots, N \tag{7.10}$$

The pseudo-code of the dynamic resource management strategy is shown in Algorithm 7.3. The pseudo-code consists of two parts. The first part (the part from Line 2 to Line 10 in Algorithm 7.3) aims to obtain SV[N][R_{total}] and the obtaining method is shown in the expression (7.10). The second part of the pseudo-code (the part from Line 12 to Line 22 in Algorithm 7.3) aims to find the corresponding protected node number based in the obtained SV[N][R_{total}]. The idea is to search in the reverse order. For the currently being searched node i, if the current remaining resource consumption R can be used to protect the node i and the difference between the maximum security value SV[i][R] that can be obtained by the protected nodes 1 to i with the resource consumption R and the maximum security value SV [i − 1][R − R_i] that can be obtained by the protected nodes 1 to (i − 1) with the resource consumption R − R_i is just the node security value NSV_i of Node i, it is indicated that Node i is the node that needs to be protected. It can be seen from Algorithm 7.3 that the time complexity of the dynamic resource management strategy is determined by the first half of the double cycle, the loop variable of the outer loop changes from 1 to N and the loop variable of the inner loop changes from R_i to R_{total}. Therefore, the time complexity of the dynamic resource management strategy is $O(N \times R_{\text{total}})$.

Algorithm 7.3 Pseudo-Code of the dynamic resource management strategy

Input: N; R_{total}; NSV_i, $i=1, 2, \ldots, N$

Output: SV[N][R_{total}]; Set of protected nodes *Combination*

```
// Get SV[N][Rtotal]
for i ← 1 to N do
    for R←Ri to Rtotal do
        if (SV[i−1][R]<SV[i−1][R−Ri]+NSVi) then
            SV[i][R]←SV[i−1][R−Ri]+NSVi
        else
            SV[i][R]←SV[i−1][R]
        end if
    end for
end for
//Obtain the set of the protected nodes Combination
R←Rtotal
for i ← 1 to N do
    Combination [i] ← 0
end for
for i ← N down to 1 do
    if (R≥Ri and SV[i][R]==SV[i−1][R−Ri]+NSVi) then
        //Add Node i to the protected set
        Combination [i] ← 1
        R←R−Ri
    end if
end for
```

Figure 7.18 shows an example of a dynamic resource management strategy of the DFG operation with five nodes under different resource constraints. n_i is determined by the number of the outputs connected to Node i. Therefore, $n_1 = 3$, $n_2 = 2$, $n_3 = n_4 = n_5 = 1$ can be obtained. Figure 7.18 does not involve any specific operations. Figure 7.18 gives p_i and R_i of Node i, where p_i is normalized and the sum of all the nodes p_i is 1. The node security value is calculated by using the expression (7.7). In Fig. 7.18, the SV is also processed accordingly. Set the SV value to 1 when all the DFG nodes are protected. For different R_{total}, the SV values obtained after the dynamic resource management strategy is implemented are also different. Taking $R_{\text{total}} = 3$ for example. In this case, Node 1 and Node 2 are protected, and the SV value of the protected node is 1.4 (NSV_1+NSV_2). If the sum

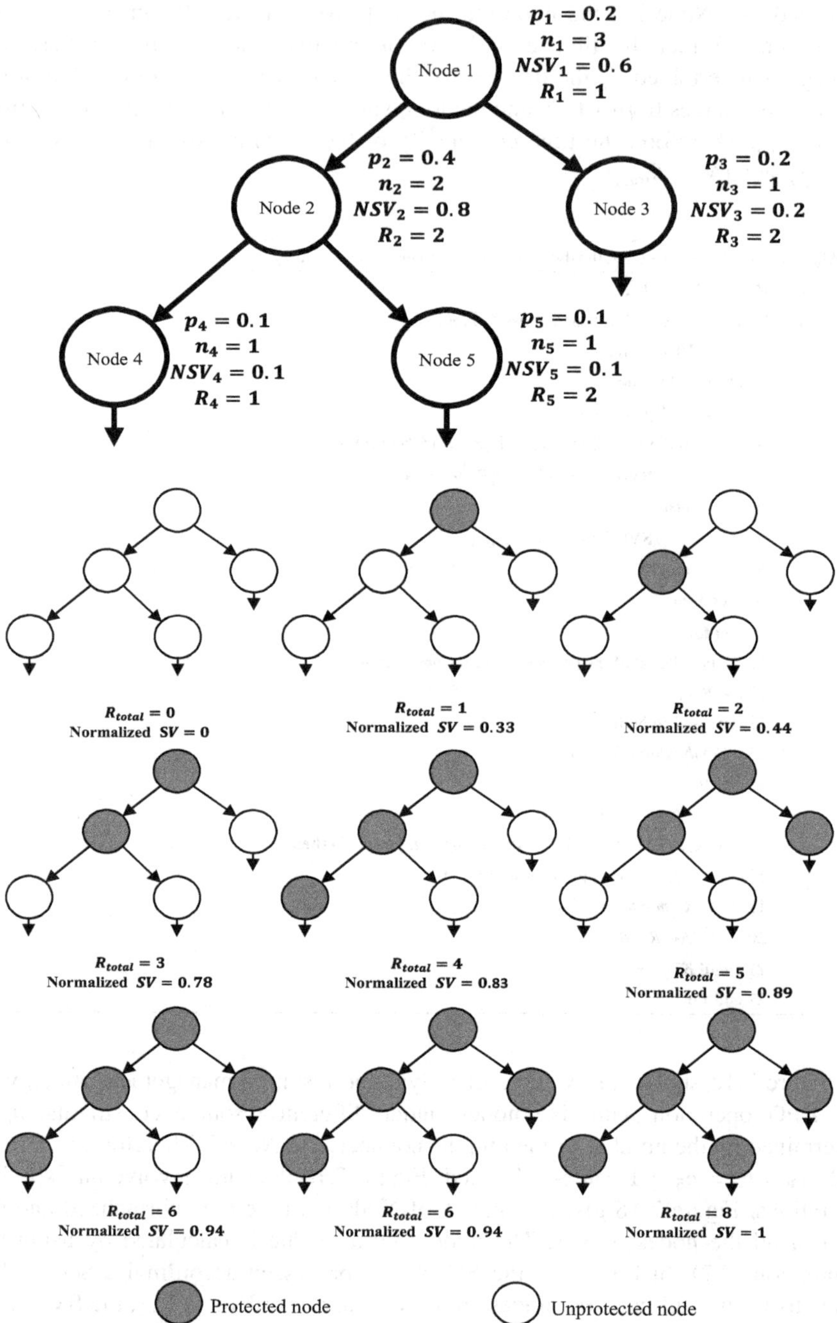

Fig. 7.18 Example of the dynamic resource management strategy

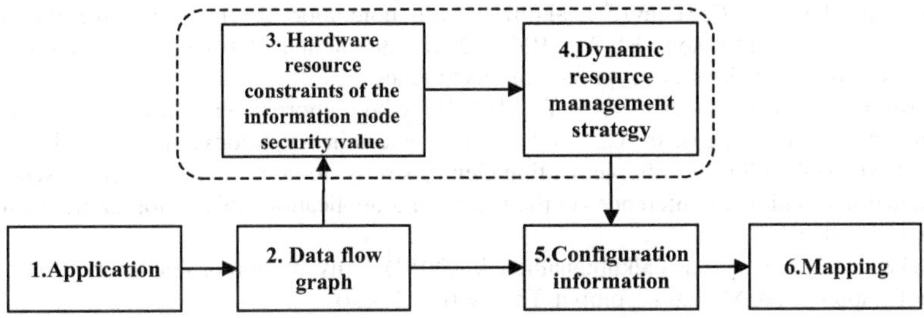

Fig. 7.19 DRMaSV: security mapping method

of the node security values for the entire DFG is 1.8 ($\sum_{i=1}^{5} NSV_i$), then the normalized SV value is 0.78 (1.4/1.8).

After the security value metric and the dynamic resource management strategy are described, the following will describe DRMaSV. DRMaSV is a secure mapping approach based on the dynamic resource management strategy (Fig. 7.19). It is implemented by adding extra steps to the mapping flow of the original reconfigurable cryptographic processor, with the additional steps shown by the dashed box in Fig. 7.19. The detailed steps of DRMaSV are described as follows.

① Transform the applications that are mapped to reconfigurable cryptographic processors into the DFG.

② Based on the hardware Trojan type and the configured functions in each node, we infer the distribution of hardware Trojans, calculate the node security value of each node, and obtain the resource consumption of each node.

③ Under given resource constraints, obtain the optimal set of the nodes to be protected based on the dynamic resource management strategy.

④ Generate a series of configuration information for the original DFG and additional protection tasks and map them to a PEA for execution.

References

1. Shor PW (1994) Algorithms for quantum computation: discrete logarithms and factoring. In: Proceedings 35th annual symposium on foundations of computer science, pp 124–134
2. Monz T, Nigg D, Martinez EA et al (2015) Realization of a scalable Shor algorithm. Science 351(6277):1068–1070
3. Howe J, Moore C, O'Neill M et al (2016) Lattice-based encryption over standard lattices in hardware. In: Proceedings of the 53rd annual design automation conference
4. Gentry C (2009) Fully homomorphic encryption using ideal lattices. In: Proceedings of the 2009 ACM symposium on theory of computing, pp 169–178
5. Rivest RL, Adleman L, Dertouzos ML (1978) On data banks and privacy homomorphisms. Academic Press, New York, pp 169–180

6. van Dijk M, Gentry C, Halevi S et al (2010) Fully homomorphic encryption over the integers. In: Advances in cryptology-EUROCRYPT 2010: 29th annual international conference on the theory and applications of cryptographic techniques
7. Coron J, Mandal A, Naccache D et al (2011) Fully homomorphic encryption over the integers with shorter public keys. In: Conference on advances in cryptology, pp 487–504
8. Gentry C, Halevi S (2011) Implementing Gentry's fully-homomorphic encryption scheme. In: Annual international conference on the theory and applications of cryptographic techniques, pp 129–148
9. Brakerski Z, Gentry C, Vaikuntanathan V (2014) Fully homomorphic encryption without bootstrapping. ACM Trans Comput Theory 6(3):13–36
10. Gentry C, Halevi S, Smart NP (2012) Homomorphic evaluation of the AES circuit. In: The 32nd annual international cryptology conference, pp 850–867
11. Gentry C, Sahai A, Waters B (2013) Homomorphic encryption from learning with errors: conceptually-simpler, asymptotically-faster, attribute-based. In: The 33rd annual international cryptology conference, pp 75–92
12. Fan J, Vercauteren F (2012) Somewhat practical fully homomorphic encryption. IACR Cryptology ePrint Archive 144
13. Bos JW, Lauter K, Loftus J et al (2013) Improved security for a ring based fully homomorphic encryption scheme. In: The 14th IMA international conference on cryptography and coding, pp 45–64
14. López-Alt A, Tromer E, Vaikuntanathan V (2012) On-the-fly multiparty computation on the cloud via multikey fully homomorphic encryption. In: Proceedings of the forty-fourth annual ACM symposium on theory of computing. ACM, pp 1219–1234
15. Stehl D, Steinfeld R (2011) Making NTRU as secure as worst-case problems over ideal lattices. In: EUROCRYPT 2011 international conference on the theory and applications of cryptographic techniques, pp 27–47
16. Erdinç Öztürk et al (2017) A custom accelerator for homomorphic encryption applications. IEEE Trans Comput 66(1):316
17. Aguilar-Melchor C, Barrier J, Guelton S et al (2016) NFLlib: NTT-based fast lattice library. In: Cryptographers track at the RSA conference, pp 341–356
18. Lepoint T, Naehrig M (2014) A comparison of the homomorphic encryption schemes FV and YASHE. In: The 7th international conference on cryptology in Africa, pp 318–335
19. Khedr A, Gulak G, Vaikuntanathan V (2016) SHIELD: scalable homomorphic implementation of encrypted data-classifiers. IEEE Trans Comput 65(9):2848–2858
20. Dai W, Doroz Y, Sunar B (2014) Accelerating NTRU based homomorphic encryption using GPUs. In: 2014 IEEE high performance extreme computing conference
21. Roy SS et al (2017) Hardware assisted fully homomorphic function evaluation and encrypted search. IEEE Trans Comput PP(99):1
22. Doroz Y, Ozturk E, Sunar B (2015) Accelerating fully homomorphic encryption in hardware. IEEE Trans Comput 64(6):1509–1521
23. Wang W, Huang XM, Emmart N et al (2014) VLSI design of a large-number multiplier for fully homomorphic encryption. IEEE Trans Very Large Scale Integr Syst 22(9):1879–1887
24. Doroz Y, Hu Y, Sunar B (2016) Homomorphic AES evaluation using the modified LTV scheme. Des Codes Crypt 80(2):333–358
25. Shoup V.NTL: a library for doing number theory[EB/OL]. http://www.shoup.net/ntl/[201731]
26. Hart W.FLINT: fast library for number Theory[EB/OL]. http://www.flintlib.org[201731]
27. Halevi S, Shoup V (2014) Algorithms in HElib. In: The 34th annual international cryptology conference, pp 554–571
28. Halevi S, Shoup V (2015) Bootstrapping for helib. In: Annual International conference on the theory and applications of cryptographic techniques, pp 641–670
29. Chen H, Laine K, Player R (2016) Simple encrypted arithmetic library-SEAL (v2.1) [EB/OL]. https://www.microsoft.com/enus/research/wpcontent/uploads/2016/09/sealmanual2.pdf [20160901]

30. Wang W, Hu Y, Chen LM et al (2015) Exploring the feasibility of fully homomorphic encryption. IEEE Trans Comput 64(3):698–706
31. Wei D, Sunar B (2016) cuHE: a homomorphic encryption accelerator library. In: The second international conference on cryptography and information security in the Balkans, pp 169–186
32. Wang W, Huang X M (2013) FPGA implementation of a large-number multiplier for fully homomorphic encryption. In: IEEE international symposium on circuits and systems, pp 2589–2592
33. Poppelmann T, Naehrig M, Putnam A et al (2015) Accelerating homomorphic evaluation on reconfigurable hardware. In: The 17th international workshop on cryptographic hardware and embedded systems, pp 143–163
34. Sinha Roy S, Järvinen K, Vercauteren F et al (2015) Modular hardware architecture for somewhat homomorphic function evaluation. In: The 17th international workshop on cryptographic hardware and embedded systems, pp 164–184
35. Doroz Y, Ozturb E, Sunar B (2014) A million-bit multiplier architecture for fully homomorphic encryption. Microprocess Microsyst 38(8):766–775
36. Wang W, Hu Y, Chen L et al (2012) Accelerating fully homomorphic encryption using GPU. In: IEEE conference on high performance extreme computing. IEEE, pp 1–5
37. Moore C, O'Neill M, O'Sullivan E, et al (2014) Practical homomorphic encryption: a survey. In: IEEE international symposium on circuits and systems, pp 2792–2795
38. Cooley JW, Tukey JW (1965) An algorithm for the machine calculation of complex Fourier series. Math Comput 19(90):297–301
39. Bhunia S, Hsiao MS, Banga M et al (2014) Hardware trojan attacks: threat analysis and countermeasures. Proc IEEE 102(8):1229–1247
40. Chakraborty RS, Narasimhan S, Bhunia S (2009) Hardware Trojan: threats and emerging solutions. In: IEEE international high-level design validation and test workshop. IEEE, pp 166–171
41. Chen Z, Guo X, Nagesh R et al (2008) Hardware Trojan designs on BASYS FPGA board. Embed Syst Chall Contest Cyber Secur Aware Week
42. Lin L, Kasper M, Güneysu T et al (2009) Trojan side-channels: lightweight hardware trojans through side-channel engineering. Springer, New York, pp 382–395
43. Lin L, Burleson W, Paar C (2009) MOLES: malicious off-chip leakage enabled by side-channels. In: Proceedings of the 2009 international conference on computer-aided design, pp 117–122
44. Bhasin S, Danger J, Guilley S et al (2013) Hardware Trojan horses in cryptographic IPcores. In: Workshop on fault diagnosis and tolerance in cryptography, pp 15–29
45. Piret G, Quisquater J (2003) A differential fault attack technique against SPN structures, with application to the AES and KHAZAD. In: International workshop on cryptographic hardware and embedded systems, pp 77–88
46. Banga M, Hsiao MS (2010) Trusted RTL: Trojan detection methodology in pre-silicon designs. In: 2010 IEEE international symposium on hardware-oriented security and trust (HOST). IEEE, pp 56–59
47. Love E, Jin Y, Makris Y (2012) Proofcarrying hardware intellectual property: a pathway to trusted module acquisition. IEEE Trans Inf Forensics Secur 7(1):25–40
48. Singh B, Shankar A, Wolff F et al (2014) Cross-correlation of specification and RTL for soft IP analysis. In: Proceedings of the conference on design, automation and test in Europe, European design and automation association, p 290
49. Chakraborty RS, Wolff F, Paul S et al (2009) MERO: A statistical approach for hardware Trojan detection. Springer, New York, pp 396–410
50. Narasimhan S, Du D, Chakraborty RS et al (2013) Hardware Trojan detection by multiple-parameter side-channel analysis. IEEE Trans Comput 62(11):2183–2195
51. Rai D, Lach J (2009) Performance of delay-based Trojan detection techniques under parameter variations. In: IEEE international workshop on hardware-oriented security and trust, pp 58–65

52. Chakraborty RS, Bhunia S (2011) Security against hardware trojan attacks using key-based design obfuscation. J Electron Test 27(6):767–785
53. Xiao K, Tehranipoor M (2013) BISA: Built-in self-verification for preventing hardware trojan insertion. In: IEEE international workshop on hardware-oriented security and trust, pp 45–50
54. Salmani H, Tehranipoor M, Plusquellic J (2012) A novel technique for improving hardware Trojan detection and reducing trojan activation time. IEEE Trans Very Large Scale Integr VLSI Syst 20(1):112–125
55. Waksman A, Sethumadhavan S (2010) Tamper evident microprocessors. In: IEEE symposium on security and privacy (SP), pp 173–188
56. MalSarkar S, Krishna A, Ghosh A et al (2014) Hardware trojan attacks in FPGA devices: threat analysis and effective counter measures. In: Proceedings of the 24th edition of the Great Lakes symposium on VLSI, pp 287–292
57. Zhang J, Su G, Liu Y et al (2014) On Trojan side channel design and identification. In: 2014 IEEE/ACM international conference on computer-aided design, pp 278–285
58. Voitsechov D, Etsion Y (2014) Single-graph multiple flows: energy efficient design alternative for GPGPUs. In: ACM SIGARCH computer architecture news, pp 205–216

Afterword

We have been researching reconfigurable computing processors since 2006. At the beginning, our attention only focused on its hardware architecture, mapping mechanism, and programming paradigms, and no additional work was done on cryptographic application. With the constant progress of the research, we found that reconfigurable computing technology could possibly be used to solve many difficult problems that could not be solved by other traditional technologies if the reconfigurable computing technology was applied to cryptographic processors. Reconfigurable computing technology was probably the brand-new technology very much needed by cryptographic processors, which surprised us and made us feel excited! Then we began to establish our team from scratch, established a platform, and we mainly concentrated on this research direction. Supported by the Key Project of the National 863 Program and the National Natural Science Foundation of China (NSFC), with the continuous efforts of several-year postgraduates and postdoctoral researchers such as Min Zhu, Bo Wang, Hai Huang, Chenchen Deng, Yansong Guo, Xing Wang, Le Chang, Kai Luo, Shengyang Mao, Qihuan Huang, Neng Zhang, Zhuoquan Zhou, Dongxing Wang, and Ao Li, our team has successively published over ten high-quality academic papers on the cryptographic chip field and the hardware security field, authorized dozens of invention patents, implemented a series of practical technologies and solved some practical difficult problems of industrial production. Looking back at the past, we just happened to pay our attention to the application of the reconfigurable computing technology to cryptographic chips. However, it was deliberate that we then concentrated on the research direction and our team did the research with great efforts!

In 2014, we summarized the technological breakthroughs in the reconfigurable computing field we had achieved in the past ten years, wrote the book of *Reconfigurable Computing* and had it published by Science Press. On this basis, we have summarized the accumulated research results that we have achieved on reconfigurable cryptographic processors in recent seven to eight years in the same way and have written this book about the results to share with readers. We have had

L. Liu et al., *Reconfigurable Cryptographic Processor*,
https://doi.org/10.1007/978-981-10-8899-5

our book published by the same press. If you find any improper description or anything not precise enough in the book, please let us know and we will appreciate it very much.

Tsinghua University, Beijing, P. R. China Leibo Liu